517.944:535.355.12

MATHEMATICAL
METHODS FOR
WAVE PHENOMENA

This is a volume in
COMPUTER SCIENCE AND APPLIED MATHEMATICS
A Series of Monographs and Textbooks

Editor: WERNER RHEINBOLDT

A complete list of titles in this series appears at the end of this volume.

MATHEMATICAL METHODS FOR WAVE PHENOMENA

NORMAN BLEISTEIN

Department of Mathematics
Colorado School of Mines
Golden, Colorado

 1984

ACADEMIC PRESS, INC.

(Harcourt Brace Jovanovich, Publishers)

Orlando   San Diego   San Francisco   New York
London   Toronto   Montreal   Sydney   Tokyo

ACADEMIC PRESS, INC.
Orlando, Florida 32887

United Kingdom Edition published by
ACADEMIC PRESS, INC. (LONDON) LTD.
24/28 Oval Road, London NW1 7DX

Library of Congress Cataloging in Publication Data

Bleistein, Norman.
    Mathematical methods for wave phenomena.

    (Computer science and applied mathematics)
    Includes index.
    1. Wave motion, Theory of. 2. Wave equation.
I. Title. II. Series.

QA927.B54 1984   531'.1133   83−21345
ISBN 0−12−105650−3 (alk. paper)

PRINTED IN THE UNITED STATES OF AMERICA

84 85 86 87   9 8 7 6 5 4 3 2 1

For my wife, Sandra, and my children, Steven and Abby,
for their love, patience, and support
and
Dedicated to the memory of my brother Eddie,
who always knew I could,
and to my teacher Bob Lewis,
who showed me how

CONTENTS

Preface   xi

1   FIRST-ORDER PARTIAL DIFFERENTIAL EQUATIONS

1.1   First-Order Quasi-Linear Differential Equations   1
1.2   An Illustrative Example   7
1.3   First-Order Nonlinear Differential Equations   12
1.4   Examples—The Eikonal Equation—and More Theory   18
1.5   Propagation of Wave Fronts   27
1.6   Variable Index of Refraction   38
1.7   Higher Dimensions   42
      References   44

2   THE DIRAC DELTA FUNCTION, FOURIER TRANSFORMS, AND ASYMPTOTICS

2.1   The Dirac Delta Function and Related Distributions   45
2.2   Fourier Transforms   52
2.3   Fourier Transforms of Distributions   58
2.4   Multidimensional Fourier Transforms   61
2.5   Asymptotic Expansions   66
2.6   Asymptotic Expansions of Fourier Integrals with Monotonic Phase   73
2.7   The Method of Stationary Phase   77
2.8   Multidimensional Fourier Integrals   82
      References   89

3 SECOND-ORDER PARTIAL DIFFERENTIAL EQUATIONS

3.1 Prototype Second-Order Equations   92
3.2 Some Simple Examples   94
    References   99

4 THE WAVE EQUATION IN ONE SPACE DIMENSION

4.1 Characteristics for the Wave Equation in One Space Dimension   100
4.2 The Initial Boundary Value Problem   105
4.3 The Initial Boundary Value Problem Continued   109
4.4 The Adjoint Equation and the Riemann Function   120
4.5 The Green's Function   130
4.6 Asymptotic Solution of the Klein–Gordon Equation   133
4.7 More on Asymptotic Solutions   136
    References   144

5 THE WAVE EQUATION IN TWO AND THREE DIMENSIONS

5.1 Characteristics and Ill-Posed Cauchy Problems   145
5.2 The Energy Integral, Domain of Dependence, and Uniqueness   150
5.3 The Green's Function   153
5.4 Scattering Problems   158
    References   163

6 THE HELMHOLTZ EQUATION AND OTHER ELLIPTIC EQUATIONS

6.1 Green's Identities and Uniqueness Results   165
6.2 Some Special Features of Laplace's Equation   170
6.3 Green's Functions   174
6.4 Problems in Unbounded Domains and the Sommerfeld Radiation Condition   180
6.5 Some Exact Solutions   192
    References   202

7 MORE ON ASYMPTOTICS

7.1 Watson's Lemma   204
7.2 The Method of Steepest Descents: Preliminary Results   211
7.3 Formulas for the Method of Steepest Descents   225
7.4 The Method of Steepest Descents: Implementation   229
    References   240

8 ASYMPTOTIC TECHNIQUES FOR DIRECT SCATTERING PROBLEMS

8.1 Scattering by a Half-Space: Analysis by Steepest Descents   241
8.2 Introduction to Ray Methods   257
8.3 Determination of Ray Data   268
8.4 The Kirchhoff Approximation   281
    References   299

9 INVERSE METHODS FOR REFLECTOR IMAGING

9.1 The Singular Function and the Characteristic Function   302
9.2 Physical Optics Far-Field Inverse Scattering (POFFIS)   312
9.3 The Seismic Inverse Problem   321
    References   337

Index   339

PREFACE

I decided to write this book when I began looking for a textbook for a methods of applied mathematics course that I was to teach during the 1981–1982 academic year. There were many fine textbooks available—with some of the recent ones having been written by friends and colleagues for whom I have a great deal of respect. However, I felt that over the fifteen years that had elapsed since I left the Courant Institute, my research and point of view about the problems of interest to me had evolved in a manner unique to the combination of scientific experiences that have made up my career to date. It was my own point of view that I wanted to communicate in the course I was to teach. Certainly, the reader will find common ground here with other texts and references. However, it is my hope that I have communicated enough of the ideas that comprise my approach to direct and inverse scattering problems to have made this project worthwhile.

Much of my success in research is based on a fundamental education in *ray methods*, in particular, and *asymptotic methods*, in general, to which I was introduced at the Courant Institute. I take the point of view that an exact solution to a problem in wave phenomena is not an end in itself. Rather, it is the *asymptotic solution* that provides a means of interpretation and a basis for understanding. The exact solution, then, only provides a point of departure for obtaining a *meaningful* solution. This point of view can be seen in the contents of this book, where I have made relatively short shrift of exact solutions on the road to asymptotic techniques for wave problems.

The goal I was trying to reach appears in Chapters 8 and 9. In the former I discuss some asymptotic techniques for direct scattering problems. In the latter, I discuss the class of inverse problems in which one seeks to image reflecting surfaces from "backscattered" data. In order to reach this material in what I consider a one-year course, I have had to make some compromises with more traditional subject matter in a first course in applied mathematics. Certainly my choices are open to criticism. However, after having gone through this process myself, I retract all such casual criticism of others that I have ever made in the past; the authors who are my predecessors now have my profound respect for having made the hard choices that were necessary.

In those last two chapters, I have referenced material in all of the preceding chapters. I believe therefore that the material "hangs together," as I hoped it would.

Throughout the book, I have used the exercises to teach additional material. I recommend perusal of the exercises to even the most casual reader.

The student taking a course that uses this book should have a good basic education in classical analysis and differential equations. I make extensive use of complex function theory techniques, especially contour integration and analysis of singularities of functions of a complex variable. I also assume that the student has had a course in ordinary differential equations and one in linear algebra.

The prerequisite that is harder to define is "some experience in applied mathematics." By its very nature, applied mathematics is interdisciplinary. Unfortunately, it is not often practiced or taught in mathematics departments but as an adjunct in engineering, physics, and geophysics departments, often under titles that do not necessarily indicate to the uninitiated that the course is really one in applied mathematics. A course in electromagnetics using the text by Jackson is a perfect example. I can only suggest that the student who has taken courses in complementary disciplines or who has on-the-job experience with solutions to real-world problems has an advantage over the student lacking that background.

Richard Courant said[†] that the applied mathematician stands "in active and reciprocal relation" to the other sciences and to society. I am an adherent to that point of view. I have made some attempt in the text to draw on my experience in applications to give meaning to some of the text material. However, it is difficult to do much of that in the press of space on the mathematical material itself. It is my hope that each instructor using this textbook will have an equivalent store of experience to bring to the classroom or that the self-taught person has some resource of experience to provide a context for the material.

My objective in Chapter 1, on nonlinear first-order partial differential equations, is to discuss the eikonal equation—rays and wave front propagation. In particular, I wanted to present all of the anomalous cases that lead

---

[†] Constance Reid, "Courant in Göttingen and New York, The Story of an Improbable Mathematician," Springer-Verlag, New York, 1976.

to focusing, caustics, and various diffraction phenomena. My interest in these more exotic aspects of ray theory first arose from a research project at the Courant Institute with R. M. Lewis and D. Ludwig on smooth-body diffraction. I first wrote about this material from the point of view presented here in a set of lecture notes as part of the North British Symposium on Differential Equations in Dundee, Scotland, in 1972. It is my hope that this chapter has profited from the additional insight I have gained over the intervening nine years.

Chapter 2 contains some topics covering necessary machinery. I have tried to present coherent developments of distribution theory, (one- and multidimensional) stationary phase, along with a synthesis of all three of these subjects. In my opinion, stationary phase is the method of choice for analyzing Fourier representations for multidimensional wave fields.

The discussion of the one-dimensional wave equation in Chapter 4 provides an opportunity to present some theoretical results about the wave equation in a relatively simple setting while also allowing—through the device of studying the vibrating string in great detail—an introduction to complex variable methods for analysis of Fourier representations, eigenfunction expansions, and the WKB method. It also provides an opportunity to use the method of stationary phase and to develop some simple ideas about superposition of waves and dispersion relations.

In Chapter 5, on the wave equation in higher dimensions, the ideas of domains of dependence and influence and energy conservation are introduced. It is apparent through the sparsity of discussion of exact solutions in this chapter that my interests and objectives in this book are to get on to the Helmholtz equation, where I can address asymptotic "high-frequency" solutions. I have gotten to that equation in Chapter 6. I have also taken some pains there to discuss uniqueness of solutions for the frequency variable in an upper or lower half plane. This again reflects my own interests in "causal" solutions to wave problems. Also in this chapter I have addressed the equivalence between the Sommerfeld radiation condition for the Helmholtz equation and the outward propagation of energy for the wave equation.

Chapter 7 discusses the method of steepest descents, which I have brought to bear with great success on many problems in wave theory. This method requires a strong grounding in certain aspects of complex function theory. However, the gain is well worth the pain!

The difficulty here and in the final two chapters was to filter from all of the ideas I would have liked to address those that would make up a complete introduction to the subject matter. The size of these last chapters attests to the difficulty I had in achieving that goal!

Finally, in Chapters 8 and 9, I have tried to give the reader a basic introduction to the methods that have dominated my research career and will probably continue to do so. To me there is a vitality and richness of experience associated with the implementation of these methods on problems in seismic exploration and nondestructive testing. In both areas, I have found that we with labels reading "applied mathematician" are late to the area; we have much catching up to do on methods that work and also have much to contribute to help make these methods work better.

A particular bias of mine will become apparent when the reader discovers that one-dimensional inverse problems are relegated to the exercises in Chapter 9. In my own research, I tend to examine only those one-dimensional inverse problems that arise as specializations of three-dimensional real-world inverse problems. I believe that the alternative—the development of sophisticated methods that do not extend from one to higher dimensions— teaches the student *and the researcher* the wrong lessons. A standard joke in our research group is that "only graduate students study the one-dimensional inverse problem!"

I owe a great debt of gratitude to the students in that first course—Mourad Lahlou, Dennis Nesser, Michael Shields, Marc Thuillard, Shelby Worley, and Bruce Zuver—for suffering through the first draft of much of this material. Often, these people would return to class after one week and explain to me how I *meant* to write the exercises. Shelby Worley and Mourad Lahlou were especially helpful in finding typos and errors.

I also recognize the important contribution to my growth as an applied mathematician made by the various contract monitors who have supported my research over the years: Stuart Brodsky at ONR and Milton E. Rose at DOE/ERDA, who supported my early efforts in inverse problems—work that was obviously primitive when viewed from the present perspective, ten years later; Hugo Bezdek and J. Michael McKisic at ONR, who supported our group's work in an ocean acoustics program even though we were mathematicians; Charles Holland at ONR, who has continued the support program begun under Stuart Brodsky. The encouragement, recognition, and financial support provided through these people have been essential elements in our continued success.

I owe a special debt of thanks to my colleagues Jack K. Cohen, Frank G. Hagin, and John A. DeSanto, all of whom have supported and encouraged me throughout this project and all of whom have taken up some of the slack in joint efforts caused by my distraction from those efforts during the writing of this book. In addition, Jack Cohen provided many suggestions to improve the exposition in Chapters 1–4.

I entered the entire text on a word processor, using the Interactive Corporation Unix System on a PDP/11. I used NROFF-based macros called -ms and -neqn. I am extremely grateful to Mourad Lahlou for writing a filter to allow the commands of these files to be reinterpreted as print commands for an NEC Spinwriter. I created large amounts of the text off-line on my own Northstar Advantage Micro-Computer. I then transmitted the material to the mainframe with a smart terminal program. Near the end of this project, the University of Denver, where I had been printing, changed operating systems. The National Center for Atmospheric Research in Boulder allowed me time on their PDP running the Interactive Unix System. I shall be eternally grateful for their assistance, which was crucial to the timely completion of this project.

I shall probably never write by hand again!

1 FIRST-ORDER PARTIAL DIFFERENTIAL EQUATIONS

In this chapter, we shall develop the theory of first-order linear and nonlinear partial differential equations. An important example of the latter is the eikonal equation, which characterizes the propagation of wave fronts and discontinuities for the acoustic wave equation, Maxwell's equations, and the equations of elastic wave propagation. In this development, the calculus of curves and surfaces will play an important role. Thus, a review of this material might be in order for the reader. Two possible sources [Kreyszig, 1959; Lipschutz, 1969] for this material are listed in the references at the end of the chapter.

1.1 FIRST-ORDER QUASI-LINEAR DIFFERENTIAL EQUATIONS

We shall begin our discussion with a study of a function of two variables $u(x, y)$. This will allow us to represent the solution as a surface in three-dimensional space or as a family of level curves in two-space. Both representations will prove useful to us. Thus, we shall consider the equation

$$a(x, y, u)u_x + b(x, y, u)u_y = c(x, y, u). \tag{1.1.1}$$

When a and b are independent of u and c is linear in u, this equation is a *linear first-order partial differential equation* for u. Otherwise, the equation is called *quasi-linear.*

For increments along the solution surface, we recall that

$$u_x \, dx + u_y \, dy = du. \tag{1.1.2}$$

This equation states that the *normal vector* $(u_x, u_y, -1)$ is perpendicular at each point to any tangent vector (dx, dy, du) to the surface. Comparison of (1.1.1) and (1.1.2) suggests that we view the solution surface as being made up of a family of curves on each of which the direction of the tangent to the curve at each point is given by the set of direction numbers (a, b, c). These curves are called the *characteristic curves*. See Fig. 1.1. Thus, the solution surface is made up of characteristic curves on which

$$\frac{dx}{a} = \frac{dy}{b} = \frac{du}{c}. \tag{1.1.3}$$

When we think of the solution u as a family of level curves in the (x, y) plane, the corresponding curves of interest are the *projections* of the characteristics onto that plane. These curves are called the *characteristic base curves* and sometimes (confusingly) are also called characteristic curves.

There are really two equations in (1.1.3). In order to understand the nature of the solution to these two equations, let us for the moment rewrite them in the form

$$dy/dx = b/a, \qquad du/dx = c/a. \tag{1.1.4}$$

If we were now given two *initial values* for y and u, we could obtain a *solution curve* with y and u on that curve prescribed in terms of x and the two initial values. The two initial values are parameters. By varying them, we would obtain other solution curves to the pair of equations (1.1.4). A *solution surface* would then consist of a one-parameter subset of this two-parameter family of curves.

How does one obtain this one-parameter subset? We have already characterized the solution in terms of initial values of y and u, that is, in terms of the value of y and u in the plane $x = 0$. Let us suppose, then, that we are interested in a solution surface that passes through a prescribed curve in the

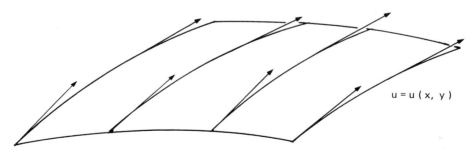

Fig. 1.1. A solution surface $u = u(x, y)$ made up of characteristic curves, each of which is tangent to the vector with direction numbers (a, b, c) at each point on the surface u.

$x = 0$ plane. Thus, we would define a relationship between the initial values of y and u. This relationship reduces our two-parameter family to a one-parameter family.

There is no a priori reason to single out x, or y or u for that matter. Let us suppose more generally that we seek a solution surface that passes through a prescribed curve

$$x = x_0(\tau), \qquad y = y_0(\tau), \qquad u = u_0(\tau). \tag{1.1.5}$$

Thus, the solution surface will consist of a one-parameter (τ) family of characteristic curves passing through the prescribed curve (1.1.5).

In order to facilitate the solution of (1.1.3), we introduce a *second parameter* σ along the characteristic curves. We will then require that the derivative of the vector (x, y, u) along the characteristic be proportional to the vector (a, b, c). (We remark that in our momentary digression above x played the role of the parameter along the characteristics.) Then we rewrite (1.1.3) as

$$dx/d\sigma = \lambda a, \qquad dy/d\sigma = \lambda b, \qquad du/d\sigma = \lambda c, \tag{1.1.6}$$

subject to the *initial data* (1.1.5) when $\sigma = 0$.

The choice of λ is at our disposal. If we set

$$\lambda = 1/\sqrt{a^2 + b^2 + c^2}, \tag{1.1.7}$$

then the parameter σ is arc length along the characteristic curves. If we set

$$\lambda = 1/\sqrt{a^2 + b^2}, \tag{1.1.8}$$

then σ is arc length along the *characteristic base curves*. While one or the other of these may prove useful for analysis of the properties of a solution, when one is actually *solving* the equation, it is usually most convenient to set

$$\lambda = 1, \qquad dx/d\sigma = a, \qquad dy/d\sigma = b, \qquad du/d\sigma = c. \tag{1.1.9}$$

We shall continue with the choice (1.1.9) below.

A solution surface now consists of a two-parameter family of points

$$x = x(\sigma, \tau), \qquad y = y(\sigma, \tau), \qquad u = u(\sigma, \tau). \tag{1.1.10}$$

Each choice of τ "labels" a characteristic curve, while σ varies along the characteristic. We rely on the existence and uniqueness theory for ordinary differential equations to assure us that we do indeed have a solution to (1.1.6). The method we have proposed here is known as the *method of characteristics*. We remark that the method, as presented, would seem to develop a "one-sided" solution surface with the prescribed "initial curve" as an edge. The method could as easily have been developed in terms of a "final value" problem, in which case the prescribed curve contained in the solution surface

u would still be an edge. Of course, we do not have to redevelop the theory for this case but only take the attitude that σ may be positive or negative. Then, the prescribed curve is truly embedded in the solution surface.

It should be noted that this still does not produce a solution surface $u(x, y)$. To do this, we must invert the first two equations in (1.1.10) to yield σ and τ as functions of x and y and then insert that solution into the third equation of (1.1.10). Let us suppose that at some point on the initial curve (1.1.5) we have such a solution for σ and τ and we wish to extend the solution off the initial curve. A *sufficient* condition that we be allowed to do so is that the Jacobian

$$J = \det \begin{bmatrix} x_\sigma & y_\sigma \\ x_\tau & y_\tau \end{bmatrix} \tag{1.1.11}$$

is not zero. Later, when studying nonlinear first-order equations, we will confirm that for $J \neq 0$ (1.1.6) does yield a solution of the equation for $u(x, y)$ in some neighborhood of the initial curve. Since the nonlinear case subsumes the linear, we will not stop to confirm this fact now.

Let us suppose now that $J = 0$ at some point on the initial curve. In this case, (1.1.9) and (1.1.11) imply that, at that point,

$$\frac{x_0'}{y_0'} = \frac{x_\sigma}{y_\sigma} = \frac{a}{b}, \qquad \sigma = 0, \qquad ' = \frac{d}{d\tau}. \tag{1.1.12}$$

This means that the projection of the initial curve in the (x, y) plane is parallel to the characteristic base curve at that point. Let us suppose that, nonetheless, in a neighborhood of this point a solution with well-behaved partial derivatives exists. Then

$$u_\tau = u_x x_\tau + u_y y_\tau \rightarrow u_0'(\tau) = u_x x_0' + u_y y_0', \tag{1.1.13}$$

from which we calculate

$$u_0' = y_0' \left[\frac{x_0'}{y_0'} u_x + u_y \right] = y_0' \left[\frac{x_\sigma}{y_\sigma} u_x + u_y \right]$$

$$= y_0' \left[\frac{a}{b} u_x + u_y \right] = \frac{y_0'}{b} \{ a u_x + b u_y \} = \frac{y_0'}{b} c. \tag{1.1.14}$$

Thus, we conclude that

$$\frac{x_0'}{a} = \frac{y_0'}{b} = \frac{u_0'}{c}. \tag{1.1.15}$$

That is, if $J = 0$ at some point on the initial curve and u_x and u_y are defined and satisfy (1.1.1), then *necessarily* the initial data must satisfy the differential equations of the characteristics at that point. The data are called *characteristic initial data* at that point, and the point itself is called a *characteristic point*. If $J = 0$ at *every* point on the initial curve, then the initial curve must be a characteristic curve, and the data satisfying (1.1.15) are called *characteristic initial data*. In this case, however, we can obtain many solutions containing the initial curve. To do so, pass any noncharacteristic curve through the given initial curve. Solve the problem with this new curve as the initial curve. Of necessity, it contains the old initial curve as the characteristic curve in the solution surface emanating from the intersection of the two curves.

Let us now suppose that the initial curve was "broken" as at the points A and B in Fig. 1.2 but that this curve is not characteristic. The discontinuity of the initial curve produces a discontinuous surface whose edges are characteristics. The initial value problems for the characteristics emanating from each point on the initial curve still have solutions. Thus, from this initial data we develop a solution as a two-sheeted surface. Of necessity, the edges of the two sheets are characteristics. That is, the discontinuity in the initial data *propagates* along characteristics. Indeed, for every segment of an initial curve, we can trace on characteristics how that segment affects the development of the solution surface. Therefore, we define the *range of influence* of a section of the initial curve as the part of the surface traced out by the characteristics emanating from that section of the initial curve. Correspondingly, we define the *domain of dependence* of a noncharacteristic curve in the solution surface as the piece of initial curve whose characteristics pass through the prescribed noncharacteristic curve. See Fig. 1.3.

The case in which $J = 0$ at an isolated noncharacteristic point on the initial curve can now be addressed. Let us consider the initial curve obtained by deleting a neighborhood of this critical point and also consider the two ranges of influence of the separate pieces of the initial curve. For each of these surfaces, we expect a solution surface that is well behaved in some region near the initial curve. When we "shrink" the excluded neighborhood of initial data about the critical point, we expect to develop some sort of limiting solution

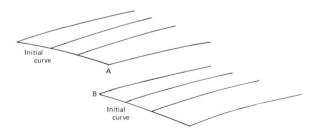

Fig. 1.2. A solution surface *u* emanating from a discontinuous initial curve.

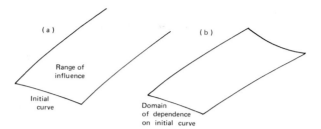

Fig. 1.3. (a) The range of influence of a segment of the initial curve—that part of the surface generated by characteristics starting from the given section of the initial curve; (b) the domain of dependence of a curve in the solution surface—that section of initial curve obtained by tracing back along the characteristics from the given curve to the initial curve.

surface. When both equations of (1.1.15) are satisfied (characteristic initial data), we expect this limiting process to yield a continuous solution surface. Indeed, the vanishing of J might only signal that the transformation from (σ, τ) to (x, y) has some pathology, perhaps only multivaluedness at the critical point. If the second equation in (1.1.15) is not satisfied (noncharacteristic initial data), we might well expect some sort of singular behavior in the solution surface all along the base characteristic curve through the critical point. It is even dangerous to speak of the full characteristic curves here because we do not even know in this case whether the method has any validity at all near the critical point. Whatever this singular behavior might be, we can expect that it will propagate on the characteristics. We shall see examples of this type of behavior in the following sections. Such examples are not to be avoided. They provide a richness to the theory that expands the class of physical phenomena that can be modeled mathematically.

To recapitulate the major features of this development:

Any curve solving (1.1.3) is called a characteristic curve for (1.1.1). Every solution of (1.1.1) consists of a one-parameter family of such characteristics. These can be chosen uniquely (in the "small") to pass through an initial curve so long as

$$ay_0'(\tau) - bx_0'(\tau) \neq 0$$

on that curve as parameterized by τ. When this condition fails, there will not be a solution in the ordinary sense of an invertible, differentiable transformation from parameters (σ, τ) to (x, y) unless the initial curve is itself a characteristic, in which case the solution exists but is nonunique (has "many" solutions).

We remark that a solution can be continued away from the initial curve until the procedure breaks down, that is, until some anomaly—such as those described earlier in the context of the initial curve—arises.

1.2   AN ILLUSTRATIVE EXAMPLE

We shall consider here the following linear partial differential equation

$$u_x + u_y = u. \tag{1.2.1}$$

For this problem, we deduce, following (1.1.3), that the equations for the characteristics are

$$\frac{dx}{1} = \frac{dy}{1} = \frac{du}{u}. \tag{1.2.2}$$

These equations have the general solutions

$$y = x + c_1, \qquad u = c_2 e^x. \tag{1.2.3}$$

Each choice of the constants c_1, c_2 produces a characteristic curve. That is, (1.2.3) is a two-parameter family of characteristic curves. A solution surface could be determined by prescribing a functional relationship between c_1 and c_2, that is, by prescribing, for example,

$$c_2 = f(c_1) \tag{1.2.4}$$

or, equivalently,

$$ue^{-x} = f(y - x). \tag{1.2.5}$$

Solving for u, we conclude that

$$u = e^x f(y - x). \tag{1.2.6}$$

By direct substitution one can verify that this is a solution to the differential equation (1.2.1) for any differentiable function f. This solution is an *envelope* of solution curves (1.2.3) to the equation (1.2.2). We shall refer to such a solution as an *envelope solution*. We could as easily characterize the envelope solution by setting c_1 equal to an arbitrary function of c_2 or by setting an arbitrary function of the two variables c_1 and c_2 equal to zero. Our particular choice here was motivated by the tidy form of the solution (1.2.6) expressing u as a function of x and y explicitly.

In the preceding section, we considered first a problem in which a curve in the plane $x = 0$ was prescribed through which the solution curve was required to pass. From (1.2.6), we see that

$$u = f(y), \qquad x = 0. \tag{1.2.7}$$

This is, the functional relationship between u and y in the $x = 0$ plane determines the arbitrary function f in the envelope solution.

For example, suppose that we seek a solution to (1.2.1) that passes through the curve

$$u = y^2, \qquad x = 0.$$

The solution is

$$u = e^x(y - x)^2. \tag{1.2.8}$$

Thus, we have demonstrated an envelope solution and the solution to a problem in which u is required to pass through a specific curve in the $x = 0$ plane.

Let us turn now to solution by the method of characteristics. For (1.2.1), we conclude from (1.1.9) that the differential equations for the characteristic curves are

$$dx/d\sigma = 1, \quad dy/d\sigma = 1, \quad du/d\sigma = u. \tag{1.2.9}$$

The general solution to this system of equations is

$$x = \sigma + x_0(\tau), \quad y = \sigma + y_0(\tau), \quad u = u_0(\tau)e^\sigma. \tag{1.2.10}$$

Here the initial curve is defined by

$$x = x_0(\tau), \quad y = y_0(\tau), \quad u = u_0(\tau), \quad \sigma = 0. \tag{1.2.11}$$

We will be able to produce a solution $u(x, y)$ in some neighborhood of the initial curve if we can solve for σ and τ on the initial curve (that is, we can solve for τ when $\sigma = 0$) and also the Jacobian (1.1.11)

$$J = y_0'(\tau) - x_0'(\tau) \neq 0. \tag{1.2.12}$$

As a simple example, let us suppose that the initial curve is given by

$$x_0(\tau) = \tau, \quad y_0(\tau) = -\tau, \quad u_0(\tau) = \tau. \tag{1.2.13}$$

Then from (1.2.10)

$$x = \sigma + \tau, \quad y = \sigma - \tau, \quad u = \tau e^\sigma. \tag{1.2.14}$$

It is now straightforward to solve for σ and τ in terms of x and y and to substitute these values in the expression for u. The result is

$$u = \tfrac{1}{2}[x - y]e^{(x+y)/2}. \tag{1.2.15}$$

We leave as an exercise that this solution is really of the form (1.2.6). Now let us suppose that

$$x_0(\tau) = y_0(\tau) = \tau. \tag{1.2.16}$$

In this case, J is identically zero on the initial curve. Thus, a solution will only exist if u_0 satisfies the characteristic differential equations (1.1.15). For example, we may take

$$u_0 = e^\tau, \tag{1.2.17}$$

and then we obtain a solution by passing an arbitrary noncharacteristic curve through a point of the prescribed initial curve. Alternatively, we can determine all solutions containing this characteristic initial data by exploiting our envelope solution (1.2.6) and restricting the function f to contain the characteristic curve defined by (1.2.16) and (1.2.17). Substitution of the latter two equations into the former yields

$$f(0) = 1. \tag{1.2.18}$$

Thus, any differentiable function satisfying this condition will produce a solution (1.2.6) of (1.2.1). Clearly, the solution is nonunique.

Now we shall create a problem in which $J = 0$ at only one point. Then we shall choose data for u so that the full set of data is or is not characteristic at that point. We begin by setting

$$x_0(\tau) = \tau, \qquad y_0(\tau) = \tfrac{1}{2}\tau^2, \tag{1.2.19}$$

so that

$$x_0' = y_0' = 1, \qquad \tau = 1, \tag{1.2.20}$$

and J (1.2.12) is zero at this point. There are many functions $u_0(\tau)$ that will make the initial data characteristic at one point. As an example, we choose the function

$$u_0(\tau) = \tfrac{1}{3}(2 + \tau^3), \tag{1.2.21}$$

which, along with (1.2.19), can be seen to satisfy the characteristic equations (1.1.15) at the point $\tau = 1$.

For these initial data, the solution (1.2.10) is

$$x = \sigma + \tau, \qquad y = \sigma + \tfrac{1}{2}\tau^2, \qquad u = \tfrac{1}{3}[2 + \tau^3]e^\sigma. \tag{1.2.22}$$

In Fig. 1.4, we depict the initial base curve (a parabola) and the base characteristics, which are straight lines at a slope equal to unity. It is seen that the base characteristics form a *double covering* of a portion of the (x, y) plane, with the boundary of that double covering passing tangentially through the initial curve at the point where $J = 0$.

Solving for σ and τ in (1.2.22), we find

$$\sigma = x - (1 \pm \sqrt{1 - 2(x - y)}), \qquad \tau = 1 \pm \sqrt{1 - 2(x - y)}, \tag{1.2.23}$$

and then

$$u = \tfrac{1}{3}[2 + \{1 \pm \sqrt{1 - 2(x - y)}\}^3] \exp\{x - (1 \pm \sqrt{1 - 2(x - y)})\}. \tag{1.2.24}$$

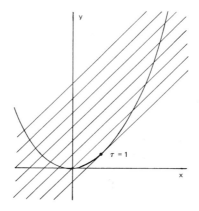

Fig. 1.4. Base initial curve (1.2.19) and base characteristics from (1.2.22).

From this expression, we calculate that

$$u_y = \pm \frac{2}{\{1 - 2(x - y)\}^{1/2}} \exp\{x - (1 \pm \sqrt{1 - 2(x - y)})\}$$

$$\cdot [(1 \pm \sqrt{1 - 2(x - y)})^2 - \tfrac{1}{3}[(1 \pm \sqrt{1 - 2(x - y)})^3 + 2]]. \quad (1.2.25)$$

One can see that this partial derivative is defined everywhere except on the straight line

$$2(x - y) = 1. \quad (1.2.26)$$

This is the base characteristic that passes through the special point $\tau = 1$. On examining the expression for u_y in the limit as one approaches this characteristic, one finds that it has a finite limit. Furthermore, u_x has the same pathology and the partial differential equation (1.2.1) is satisfied, even in this limit. Therefore, the characteristic data have produced a *double* surface over a portion of the (x, y) plane on which the partial derivatives remain finite in the limit as one approaches the edge of the two surfaces.

On the other hand, let us change the data for u in such a manner that they are no longer characteristic data. A simple way to do this is to change the constant 2 in (1.2.21) to any other value. The effect of this will be to change the constant 2 in the second line of (1.2.25) to this same new value. In this case, that second line no longer has a zero limit as the base characteristic (1.2.26) is approached, and u_y becomes infinite in this limit, as does u_x. This is really not so severe as one might expect. Actually, all that has happened is that the surface has become *vertical* above this base characteristic in the sense that the normal to the surface is horizontal. By setting $x = y$ in (1.2.24), we see that in three-space, the singularity lies on the characteristic through the singular point. That is, the singularity of the data propagates on the characteristic curve through the singular point whether or not the value of u at the critical point on the initial curve is such that the data are characteristic.

The purpose of these last cases was to demonstrate that failure of the conditions leading to a straightforward solution by the method of characteristics does not mean that there is no solution. On the contrary, the situation in which $J = 0$ at one or more points enriches the class of solutions. The reader is cautioned not to abandon a problem because of the presence of a point at which the data are such that $J = 0$.

Exercises

1.1 (a) Verify by direct substitution that (1.2.6) is a solution of (1.2.1).

(b) Solve (1.2.2) by finding x and u as functions of y and constants and obtain the envelope solution

$$u = e^y g(x - y).$$

Reconcile this solution with the solution (1.2.6).

(c) Show that the solution (1.2.15) is of the form (1.2.6).

(d) Find the most general choice of u_0 to combine with (1.2.16) to create characteristic data for (1.2.1).

1.2 Repeat the steps of this section for the equation

$$u_x + u_y = 1.$$

That is:

(a) Find the envelope solution.

(b) Find the solution by the method of characteristics, analogous to the form (1.2.10), for arbitrary initial data.

(c) Obtain a solution for $u(x, y)$ for the data (1.2.13).

(d) For the initial base curve (1.2.16), find u_0 to make the data characteristic and characterize the class of nonunique solutions to this problem.

(e) Now take the data (1.2.19) and

$$u_0(\tau) = \tfrac{1}{2}\tau^2$$

and solve for u.

1.3 Find u satisfying

$$(x + 1)u_x + yu_y = u$$

passing through the curve

$$x = 0, \qquad y^2 - u^2 = 1.$$

1.4   (a)   Show that the envelope solution of the equation

$$uu_x + yu_y = u^2$$

can be written as

$$y = e^{-1/u}f(ue^{-x}).$$

(b)   Show that for the initial curve $x_0(\tau)$, $y_0(\tau)$, $u_0(\tau)$ the solution can be written as

$$x = x_0(\tau) - \ln[1 - \sigma u_0(\tau)], \qquad y = y_0(\tau)e^{\sigma}, \qquad u = \frac{u_0(\tau)}{1 - \sigma u_0(\tau)}.$$

1.3   FIRST-ORDER NONLINEAR DIFFERENTIAL EQUATIONS

We now consider the equation

$$F(x, y, u, p, q) = 0, \qquad p = u_x, \quad q = u_y. \qquad (1.3.1)$$

To ensure that this is, in fact, a partial differential equation, we require that either or both

$$F_p \neq 0, \qquad F_q \neq 0. \qquad (1.3.2)$$

We shall develop a solution technique for (1.3.1) by replacing this equation with quasi-linear equations to which the theory of Section 1.1 applies. We note that for a solution to (1.3.1), small increments in x and y produce no change in the value of F; that is,

$$\begin{aligned}
0 = \Delta F = &[F_x + pF_u + F_p p_x + F_q p_y]\,\Delta x \\
&+ [F_y + qF_u + F_p q_x + F_q q_y]\,\Delta y.
\end{aligned} \qquad (1.3.3)$$

Here we have assumed that the solution is sufficiently smooth to allow the interchange of orders of differentiation

$$p_y = q_x. \qquad (1.3.4)$$

Since the increments in x and y are independent, their coefficients must each be zero. Thus,

$$F_p p_x + F_q p_y = -F_x - pF_u, \qquad F_p q_x + F_q q_y = -F_y - qF_u. \quad (1.3.5)$$

If (1.3.1) does not depend explicitly on u, then neither does (1.3.5). If (1.3.1) does depend on u, we can think of solving for u and substituting into (1.3.5). In either case, we obtain a simultaneous system of *quasi-linear* partial differential equations for p and q. Significantly, the *base characteristics* for these two equations are the same, and we can view the equations as describing the simultaneous propagation of p and q along these base characteristics. Thus, by exploiting our linear theory,

$$\frac{dx}{F_p} = \frac{dy}{F_q} = -\frac{dp}{F_x + pF_u} = -\frac{dq}{F_y + qF_u}. \tag{1.3.6}$$

We calculate the change in u along these base characteristics as

$$du = p\,dx + q\,dy = [pF_p + qF_q](dx/F_p). \tag{1.3.7}$$

By dividing here, we obtain an equation for du that completes the set of equations in (1.3.6). The result is

$$\frac{dx}{F_p} = \frac{dy}{F_q} = \frac{du}{pF_p + qF_q} = -\frac{dp}{F_x + pF_u} = -\frac{dq}{F_y + qF_u}. \tag{1.3.8}$$

Having deduced this full set of equations, we can now abandon the intermediary step of eliminating u in order to obtain (1.3.6).

The determination of u for each (x, y) on a base characteristic defines a characteristic curve in three-space. Here, however, we also solve for p and q. That is, a solution of this system of equations yields at each point on the characteristic curve a normal direction of the solution surface as well. Alternatively, we can think of the solution as a characteristic curve and a differential element of tangent plane at each point. Together, the characteristic curve and planar element yield a *characteristic strip*. See Fig. 1.5.

As in the quasi-linear case, it will be helpful for us to introduce a parameter σ along the characteristic strips and rewrite (1.3.8) as

$$dx/d\sigma = \lambda F_p, \qquad dy/d\sigma = \lambda F_q, \qquad du/d\sigma = \lambda[pF_p + qF_q],$$
$$dp/d\sigma = -\lambda[F_x + pF_u], \qquad dq/d\sigma = -\lambda[F_y + qF_u]. \tag{1.3.9}$$

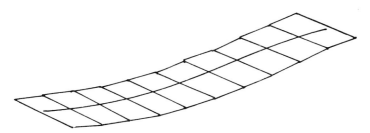

Fig. 1.5. A characteristic strip consisting of a characteristic curve and differential elements of tangent planes.

We have the same options for λ as discussed in Section 1.1 and proceed with our discussion here with $\lambda = 1$:

$$dx/d\sigma = F_p, \qquad dy/d\sigma = F_q, \qquad du/d\sigma = pF_p + qF_q,$$
$$dp/d\sigma = -F_x - pF_u, \qquad dq/d\sigma = -F_y - qF_u. \tag{1.3.10}$$

Since the derivation of this system of equations started from (1.3.3), it follows that $F = $ const for any solution of (1.3.10). That is, $F = $ const is an integral of this system. The condition that this constant is zero is simply a constraint on the constants of integration of the solution to (1.3.10). Thus, (1.3.1) will be satisfied everywhere on the characteristic strip solving (1.3.10) if it is satisfied at one point on the strip.

The reader is cautioned that we have not yet confirmed that the system (1.3.10) yields a solution surface. Indeed, we do not have a solution surface at all (we have only a characteristic strip) and therefore no way to relate p and q to u_x and u_y. A solution surface will consist of a one-parameter family of characteristic strips. We shall now demonstrate how one might find this family.

Let us suppose that we seek a solution passing through the curve

$$x = x_0(\tau), \qquad y = y_0(\tau), \qquad u = u_0(\tau). \tag{1.3.11}$$

This one-parameter (τ) family of values of x, y, and u provides initial data (at $\sigma = 0$) for the first three unknowns in (1.3.10). Now data for p and q, say, $p_0(\tau)$ and $q_0(\tau)$, are required. We obtain one equation for these functions by imposing (1.3.1) and a second equation from the data (1.3.11) themselves through differentiation. That is,

$$F(x_0(\tau), y_0(\tau), u_0(\tau), p_0(\tau), q_0(\tau)) = 0,$$
$$u_0'(\tau) = p_0(\tau)x_0'(\tau) + q_0(\tau)y_0'(\tau). \tag{1.3.12}$$

We use these equations to obtain initial values

$$p = p_0(\tau), \qquad q = q_0(\tau) \tag{1.3.13}$$

that, together with (1.3.11), provide initial data for the system of equations (1.3.10).

If we have a solution to (1.3.12) at one point, say, τ_0, then a sufficient condition that we be able to solve for these functions in a neighborhood of this point is that

$$J = \frac{\partial(F, u_0')}{\partial(p, q)} = \det \begin{bmatrix} F_p & F_q \\ x_0'(\tau) & y_0'(\tau) \end{bmatrix} \neq 0. \tag{1.3.14}$$

If this condition holds everywhere on the initial curve (1.3.11), then we can solve for the functions p_0 and q_0 all along the curve. We will assume this. In view of the differential equations (1.3.10), this equation can be rewritten as

$$J = \det \begin{bmatrix} x_\sigma & y_\sigma \\ x_0'(\tau) & y_0'(\tau) \end{bmatrix} \neq 0. \tag{1.3.15}$$

Thus, in order that a solution to (1.3.12) at one point be extendible to a solution everywhere on the initial curve, it is sufficient that the projection of the initial curve on the (x, y) plane not be a base characteristic. We remark that for (1.3.1) being truly nonlinear there may be more than one solution for p_0 and q_0 at τ_0. Each such pair for which (1.3.14) holds provides a different set of initial data for (1.3.10) and hence a different solution surface.

We solve (1.3.10) for functions

$$x = x(\sigma, \tau), \qquad y = y(\sigma, \tau), \qquad u = u(\sigma, \tau),$$
$$p = p(\sigma, \tau), \qquad q = q(\sigma, \tau). \tag{1.3.16}$$

Next, we show that at least in some neighborhood of the initial curve the values of p and q obtained here are the partial derivatives of u. To this end, we introduce the functions

$$U = \frac{\partial u}{\partial \sigma} - p \frac{\partial x}{\partial \sigma} - q \frac{\partial y}{\partial \sigma}, \qquad V = \frac{\partial u}{\partial \tau} - p \frac{\partial x}{\partial \tau} - q \frac{\partial y}{\partial \tau}. \tag{1.3.17}$$

For $J \neq 0$, if we could show that $U \equiv 0$ and $V \equiv 0$, then we could conclude that $p = u_x$ and $q = u_y$. The former of these, i.e., $U \equiv 0$, follows from (1.3.10). To check that $V \equiv 0$, we differentiate with respect to σ and calculate

$$
\begin{aligned}
\frac{\partial V}{\partial \sigma} &= \frac{\partial^2 u}{\partial \sigma\, \partial \tau} - p \frac{\partial^2 x}{\partial \sigma\, \partial \tau} - \frac{\partial p}{\partial \sigma} \frac{\partial x}{\partial \tau} - q \frac{\partial^2 y}{\partial \sigma\, \partial \tau} - \frac{\partial q}{\partial \sigma} \frac{\partial y}{\partial \tau} \\
&= \frac{\partial}{\partial \tau}\left[\frac{\partial u}{\partial \sigma} - p \frac{\partial x}{\partial \sigma} - q \frac{\partial y}{\partial \sigma} \right] + \frac{\partial p}{\partial \tau} \frac{\partial x}{\partial \sigma} + \frac{\partial q}{\partial \tau} \frac{\partial y}{\partial \sigma} - \frac{\partial p}{\partial \sigma} \frac{\partial x}{\partial \tau} - \frac{\partial q}{\partial \sigma} \frac{\partial y}{\partial \tau} \\
&= \frac{\partial U}{\partial \tau} + \frac{\partial p}{\partial \tau} F_p + \frac{\partial q}{\partial \tau} F_q + \frac{\partial x}{\partial \tau}[F_x + p F_u] + \frac{\partial y}{\partial \tau}[F_y + q F_u] \\
&= 0 + \frac{\partial x}{\partial \tau} F_x + \frac{\partial y}{\partial \tau} F_y + \frac{\partial u}{\partial \tau} F_u + \frac{\partial p}{\partial \tau} F_p + \frac{\partial q}{\partial \tau} F_q + \left[p \frac{\partial x}{\partial \tau} + q \frac{\partial y}{\partial \tau} - \frac{\partial u}{\partial \tau} \right] F_u \\
&= \frac{\partial F}{\partial \tau} - V F_u = -V F_u.
\end{aligned}
$$

The solution to this equation is

$$V(\sigma, \tau) = V(0, \tau) \exp\left[-\int_0^\sigma F_u(s, \tau)\, ds \right].$$

However, $V(0, \tau) = 0$ by the second equation in (1.3.12). Thus, $V \equiv 0$ and p and q are indeed u_x and u_y, respectively, at least in some neighborhood of the initial curve. The same condition, that $J \neq 0$ on the initial curve, also allows us to invert the first two equations in (1.3.16) to obtain both σ and τ as functions of x and y and to substitute into the third equation in (1.3.16) to obtain $u(x, y)$, again in some neighborhood of the initial curve. We remark that this solution can be continued so long as

$$J(\sigma, \tau) = \det \begin{bmatrix} x_\sigma & y_\sigma \\ x_\tau & y_\tau \end{bmatrix} \neq 0. \qquad (1.3.18)$$

Just as for the case of linear equations in Section 1.1, the method we have described is called *method of characteristics*. To recapitulate:

In order to find a solution surface u satisfying (1.3.1) and passing through the initial curve (1.3.11), we solve the system of ordinary differential equations (1.3.10) subject to the initial conditions (1.3.11) and (1.3.13), the latter being determined as a solution of (1.3.12). We obtain a solution (1.3.16). By solving for σ and τ in terms of x and y, we then find a solution surface $u(x, y)$. This process can be carried out in some neighborhood of the initial curve so long as (1.3.14) holds on the initial curve.

As with the quasi-linear case, $J = 0$ in (1.3.14) does not necessarily imply that there is no solution. Rather, it signals that the method as proposed may break down in some (perhaps nonfatal) manner. As in Section 1.1, let us first examine the situation in which $J = 0$ everywhere on the initial curve and in which the method of characteristics is to produce a well-behaved solution anyway. As a first consequence of $J = 0$ in (1.3.14),

$$\frac{x_0'(\tau)}{F_p} = \frac{y_0'(\tau)}{F_q}. \qquad (1.3.19)$$

Now we calculate

$$u_0' = p_0 x_0' + q_0 y_0' = (x_0'/F_p)[p_0 F_p + q_0 F_q].$$

Here (1.3.19) was used. We conclude from this result that

$$\frac{x_0'(\tau)}{F_p} = \frac{y_0'(\tau)}{F_q} = \frac{u_0'(\tau)}{p_0 F_p + q_0 F_q}. \qquad (1.3.20)$$

That is, just as in the quasi-linear case, the initial curve must be a characteristic curve. However, for the nonlinear equation, the data p_0 and q_0 are deduced as part of the method. Therefore, we should examine these values in this characteristic case as well. We calculate p_0' and q_0' exactly as we calculated

u_0' and conclude that these functions must satisfy the characteristic equations as well:

$$\frac{x_0'(\tau)}{F_p} = \frac{y_0'(\tau)}{F_q} = \frac{u_0'(\tau)}{p_0 F_p + q_0 F_q} = -\frac{p_0'(\tau)}{F_x + p_0 F_u} = -\frac{q_0'(\tau)}{F_y + q_0 F_u}. \quad (1.3.21)$$

Therefore, it is necessary that the initial data form a characteristic strip.

Now suppose that we pass a noncharacteristic curve through this characteristic initial strip. This is to be a new initial curve. However, at the point of intersection of the two curves, the deduced initial data for p and q must agree with the data p_0 and q_0. Then the solution surface will contain the characteristic initial strip. Thus, as in the quasi-linear case, there are still many solutions. However, now these solutions not only contain a curve, they contain a strip and therefore all meet tangentially along the original characteristic initial strip. See Fig. 1.6.

If we give up the continuity of second partial derivatives, then (1.3.4) need not be true and we cannot deduce (1.3.20). If we give up the second equation in (1.3.20), then the initial curve is not a characteristic curve, although the initial base curve is still a base characteristic curve. In this case, the normal to the surface, being normal both to the initial curve and to the characteristic curves, must be horizontal. Thus, the normal has third component zero and is not well described by a set of direction numbers with third component equal to unity. Of course, each of these anomalies can exist at a single point as well as on a segment of the initial curve. We expand our idea of solution to include all of them.

Exercises

1.5 Suppose that $J = 0$ at a point on the initial curve and the second equation in (1.3.20) is not satisfied but that there is a solution surface anyway.

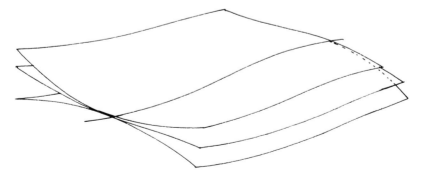

Fig. 1.6. A family of solution surfaces through a characteristic initial strip.

Show that necessarily the normal to the solution surface must be horizontal at that point.

1.6 Derive the last two equations in (1.3.21). Note the point at which (1.3.4) is used in this derivation.

1.4 EXAMPLES—THE EIKONAL EQUATION— AND MORE THEORY

We shall discuss the equation

$$p^2 + q^2 = 1, \qquad p = u_x, \quad q = u_y. \tag{1.4.1}$$

This is the eikonal equation. It arises in the analysis of the propagation of waves in many problems in mathematical physics. Perhaps the easiest to visualize is the case of shallow water waves. In that theory, the wave crests or troughs or any curves of constant phase are the level curves of

$$u(x, y) - c_0 t = \text{const}, \tag{1.4.2}$$

with t being time. As time progresses, the waves are seen to propagate in the (x, y) plane.

The eikonal equation also describes the propagation of cylindrical waves [there is no z dependence in (1.4.1)] in optics, acoustics, electromagnetics, and elasticity. The results of the analysis of this equation are usually meaningful only when the *wavelength*, the shortest distance between wave crests, is small compared to the other dimensions of the problem, such as the radius of curvature of the level curve or the distance of propagation since initiation of the wave. If a is a "typical dimension" and Λ a "typical wavelength," then this condition is

$$a/\Lambda \gg 1.^\dagger \tag{1.4.3}$$

From this description it is seen that this is a situation in which the solution is more meaningfully depicted in terms of the level curves in the (x, y) plane and therefore in terms of the *base characteristics* rather than the full characteristics. Indeed, in geometrical optics, the base characteristics are just the *rays* along which light propagates. Consequently, the terminology of optics is often borrowed in the other physical examples cited.

With the right side equal to unity in (1.4.1), the "medium" in which the waves propagate is homogeneous or uniform. In an inhomogeneous medium, the right side should be replaced by the *index of refraction*

$$n^2(x, y) = c_0^2/c^2(x, y). \tag{1.4.4}$$

---

† Read $\gg$ as much greater than; in practice, typically at least 3.

Here c_0 is a constant "reference speed" (e.g., the speed of light in vacuum) and $c(x, y)$ the variable speed of the medium.

For (1.4.1),

$$F(x, y, u, p, q) = p^2 + q^2 - 1. \tag{1.4.5}$$

The equations of the characteristics (1.3.10) are

$$\frac{dx}{d\sigma} = 2p, \quad \frac{dy}{d\sigma} = 2q, \quad \frac{du}{d\sigma} = 2, \quad \frac{dp}{d\sigma} = 0, \quad \frac{dq}{d\sigma} = 0. \tag{1.4.6}$$

From the first two equations here we conclude that the tangent direction to the base characteristic at each point is the direction of the normal (p, q) to the level curve of u. That is, the base characteristics, or rays, are the orthogonal trajectories to the wave fronts. This is true even for a variable index of refraction. From the last two equations here we conclude that p and q are constant on the rays (this is true only for the homogeneous case) and that therefore the rays are straight lines.

The solution to the system of equations (1.4.6) is

$$x = x_0 + 2p_0\sigma, \quad y = y_0 + 2q_0\sigma, \quad u = u_0 + 2\sigma,$$
$$p = p_0, \quad q = q_0, \quad p_0^2 + q_0^2 = 1. \tag{1.4.7}$$

We remark that along each ray the equation for u here is exactly of the form of (1.4.2), with σ playing the role of a scaled time and u_0 independent of that time like variable. Thus, the concept of propagation of a wave may be identified with movement along rays from one level curve to the next. By eliminating p_0, q_0, and σ, we can rewrite this solution as

$$(x - x_0)^2 + (y - y_0)^2 = (u - u_0)^2. \tag{1.4.8}$$

This is a three-parameter family of solutions to (1.4.1). In fact, it is the *general solution* to (1.4.1).

Digression (the General Solution)   With this problem as a model, it is perhaps easier to see that (1.3.1) will always have a general solution with three parameters. The five differential equations (1.3.9) will have, as a general solution in terms of σ, five functions of σ and initial values, that is, five constants of integration. These five equations and the original partial differential equation (1.3.1) are six equations relating the eleven variables, x, y, u, p, q, x_0, y_0, u_0, p_0, q_0, and σ. We use five of these equations to evaluate five of the variables in terms of the other six and substitute into the sixth equation. This is then one equation in six unknowns, preferably, x, y, u, and any three of the parameters. This is a three-parameter family of solutions. Thus, a general solution to the set of characteristic equations leads to a three-parameter family of solutions of the original differential equation.

Digression (the Conoidal Solution)   If the parameters x_0, y_0, and u_0 in (1.4.7) are taken to be fixed constants, then the solution we obtain is a right circular cone. That a conic surface solution exists is also a typical feature of the general equation (1.3.1). At a fixed point, (1.3.1) is a relation between p and q. All of the normals satisfying this relationship at that point form a cone. The tangent planes associated with these normals have an envelope that is also a cone whose generators are at right angles to the generators of the previous cone. See Fig. 1.7. The continuation of this cone forms a solution that is called the *conoidal solution*.

The question arises as to how one finds this conoidal solution for the more general problem. One starts from the general solution to (1.3.10), fixing the initial values of x, y, and u so that the cone starts at the desired point. One then imposes the original equation (1.3.1) on the initial data to obtain a functional relationship between p_0 and q_0 and thus defines these variables as a function of a single parameter (which may be one of them). The representation of x, y, and u in terms of this parameter and σ is the conoidal solution. It is to be expected that this solution will not have first partial derivatives at the apex of the cone. Indeed, this can be seen for the specific example (1.4.8).

Now let us suppose that we seek a solution passing through the initial curve (1.3.11), x_0, y_0, u_0. The equations (1.3.12) for p_0 and q_0 are

$$p_0^2 + q_0^2 = 1, \qquad p_0 x_0' + q_0 y_0' = u_0'. \tag{1.4.9}$$

Here the condition $J \neq 0$ in (1.3.18) is

$$\det \begin{bmatrix} p_0 & q_0 \\ x_0' & y_0' \end{bmatrix} \neq 0. \tag{1.4.10}$$

At each point on the initial curve at which we can solve (1.4.9), we find two

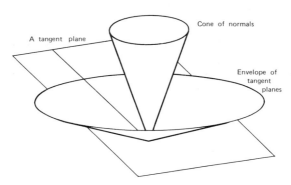

Fig. 1.7. The cone of normal directions and the cone formed as the envelope of tangent planes.

solutions, and (1.4.10) will ensure that for each of them the base characteristic (or ray) is not parallel to the base initial curve.

Let us set

$$p_0 = \cos \alpha(\tau), \qquad q_0 = \sin \alpha(\tau),$$

$$x_0'/\sqrt{x_0'^2 + y_0'^2} = \cos \theta(\tau), \qquad y_0'/\sqrt{x_0'^2 + y_0'^2} = \sin \theta(\tau). \tag{1.4.11}$$

Then

$$u_0'/\sqrt{x_0'^2 + y_0'^2} = \cos(\theta(\tau) - \alpha(\tau)). \tag{1.4.12}$$

Thus,

$$|u_0'| \leq \sqrt{x_0'^2 + y_0'^2}. \tag{1.4.13}$$

When equality holds, $\alpha = \theta$ or $\theta + \pi$; the base characteristics are tangent to the initial curve; and $J = 0$, which we have assumed, for the present is not true. If

$$|u_0'| > \sqrt{x_0'^2 + y_0'^2}, \tag{1.4.14}$$

no real solution to (1.4.9) exists and no real solution surface exists. This is not a contradiction of our theory in that there must be a solution p_0 and q_0 at a point before we can even think of extending that solution to the entire initial curve.

We now continue with strict inequality holding in (1.4.13). Now, given a solution $\alpha_1(\tau), 0 < \alpha_1 < \pi$, then $\alpha_2 = 2\theta(\tau) - \alpha_1$ is also a solution. Note that the *relative angles* $\theta(\tau) - \alpha_1$ and $\theta(\tau) - \alpha_2$ are negatives of one another. See Fig. 1.8. Thus, if one ray family is associated with waves that are *incident* on the initial curve, the other ray family is associated with the rays that are *reflected* from the initial curve.

In fact, we now contemplate the following problem. Let us suppose that u does represent some wave propagating in a region having a boundary curve $x_0(\tau), y_0(\tau)$. For this incident wave, the objective is to find the reflected wave, with u for both the incident and reflected waves being solutions of the

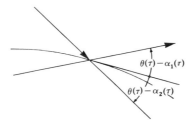

Fig. 1.8. The two ray directions at a point on the initial curve.

eikonal equation on the same side of the boundary curve. On physical grounds, one can argue that the phase of the incident and reflected waves should be the same at the boundary. That is, u_0 is taken to be the value of this incident wave at the boundary curve. Necessarily, one of the solutions for p_0 and q_0 will have to be the values associated with the incident wave. The other pair then are the values for the reflected wave. If we used the same pair for the reflected wave, then the effect would be to produce a solution that propagated through the boundary as time, or σ, increased and that would be unphysical.

The representation of the solution by level curves in the (x, y) plane is fairly straightforward. At each point on the initial curve, the base characteristic is a straight line with direction numbers p_0 and q_0, and the level curves are orthogonal to the characteristics. See Fig. 1.9. We remark that the solution surface $u(x, y)$ depicted in three-space will make an angle of $45°$ with the horizontal by virtue of (1.4.1) and the fact that the third direction number is -1. For the motivating example, this depiction is, admittedly, of less interest than the representation in Fig. 1.9.

We shall now present two examples for the eikonal equation (1.4.1). In the first an explicit inversion of the parametric representation will be practical, while in the second it will not be.

For the first example, we set

$$x_0(\tau) = u_0(\tau) = \frac{\cos \tau}{1 - \cos \tau}, \qquad y_0(\tau) = \frac{\sin \tau}{1 - \cos \tau}. \qquad (1.4.15)$$

The base initial curve here is the parabola $y^2 = 2x + 1$, which opens to the right with the focal point at the origin. The solutions to (1.4.9) for the initial data for p_0 and q_0 are

$$p_0 = 1, \qquad q_0 = 0, \qquad \text{or} \qquad p_0 = -\cos \tau, \qquad q_0 = -\sin \tau. \qquad (1.4.16)$$

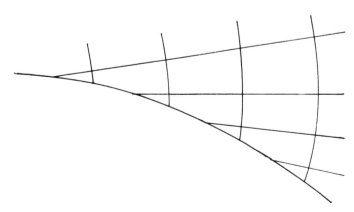

Fig. 1.9. Representation of a solution in terms of characteristics and level curves.

By using the solution formulas (1.4.7), we find for the first pair of values here that

$$x(\sigma, \tau) = u(\sigma, \tau) = \frac{\cos \tau}{1 - \cos \tau} + 2\sigma, \qquad y(\sigma, \tau) = \frac{\sin \tau}{1 - \cos \tau} \qquad (1.4.17)$$

while the second set of data produces the solution

$$x(\sigma, \tau) = \frac{\cos \tau}{1 - \cos \tau} + 2\sigma \cos \tau, \qquad y(\sigma, \tau) = \frac{\sin \tau}{1 - \cos \tau} + 2\sigma \sin \tau,$$

$$(1.4.18)$$

$$u(\sigma, \tau) = \frac{\cos \tau}{1 - \cos \tau} + 2\sigma,$$

In (1.4.17), an explicit solution for u in terms of x and y is given by the first equation. In (1.4.18), the sum of the squares of x and y yields the square of u. Thus, the two explicit solutions are

$$u = x, \qquad u = \sqrt{x^2 + y^2} - 1. \qquad (1.4.19)$$

In the (x, y) plane, the level curves of the first solution are straight lines parallel to the y axis, while the level curves of the second solution are circles. If u represents the spatial part of the phase (1.4.2), then the first solution describes a wave whose crests are straight lines while the second solution describes a wave whose crests are circles. See Fig. 1.10.

Let us consider a wave field that emanates from the origin. The wave fronts of this wave are indeed described by the second solution in (1.4.19).

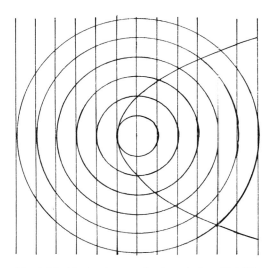

Fig. 1.10. The level curves for the solutions (1.4.19).

When the wave reflects from the parabolic boundary, it produces the first solution as the reflected wave. We have already seen that the base characteristics, or rays, for these two waves make equal angles with the reflector. In this case, we see that the reflected wave has parallel rays, all with direction numbers $(1, 0)$, the first solution in (1.4.16).

This is a well-known result for parabolic reflectors, namely, that for a point source at the focal point of the reflector, the reflected wave is a parallel beam.

Of course, we can also interpret our result in terms of the exterior problem. Here the rays of the first solution in (1.4.9) are directed toward the parabola from the left while the rays of the second solution are directed away from the exterior of the parabola. Thus, we can think here of a parallel beam incident on the exterior of the parabola producing a reflected wave that is circular.

For some applications, the explicit solutions (1.4.19) are certainly more useful. However, for others, the representations (1.4.17) and (1.4.18) in terms of rays and propagation of the phase along those rays will prove more useful. For a more complete understanding, both are desirable.

Again, we remark on the representation in three-space. The first solution surface in (1.4.19) is a plane through the y axis making an angle of $45°$ with the x axis, while the second is a right circular cone–conic angle $45°$–through the origin.

As a second example, we take the initial data

$$x_0 = \tau, \qquad y_0 = 0, \qquad u_0 = \tfrac{1}{2}\arctan \tau. \tag{1.4.20}$$

For this initial data we find the solutions of (1.4.7) to be

$$p_0 = \tfrac{1}{2}\frac{1}{1 + \tau^2}, \qquad q_0 = \pm\left[1 - \frac{1}{4(1 + \tau^2)^2}\right]^{1/2}. \tag{1.4.21}$$

In terms of these values, the solution (1.4.7) is

$$x(\sigma, \tau) = \tau + 2p_0\sigma, \qquad y(\sigma, \tau) = 2q_0\sigma,$$
$$u(\sigma, \tau) = \tfrac{1}{2}\arctan \tau + 2\sigma. \tag{1.4.22}$$

In order to eliminate σ and τ here, we must solve a quartic equation. Such a solution would not be particularly edifying. On the other hand, a qualitative depiction of the solution as level curves in the (x, y) plane is possible. We see from (1.4.21) that the ray directions are symmetric about the origin on the initial line. Furthermore, the ray making the smallest angle with the x axis passes through the origin. In the limit as $\tau \to \infty$, the ray directions become vertical. Figure 1.11 is a qualitative depiction of the solution with upper sign for q_0.

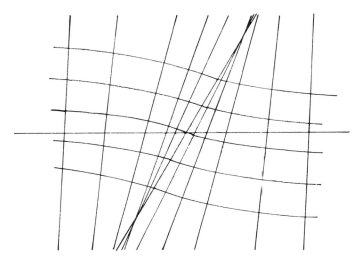

Fig. 1.11. The level curves and rays for the solution (1.4.22).

It can be seen in Fig. 1.11 that the base characteristics cross. In fact, they form one envelope above the x axis and another symmetrically placed with respect to the origin. The base characteristics initiated from the positive x axis remain to the right of the upper envelope; those from the negative x axis stay to the left of the lower envelope. The envelope is called an *edge of regression* or a *caustic*. Along this curve, $J = 0$. The verification of this is outlined in the exercises. However, the caustic is not a base characteristic in this example

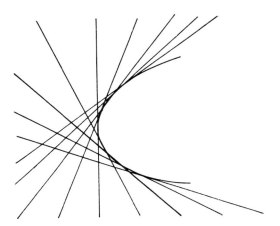

Fig. 1.12. An edge of regression or a caustic.

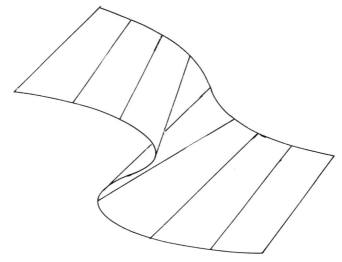

Fig. 1.13. Solution surface depiction for caustic anomaly.

because the base characteristics must be straight. Therefore, the solution surface must be horizontal above the caustic according to Exercise 1.9. Figure 1.12 is a larger depiction of an edge of regression, and Fig. 1.13 depicts a solution surface with the same anomaly.

Exercises

1.7 Suppose that U is a solution of the equation

$$U_{xx} + U_{yy} - (1/c^2)U_{tt} = 0.$$

Introduce $\xi = \phi(x, y) - c_0 t$ and any other two functions of x, y, and t as new independent variables of integration. Show that the coefficient of $U_{\xi\xi}$ is

$$(\nabla)^2 \phi - n^2,$$

with n^2 defined by (1.4.4). If ϕ satisfies the eikonal equation, what effect will discontinuities in $U_{\xi\xi}$ have on the solution?

1.8 Let U be a solution of

$$U_{xx} + U_{yy} + (\omega^2/c^2)U = 0.$$

Assume a solution of the form

$$U = A(x, y, \omega) \exp(i\omega\phi/c_0),$$

with A being a power series in inverse powers of ω whose coefficients are functions of x and y. Substitute this form into the equation for U. Conclude that the leading order term is of order ω^2 for "large" ω and that this term will be zero if ϕ satisfies the eikonal equation, with the right side n^2 as given by (1.4.4).

1.9   Suppose that a set of base characteristics is given by

$$x = f(\sigma, \tau), \qquad y = g(\sigma, \tau).$$

Then we solve for σ in the first equation and substitute into the second to obtain y as a function of x and τ. The envelope of this set of curves is defined by this equation and the equation obtained by setting its τ derivative equal to zero. Show that if this latter equation is satisfied, then $J = 0$. Here J is the Jacobian of the transformation from (x, y) to (τ, σ).

1.5   PROPAGATION OF WAVE FRONTS

We shall develop here some phenomena of wave propagation as deduced from the eikonal equation in its more general form

$$p^2 + q^2 = n^2(x, y). \tag{1.5.1}$$

Let us suppose that a wave front from the left in Fig. 1.14 is incident on the boundary curve

$$x = x_0(\tau), \qquad y = y_0(\tau) \tag{1.5.2}$$

depicted in the figure. Here it is assumed that the medium is such that

$$n^2(x, y) = \begin{cases} 1, & \text{on the left of the boundary curve,} \\ c_0^2/c_1^2, & \text{on the right of the boundary curve.} \end{cases} \tag{1.5.3}$$

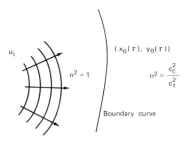

Fig. 1.14. A wave incident from the left on an interface across which n^2 changes.

Although we have already introduced the reflected wave in the preceding section, we shall repeat that discussion here as part of this more general discussion of propagation. Let us denote the wave incident at the boundary by u_I. We anticipate that this wave will give rise to two other waves, one reflected back into the left medium and denoted by u_R and another *transmitted* into the right medium and denoted by u_T.

On physical grounds, one can argue that the three functions u_I, u_R, and u_T have to be equal at the boundary. That is,

$$u_{0I}(\tau) \equiv u_I(x_0(\tau), y_0(\tau)) = u_{0R}(\tau) = u_{0T}(\tau). \tag{1.5.4}$$

Thus, the initial values of the reflected and transmitted waves are equal to the (assumed known) value of the incident wave. Differentiation of these equations provides equations for the initial values of p and q, namely,

$$p_{0R}x_0' + q_{0R}y_0' = u_{0I}', \qquad p_{0T}x_0' + q_{0T}y_0' = u_{0I}'. \tag{1.5.5}$$

The left sides here can be interpreted as dot products of the vectors (p, q) for each wave with a tangent to the boundary. The right sides have the same interpretation. That is, it also follows from the definition of u_{0I} in (1.3.12) that

$$p_{0I}x_0' + q_{0I}y_0' = u_{0I}'. \tag{1.5.6}$$

Thus, these equations state that for all three waves the projection of (p, q), the ray directions, make equal angles with the tangent to the boundary curve at each point. We introduce α_I, α_R, and α_T by setting

$$p_{0I} = \cos \alpha_I, \qquad q_{0I} = \sin \alpha_I,$$
$$p_{0R} = \cos \alpha_R, \qquad q_{0R} = \sin \alpha_R, \tag{1.5.7}$$
$$p_{0T} = n \cos \alpha_T, \qquad q_{0T} = n \sin \alpha_T.$$

Here α_I is known. Note that in the last equation, we have introduced the scale n, which by (1.5.3) is no longer equal to unity since u_T is a solution on the *right* of the boundary curve. Introduce θ as in (1.4.11) to describe the direction numbers for the boundary curve and conclude as in that discussion that

$$\alpha_R = 2\theta - \alpha_I. \tag{1.5.8}$$

Furthermore, as noted in the discussion in Section 1.4, the incident ray direction and the reflected ray direction make equal angles with the tangent to the boundary curve.

To determine α_T, we rewrite the third line in (1.5.7) as

$$n \cos(\theta - \alpha_T) = \cos(\theta - \alpha_I). \tag{1.5.9}$$

This equation is Snell's law, and the wave we are calling the transmitted wave is actually the refracted wave. Usually, Snell's law is written in terms of an angle of incidence i and an angle of refraction r with respect to the normal to the boundary, and then

$$\sin i / \sin r = n.$$

Equation (1.5.9) will have real solutions for α_T so long as

$$\cos(\theta - \alpha_1) \le n. \tag{1.5.10}$$

We proceed for the moment under the assumption that strict inequality holds here. In this case, (1.5.9) admits two solutions: one directed into the right medium, the other not; we choose the former, and identify this ray as the *transmitted ray*.

We have now determined the initial values of x_0, y_0, u_0, p_0, and q_0 for both the reflected and transmitted waves. Thus, for each of these, determination of the function describing the equiphase curves is reduced to the solution of the characteristic equations (1.3.10) with known initial data.

Let us suppose now that there is some τ_0 such that for $\tau < \tau_0$ strict inequality holds in (1.5.10) but equality holds at τ_0 and (1.5.10) fails for larger values of τ. First, for $\tau > \tau_0$ there is no transmitted wave since (1.5.9) will have no real solution. We say them that the incident wave is *totally reflected*, and the reflected and transmitted rays initiated at $(x_0(\tau_0), y_0(\tau_0))$ are called the *critically reflected* ray and the *critically transmitted* or *critically refracted* ray, respectively. See Fig. 1.15.

Right at τ_0, with equality holding in (1.5.10), the solution to (1.5.9) is that $\alpha_T = \theta$. That is, the critically transmitted ray is directed tangent to the boundary. For this case, $J = 0$ in (1.3.14). Also, by (1.5.4) and (1.5.5), u_{0T} is the proper data to make the initial curve a characteristic curve at that point.

Let us now contemplate a new set of initial data initiating from that point and remaining characteristic all along the boundary curve in the direction of

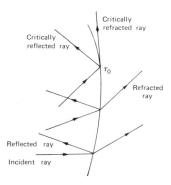

Fig. 1.15. A critically reflected ray and a critically transmitted (refracted) ray.

$(p_{0T}(\tau_0), q_{0T}(\tau_0))$. That is, let us contemplate that because this transmitted ray is initiated tangent to the boundary, it actually "splits" into two rays, one the ordinary transmitted ray and the other a ray that "clings" to the boundary. This latter ray is called the *creeping ray*, and the wave associated with it is called the *creeping wave*. For this new ray to continue along the boundary, J as defined by (1.3.14) must remain zero. Let us denote this new data set by subscript C (for creeping, critical, or characteristic). Then on the boundary curve, we require for u_{0C} that

$$p_{0C} = \frac{nx_0'}{\sqrt{x_0'^2 + y_0'^2}}, \qquad q_{0C} = \frac{ny_0'}{\sqrt{x_0'^2 + y_0'^2}}. \tag{1.5.11}$$

With these values of p_{0C} and q_{0C} defined, we find u_{0C} by setting

$$u_{0C}' = x_0' p_{0C} + y_0' q_{0C} = n\sqrt{x_0'^2 + y_0'^2}. \tag{1.5.12}$$

Integration of this equation yields the creeping wave. The initial data for this equation are prescribed at the critical point τ_0 to be

$$u_{0C}(\tau_0) = u_{0I}(\tau_0). \tag{1.5.13}$$

We now consider this creeping wave as new boundary data for the eikonal equation in both media. If the boundary curves toward the left medium as in Fig. 1.16, then at each $\tau > \tau_0$ a tangential transmitted ray is initiated and produces another wave field in the right medium. This wave phenomenon is called *smooth-body diffraction*. If the boundary is straight or curved oppositely, this family of rays and its associated wave field will not occur. However, these data also provide initial data for a new wave field in the left medium. For this wave, the rays make the same angle with the initial curve at each point as does the reflected ray at τ_0. The verification of this is left to the exercises. When the boundary curve is straight, this family of rays is therefore parallel. Hence, the wave fronts are straight lines. This wave is called the *lateral wave* or *head wave*.

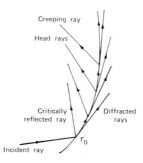

Fig. 1.16. Creeping rays, smooth-body diffracted rays, and head rays.

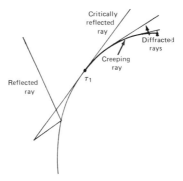

Fig. 1.17. Creeping mode and smooth-body diffracted wave in the left medium.

Let us suppose now that the curvature is the reverse of what it was in Figs. 1.14–1.16 and that there is a point on the boundary, say, with $\tau = \tau_1$, at which the incident ray is tangent to the boundary. Thus, $\alpha_1 = 0$, and from (1.5.8) $\alpha_R = 0$ as well. Now it follows from (1.5.5) and (1.5.6) that the data u_{0R} are such that the boundary curve is a characteristic curve at this point. As in the preceding discussion, we allow this tangent reflected ray to split into two rays: one the straight ray satisfying the differential equations (1.3.10) and the other the boundary curve itself. Of course, for this latter ray, the wave will propagate only if we make u along that ray be characteristic data and satisfy (1.5.12). Thus, we generate another type of creeping wave, in the right medium this time. Associated with this boundary data will be another smooth-body diffracted wave field. See Fig. 1.17.

We shall consider now the problem of the *scattering of a plane wave by a circular cylinder*. By *scattering* we generally mean everything but the incident wave. Here we shall limit the discussion to a description of the reflected, creeping, and smooth-body diffracted waves. The use of the terms *plane wave* and *circular cylinder* is a consequence of viewing this problem as a three-dimensional problem in which there is no dependence on the third variable. This is traditional, and the term is applied even when the problem being modeled really is two-dimensional, such as in the propagation of a small-amplitude surface wave in water.

We suppose that in the region of the plane exterior to a circle of radius a, u satisfies the eikonal equation. We shall not consider the transmission problem in the interior of the cylinder. All wave fields are to satisfy (1.4.1). We suppose that the incident wave is generated far away on the left and propagates parallel to the x axis toward the scatterer. Thus, within a constant, we shall take

$$u_1 = x. \qquad (1.5.14)$$

However, some care is necessary here. Those rays propagating from the left with $|y| \le a$ will be incident on the scatterer, the cylinder. On the other hand, rays for which $|y| > a$ will pass above or below the cylinder. In this manner, the geometrical optics shadow is generated.

We introduce as the parameter τ the polar angle in the (x, y) plane. Thus,

$$u_{0I} = x_0 = a \cos \tau, \qquad y_0 = a \sin \tau, \qquad \pi/2 < \tau < 3\pi/2. \quad (1.5.15)$$

We remark that for this parameterization, θ, defined by (1.4.11), is related to τ by

$$\theta = \tau - (3\pi/2). \quad (1.5.16)$$

Because of this simple relationship, we will not use θ at all.

We require that u_R satisfy (1.5.4) in the τ range prescribed in (1.5.15). The equation (1.5.5) for p_{0R} and q_{0R} now becomes

$$\sin(\tau - \alpha_R) = \sin \tau. \quad (1.5.17)$$

Here α_R is defined by (1.5.7). One solution, namely, $\alpha_R = 0$, would yield the incident wave. Therefore, we choose the other solution of (1.5.7):

$$\alpha_R = 2\tau - \pi. \quad (1.5.18)$$

Then from (1.5.7),

$$p_{0R} = -\cos 2\tau, \qquad q_{0R} = -\sin 2\tau. \quad (1.5.19)$$

This equation and (1.5.15) provide all of the data for the solution formula (1.4.7), and we obtain the parametric representation of the reflected wave

$$x = a \cos \tau - 2\sigma \cos 2\tau, \qquad y = a \sin \tau - 2\sigma \sin 2\tau,$$
$$u_R = a \cos \tau + 2\sigma. \quad (1.5.20)$$

The points of the cylinder at which $\tau = \pi/2$ or $3\pi/2$ are points at which the incident and reflected ray directions are tangent to the cylinder and the data $u_{0I} = u_{0R}$ are characteristic data. We shall develop the creeping and smooth-body diffracted waves for the critical point at which $\tau = 3\pi/2$. Using (1.5.11)–(1.5.13) and (1.5.15), we find that

$$p_{0C} = -\sin \tau, \qquad q_{0C} = \cos \tau, \qquad u_{0C} = a[\tau - (3\pi/2)], \qquad \tau > 3\pi/2. \quad (1.5.21)$$

This value becomes boundary data for a new field, the smooth-body diffracted field, which we shall call u_{0D}. We use this initial data along with the definition of the boundary curve in (1.5.15) in (1.4.7) and find that

$$x = a \cos \tau - 2\sigma \sin \tau, \qquad y = a \sin \tau + 2\sigma \cos \tau, \quad (1.5.22)$$
$$u_{0D} = a[\tau - (3\pi/2)] + 2\sigma. \quad (1.5.23)$$

We remark that for the tangent ray at $\tau = \pi/2$, the vectors (p_{0I}, q_{0I}) and (x_0', y_0'), are antiparallel. Thus, we must adjust (1.5.11) accordingly in developing the creeping wave originating at this point. Clearly, the only necessary adjustment is to reverse the signs in that equation so that (p_{0C}, q_{0C}) and (p_{0I}, q_{0I}) are colinear at the initiation of u_{0C}. We leave as an exercise the determination of u_{0C} and u_{0D} originating from $\pi/2$. Figure 1.18a depicts the rays for this problem, and Fig. 1.18b depicts the level curves of u_1, u_R, and u_D. The ray families we have described here are those that would be developed when using geometrical optics and the geometrical theory of diffraction to analyze the problem of scattering by a circular cylinder. Complete analysis of this problem is quite deep. However, an understanding of these ray families in advance is certainly an aid to that analysis.

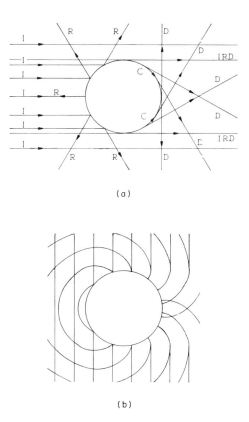

(a)

(b)

Fig. 1.18. (a) Ray families for scattering by a circular cylinder and (b) wave fronts for scattering by a circular cylinder.

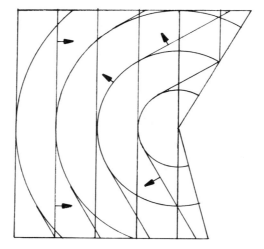

Fig. 1.19. Diffraction from a point.

Let us turn now to another type of diffraction phenomenon and consider a boundary that has a discontinuity in it, such as in Fig. 1.19. We remark that the reflected fields generated from the two sides of the discontinuity will continue to exist, as in the preceding discussion. However, they will end abruptly on the base characteristics emanating from the two edges of the discontinuity. We anticipate that this point on the interface will provide a new source point for wave generation. As a single point at which u is prescribed (to be u_1), this must be the initial point of a conoidal wave. In the diagram, one can see how this new type of diffracted wave (edge diffraction) connects the two disjoint reflected waves.

This ends our discussion of scattering phenomena. Through extended solution techniques for the eikonal equation, we have described the major primary scattering effects. Of course, each of these waves may, in turn, be rescattered and the analysis may be continued to treat those waves as well.

The rays generated by the various phenomena just described are consistent with Fermat's principle in a generalized form, which can be stated as follows. Define the travel time t between two points (x_0, y_0) and (x_1, y_1) by

$$t = \int_{s_0}^{s_1} n(x(s), y(s)) \, ds, \qquad (1.5.24)$$

where s is an arc-length variable and

$$x(s_0) = x_0, \qquad y(s_0) = y_0, \qquad x(s_1) = x_1, \qquad y(s_1) = y_1. \qquad (1.5.25)$$

Then the actual ray path connecting these two points is a *local minimum* among all ray paths connecting these two points. Thus, in a homogeneous medium:

(1)   The direct path connecting two points is a straight line.
(2)   The path of the reflected ray is the extremum among all paths that touch the boundary at one point.
(3)   Snell's law provides the minimum refracted ray path.
(4)   The head-ray path is the minimum among paths that propagate to the interface, then in the second medium and back again in the first medium.
(5)   The smooth-body diffracted-ray path is the minimum among paths that connect two points lacking a line-of-sight path.

Exercises

1.10   The purpose of this exercise is to determine the angle of the head rays with the boundary curve. Introduce the wave u_H in the left medium having as boundary data the creeping wave data u_{OC}. Introduce p_H, q_H, and α_H as in the preceding discussion. Show that the equation for α_H is

$$\cos(\theta - \alpha_H) = n$$

and that at the critical point the angle of reflection satisfies

$$\cos(\theta - \alpha_R) = n.$$

What can you conclude about the relationship between the angle of reflection at the critical point and the angle that each head ray makes with the boundary curve?

1.11   The objective of this exercise is to carry out for an explicit case the analysis of the preceding discussion to develop the representations of u_1, u_R, u_T, u_C, and u_H. (The complete ray family for this exercise is shown in Fig. 1.20.) Let us suppose in that discussion that the boundary is the line $x = 0$ and that $n < 1$. (Why is there no smooth-body diffracted wave in this example?)

(a)   Let u_1 be a circular wave (the conoidal solution) originating at the point $(-L, 0)$ in the (x, y) plane. Let τ be the angle that the incident rays make with the x axis. Derive the following parametric representation for u_1:

$$x = -L + 2\sigma \cos \tau, \qquad y = 2\sigma \sin \tau, \qquad u_1 = 2\sigma,$$

and eliminate the parameters to confirm that this is indeed a circular wave centered at $(-L, 0)$.

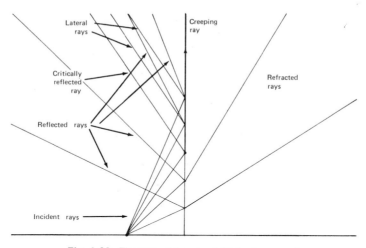

Fig. 1.20. The rays of Exercise 1.11 for $\tan \tau_0 = \frac{3}{2}$.

(b)   Show that for u_R

$$p_{OR} = -\cos \tau, \qquad q_{OR} = \sin \tau$$

and

$$x = -2\sigma \cos \tau, \qquad y = L \tan \tau + 2\sigma \sin \tau, \qquad u = L \sec \tau + 2\sigma.$$

Eliminate the parameters here to verify that this is a circular wave centered at the image point $(L, 0)$.

(c)   Show that for the transmitted wave u_T,

$$p_{OT} = \sqrt{n^2 - \sin^2 \tau}, \qquad q_{OT} = \sin \tau$$

and

$$x = 2\sigma\sqrt{n^2 - \sin^2 \tau}, \qquad y = L \tan \tau + 2\sigma \sin \tau, \qquad u = L \sec \tau + 2\sigma.$$

(d)   Verify that there are two critical values of τ for which the refracted ray is tangent to the boundary (vertical) and that they are given by

$$\sin^2 \tau_0 = n^2.$$

Because of the symmetry of the problem, continue with only the positive root. Next, show that

$$u'_{OC} = nL \sec^2 \tau$$

and that

$$u_{OC}(\tau) = nL \tan \tau + L\sqrt{1 - n^2}.$$

(e) To determine u_{0H}, first verify that

$$p_{0H} = -\sqrt{1 - n^2}, \qquad q_{0H} = n,$$

and then show that

$$x = -2\sigma\sqrt{1 - n^2}, \qquad y = L \tan \tau + 2\sigma n,$$

$$u_H = nL \tan \tau + L\sqrt{1 - n^2} + 2\sigma.$$

Finally, eliminate the parameters to obtain

$$u_H = -x\sqrt{1 - n^2} + yn + L\sqrt{1 - n^2}.$$

1.12 The objective of this exercise is to develop the wave fields for scattering by an edge. Suppose that $n = 1$ everywhere in the (x, y) plane but that there is a boundary that consists of the half line $x = 0$, $y < 0$. Suppose further that there is an incident plane wave from the upper left with direction $(p, q) = (2/\sqrt{5}, -1/\sqrt{5})$.

(a) Show that there is a ray associated with u_1 such that below this ray all rays of u_1 are incident on the boundary while above this ray all rays miss the boundary entirely. This ray defines the *shadow boundary* of the incident field.

(b) Find the reflected wave and show that it, too, has a shadow boundary.

(c) Find the edge-diffracted wave from the point $(0, 0)$. See Fig. 1.21.

1.13 Consider an ellipse with major axis along the x axis. Next, consider a source and receiver of light at one focal point of the ellipse. Assume that the medium is homogeneous. Show that there are two reflected rays that contact the ellipse on the x axis and that for the one nearer the source/receiver point, the arc length (1.5.24) is a minimum while for the other it is a maximum.

Fun and Games

Many of these waves can be generated with simple equipment. Fill a cookie pan with about 1 cm of water. Use a piece of wood across the narrower dimension of the pan to generate a wave in the pan with a short, firm push. A second board that stands higher than the depth of the water and is held in place serves as a reflector. If the scatterer is shorter than the width of the pan, an edge-diffracted wave will be visible behind it. Use a can about the size of a tuna fish can as the scatterer, and observe the reflected waves as well as the smooth-body diffracted waves in the geometrical shadow. Somewhat more difficult to see is the refracted wave. Submerge a thin board to decrease the depth of the water in half of the cookie pan. Use the source generator in the deeper part to send a wave at an angle toward the shallower part of the pan. Believers can see the refracted wave generated in the shallower part of the pan.

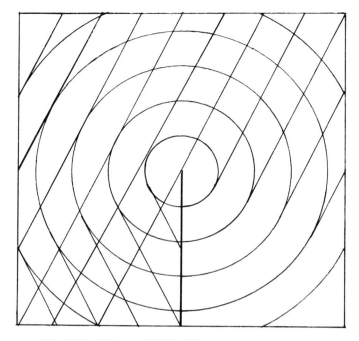

Fig. 1.21. The wave field for diffraction by a straight edge.

1.6 VARIABLE INDEX OF REFRACTION

In this section, we shall make a few remarks about the case of variable n for the eikonal equation

$$p^2 + q^2 = n^2(x, y) \tag{1.6.1}$$

and discuss one example. First, we shall write the characteristic equations with $\lambda = 1/2n$ and distinguish the independent parameter for this case by denoting it by s rather than by σ. This is the choice of λ–noted in (1.1.8)–for which the parameter s is arc length along the base characteristics, or rays. In vector form, we write the characteristic equations as

$$\dot{\mathbf{x}} = \mathbf{p}/n, \qquad \dot{u} = n, \qquad \dot{\mathbf{p}} = \nabla n,$$
$$\mathbf{x} = (x, y), \qquad \mathbf{p} = (p, q), \qquad (\dot{\ }) = d/ds. \tag{1.6.2}$$

By differentiating the equation for $\mathbf{x}$ here, we obtain the equation

$$\ddot{\mathbf{x}} = (1/n)[\nabla n - \dot{\mathbf{x}}(\nabla n \cdot \dot{\mathbf{x}})]/n. \tag{1.6.3}$$

The left side—the second derivative of the position vector with respect to arc length—is the *curvature vector* for the rays. When the initial point of this vector is at the point **x**, the terminal point is at the center of a circle making second-order contact with the ray. The right side of this equation is the gradient of n minus its projection on the tangent to the ray and hence is that part of the gradient that is perpendicular to the ray. Thus, the ray "turns" toward the normal component of the gradient of n, and the gradient of n lies in the plane of the tangent and *principal normal* of the ray.

For shallow water waves of small amplitude

$$n = \text{const}\{gh\}^{-1/2}. \tag{1.6.4}$$

Here g is the acceleration due to gravity and h the depth of the water. Thus, the direction of increasing n is the direction of decreasing depth. Consequently, the rays tend to turn toward shallower water, and the wave fronts then tend to become parallel to the level curves of h. This means that wave fronts that are propagating obliquely toward a beach will tend to turn parallel to the beach and "run up," ultimately breaking after this simple linear model breaks down. Aerial photographs demonstrating this phenomenon can be found in Stoker [1957]. To continue the fun and games from the preceding section, the reader can refer to a photograph in Stoker [1957] of an experiment with a lens-shaped obstacle submerged in a pan of water.

We remark that the vector form of the equations is equally valid in three or higher dimensions and hence so is the conclusion.

AN EXAMPLE WITH VARIABLE n

We consider now a two-dimensional example in which n is a function of x alone and monotonic, as is Fig. 1.22. We shall consider here the conoidal solution emanating from the origin. Because of the symmetry in y, we will only discuss $y \geq 0$.

We anticipate that rays will tend to turn toward the right, toward increasing n. Thus, rays starting out to the right will continue in that direction.

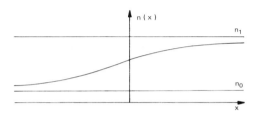

Fig. 1.22. A monotonic index of refraction with left limit n_0 and right limit n_1.

However, for rays that start out toward the left, it is not quite so straight-forward. This tendency to turn may actually cause these rays to change direction (toward the right) entirely. To see how this occurs, let us look more carefully at the characteristic equations. To do so. we will not use the representation above with arc length as parameter, but instead we choose λ in (1.3.9) to be $\frac{1}{2}$ and rewrite the characteristic equations (1.3.9) as

$$\frac{dx}{d\sigma} = p, \qquad \frac{dy}{d\sigma} = q, \qquad \frac{du}{d\sigma} = n^2, \qquad \frac{dp}{d\sigma} = n\frac{dn}{dx}, \qquad \frac{dq}{d\sigma} = 0. \quad (1.6.5)$$

Observe that the last of these equations implies that q is a constant, which allows us to quickly determine p as well from the eikonal equation itself. Thus,

$$q = q_0, \qquad p = \pm\sqrt{n^2 - q_0^2}. \quad (1.6.6)$$

For rays in the upper half plane, q_0 must be nonnegative. Furthermore, the square root in (1.6.6) must be real. Hence, q_0 is restricted to the range

$$0 \le q_0 \le n(0). \quad (1.6.7)$$

The rays are labeled by the parameter q_0. For p given by the upper sign in (1.6.6), the rays are initially inclined toward the right: $n(x)$ continually increases along the ray, as does p, while q remains constant. Thus, the tangent to the ray becomes progressively more horizontal. On the ray for which

$$q_0 = n(0), \qquad \sqrt{n^2 - q_0^2} = 0, \quad (1.6.8)$$

p must be increasing initially according to (1.6.5), and this ray turns to the right as well.

Now let us consider the lower sign in (1.6.6). This family of rays is initially directed toward the left. For a subset of these rays on which

$$n_0 < q_0 < n(0), \quad (1.6.9)$$

there is a value of x, say, x_q, such that

$$n(x_q) = q_0. \quad (1.6.10)$$

From (1.6.6), $p = 0$ at $x = x_q$ and the ray tangent is vertical at this point. Again, from (1.6.5), p must continue to increase at this point and hence must be given by the positive square root in (1.6.6) beyond this point on the ray. For those values of q_0 such that

$$0 \le q_0 \le n_0, \quad (1.6.11)$$

p is never zero on the left, and the ray continues to progress to the left. The ray picture is shown in Fig. 1.23.

For this example, it is reasonable to use x as the independent parameter on the rays and to express the solution of (1.6.5) as first integrals with respect to x

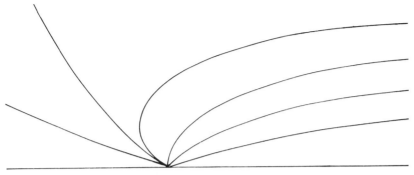

Fig. 1.23. Ray family for a monotonic $n(x)$.

with the rays labeled by the parameter q_0. Thus, the ray family and wave field are described parametrically by (1.6.6) and

$$y = \int_0^x \frac{q_0 \, d\xi}{\sqrt{n^2(\xi) - q_0^2}},$$
$$\qquad\qquad 0 \le x, \quad 0 \le q_0 \le n(0); \qquad (1.6.12)$$
$$u = \int_0^x \frac{n^2(\xi) \, d\xi}{\sqrt{n^2(\xi) - q_0^2}},$$

$$y = -\int_0^x \frac{q_0 \, d\xi}{\sqrt{n^2(\xi) - q_0^2}},$$
$$\qquad\qquad x_q \le x \le 0, \quad 0 \le q_0 < n(0); \qquad (1.6.13)$$
$$u = -\int_0^x \frac{n^2(\xi) \, d\xi}{\sqrt{n^2(\xi) - q_0^2}},$$

$$y = \int_{x_q}^0 + \int_{x_q}^x \frac{q_0 \, d\xi}{\sqrt{n^2(\xi) - q_0^2}},$$
$$\qquad\qquad x_q \le x, \quad n_1 < q_0 < n(0). \qquad (1.6.14)$$
$$u = \int_{x_q}^0 + \int_{x_q}^x \frac{n^2(\xi) \, d\xi}{\sqrt{n^2(\xi) - q_0^2}},$$

In (1.6.13), $x_q = -\infty$ for $q_0 < n_0$.

Exercises

1.14   The purpose of this exercise is to carry out the discussion in this section for a specific choice of $n(x)$. Define

$$n(x) = \begin{cases} n_0, & x < -1, \\ n_0 + \tfrac{1}{2}(n_0 + n_1)(x + 1), & -1 \le x \le 1, \\ n_1, & 1 < x. \end{cases}$$

(a)   Show that the solution (1.6.12) for this example in the region $0 \le x \le 1$ is given by

$$
y = \frac{q_0}{\tilde{n}} \log \left[\frac{n(x) + \sqrt{n^2(x) - q_0^2}}{\bar{n} + \sqrt{n^2 - q_0^2}} \right], \qquad u = q_0 x.
$$

In these equations we have set $\bar{n} = \frac{1}{2}(n_1 + n_0)$ and $\tilde{n} = \frac{1}{2}(n_1 - n_0)$.

(b)   Suppose that $n_0 < q_0 < \bar{n}$. Find x_q and verify that $-1 < x_q < 0$.

(c)   Find the solutions (1.6.13) and (1.6.14). In particular, show that (1.6.14) becomes

$$
y = \frac{q_0}{\tilde{n}} \left\{ \log \left[\frac{\bar{n} + \sqrt{n - q_0^2}}{q_0} \right] + \log \left[\frac{n(x) + \sqrt{n(x) - q_0^2}}{q_0} \right] \right\},
$$

$$
u = q_0(x - 2x_q).
$$

(d)   Find the transmitted waves in $x > 1$, $x < -1$.

(e)   Discuss the wave reflected from $x = -1$ without carrying out the details of the computation.

(f)   Discuss the wave reflected from $x = 1$, and give special emphasis to those rays for which $n_0 < q_0 < \bar{n}$.

1.7   HIGHER DIMENSIONS

We shall now consider the equation in m independent variables,

$$
F(\mathbf{x}, u, \mathbf{p}) = 0, \qquad \mathbf{x} = (x_1, \ldots, x_m), \qquad \mathbf{p} = (p_1, \ldots, p_m);
$$

$$
p_j = \frac{\partial u}{\partial x_j}, \qquad j = 1, \ldots, m. \tag{1.7.1}
$$

Although we lose our geometrical intuition here, we can derive the system analogous to (1.3.9), namely,

$$
\frac{dx_j}{d\sigma} = \lambda F_{p_j}, \qquad\qquad j = 1, \ldots, m,
$$

$$
\frac{du}{d\sigma} = \lambda \sum_{j=1}^{j=m} p_j F_{p_j}, \tag{1.7.2}
$$

$$
\frac{dp_j}{d\sigma} = -\lambda [F_{x_j} + p_j F_u], \qquad j = 1, \ldots, m.
$$

Data for u are now to be prescribed on some $(m - 1)$-dimensional *initial manifold*, which we denote by S. We represent S and the data on it by introducing $(m - 1)$ parameters and setting

$$\mathbf{x} = \mathbf{x}_0(\tau), \qquad u = u_0(\tau), \qquad \tau = (\tau_1, \ldots, \tau_{m-1}). \tag{1.7.3}$$

The vector function $\mathbf{x}_0$ describes S, and then u_0 gives the value of u at each point of S. By differentiating these equations with respect to the τ_j's, we obtain a system of $(m - 1)$ equations for the initial values of the vector $\mathbf{p}$ on S, denoted by p_{j0}:

$$\sum_{j=1}^{m} p_{j0}(\tau) \frac{\partial x_{j0}}{\partial \tau_k} = \frac{\partial u_0}{\partial \tau_k}, \qquad k = 1, \ldots, m - 1. \tag{1.7.4}$$

This system, along with the original equation (1.7.1) with $\mathbf{x}$ and u replaced by their initial values $\mathbf{x}_0(\tau)$ and $u_0(\tau)$, is a system of m equations for the m unknowns $\mathbf{p}_0(\tau)$. If this system has a solution at a point $\tau = \tau_0$ and if

$$J = \det \begin{bmatrix} \dfrac{\partial x_{1_0}}{\partial \tau_1} & \cdots & \dfrac{\partial x_{m_0}}{\partial \tau_1} \\ \cdot & \cdot & \cdot \\ \cdot & \cdot & \cdot \\ \cdot & \cdot & \cdot \\ \dfrac{\partial x_{1_0}}{\partial \tau_{m-1}} & \cdots & \dfrac{\partial x_{m_0}}{\partial \tau_{m-1}} \\ F_{p_1} & \cdots & F_{p_m} \end{bmatrix} \neq 0, \tag{1.7.5}$$

on S, we can continue this solution everywhere on S. Then we solve for $\mathbf{x}(\sigma, \tau), \mathbf{p}(\sigma, \tau)$, and $u(\sigma, \tau)$ and invert this solution in some neighborhood of the initial manifold to find a solution $u(\mathbf{x})$. This solution has continuous second partial derivatives. As in the two-dimensional discussion of the preceding sections, there is a rich set of anomalies that can occur when $J = 0$. The earlier discussions should be viewed as a basis for identifying and analyzing these anomalous cases.

When there are three independent variables, it is still possible to represent the solution pictorially in terms of the level surfaces of u and to exploit our visual conceptualization in order to understand the nature of the solution. In higher dimensions, we cannot do that and must content ourselves with analytical and numerical representations. We remark that with $\lambda = 1$ and F independent of u, the system (1.7.2) is just the set of Hamilton–Jacobi equations for a system of m particles, with Hamiltonian $F + \partial u/\partial \sigma$ and time σ. The generalized Fermat principle is the principle of least action. For this case, the hypersurface u, which is the *action*, is not as important as are the trajectories of the particles, that is, the rays.

References

Courant, R., and Hilbert, D. [1962]. "Methods of Mathematical Physics," Vol. 2. Wiley (Interscience), New York.

Garabedian, P. R. [1964]. "Partial Differential Equations." Wiley, New York.

John, F. [1982]. "Partial Differential Equations," 4th ed. Springer-Verlag, New York.

Kline, M., and Kay, I. W. [1965]. "Electromagnetic Theory and Geometrical Optics." Wiley
(Interscience), New York.

Kreyszig, E. [1959]. "Differential Geometry." Univ. of Toronto Press, Toronto.

Lipschutz, M. M. [1969]. "Differential Geometry." Shaum's Outline Series. McGraw-Hill,
New York.

Luneburg, R. K. [1964]. "Mathematical Theory of Optics." Univ. of California Press, Berkeley.

Sneddon, I. N. [1957]. "Elements of Partial Differential Equations." McGraw-Hill, New York.

Stoker, J. J. [1957]. "Water Waves." Wiley (Interscience), New York.

2 THE DIRAC DELTA FUNCTION, FOURIER TRANSFORMS, AND ASYMPTOTICS

The purpose of this chapter is to develop some tools that will aid in the discussion of solutions of partial differential equations in subsequent chapters. The introduction of the Dirac delta function and related distributions will allow us to develop Green's functions and the associated representations of solutions. Fourier transforms will help us to develop integral representations of Green's functions and through them integral equations equivalent to problems involving partial differential equations and integral representations of solutions. We shall develop enough asymptotic theory to allow us to analyze integral representations of solutions and to interpret them in terms of special functions whose wavelike character is familiar.

2.1 THE DIRAC DELTA FUNCTION
AND RELATED DISTRIBUTIONS

Distributions are an extension of the idea of ordinary functions. They were first introduced because they fill a need not met by ordinary functions in the modeling of physical phenomena. These are phenomena that occur over extremely short duration in time and/or space and would be best modeled as occurring at a point. However, despite this "point duration," it is necessary that an integral of the modeling function be finite and nonzero, representing, for example, the total intensity in a source or the total change in some physical variable across a surface. Clearly, ordinary functions cannot fulfill this requirement; thus, the need for distributions is evident.

The Dirac Delta Function

Let us consider the sequence of functions $\{s_n(x)\}$ such that for each n,

$$s_n(x) = \begin{cases} 0, & |x| > 1/2n, \\ n, & |x| \leq 1/2n. \end{cases} \tag{2.1.1}$$

Let us think of each s_n as representing a source density and then the integrals of these functions as representing the total source strength. It is apparent that the source strength remains constant, equal to one, while the density of the source is progressively increased in magnitude over a progressively narrowing interval around the origin.

The pointwise limit of this sequence everywhere but at the origin is zero. At the origin, the function increases beyond all bound. Neither of these facts is quite so important as the fact that the sequence of integrals of s_n will have limit unity on any interval containing the origin and will have limit zero on any interval not containing the origin. Although the limit of the sequence s_n as an ordinary function having this property makes little sense at all, we extend the idea of a *function* to include objects defined by sequences like this and call this new type of function a *distribution*. We denote the limit by $\delta(x)$, called the *Dirac delta function*.

The Sifting Property of the Delta Function

We define the *action* of the delta function by its *sifting property*; namely, that for an appropriate class of functions,

$$I[f(x)] = \int_{-\infty}^{\infty} f(x)\delta(x)\,dx = f(0). \tag{2.1.2}$$

With the delta function defined as the limit of a sequence of functions, this integral is to be interpreted as the limit of the sequence of integrals with δ replaced by s_n. The function $f(x)$ must certainly be continuous at the origin and then behave sufficiently well for each integral in the sequence to make sense. The class of functions that are continuous at the origin, absolutely integrable, and bounded on the whole line is an appropriate class of functions for which (2.1.2) holds. However, in the development of the theory in this section, we shall use as *test functions* those that have enough derivatives and sufficient decay at infinity to allow all of the indicated operations.

It is very tempting to think of the delta function as being a function that is zero everywhere but at the origin, where it is "so infinite" that it has finite integral. Indeed, the creation of this function was motivated by just such an idea. However, in an attempt to clothe the delta function in mathematical respectability, we find that the cloak actually came from a Pandora's box, containing, among others, sequences that should dispel the notion of the delta function as an ordinary function. For example, let us consider the sequence

$$s_n(x) = \begin{cases} -n, & |x| < 1/2n, \\ 2n, & 1/2n \leq |x| \leq 1/n, \\ 0 & \text{otherwise.} \end{cases} \tag{2.1.3}$$

At $x = 0$, $s_n(x)$ is negative, with its magnitude exceeding all bounds in the limit as $n \to \infty$. On the other hand, at $x = \pm \frac{1}{2}n$ the function is positive, with its magnitude exceeding all bounds as $n \to \infty$. Thus, this function has no limit at all at the origin (even if we allow $\pm \infty$ as a limit), while the pointwise limit anywhere but at the origin is zero. As another myth-dispelling example, we consider the sequence

$$s_n(x) = \begin{cases} (\sin n\pi x)/\pi x, & x \neq 0, \\ n, & x = 0. \end{cases} \tag{2.1.4}$$

Let us consider a punctured neighborhood of the origin, say, $0 < |x| < \frac{1}{2}$. In this interval, there are no pointwise limits at all! Despite the strange pointwise nature of these two functions, both have the property that as for the original sequence,

$$\lim_{n \to \infty} \int s_n(x)\, dx = 1 \tag{2.1.5}$$

on any interval containing the origin but the limit is zero on any interval not containing the origin. Furthermore, using this property, we can show that these sequences exhibit the sifting property [(2.1.2)] on a class of test functions.

We now derive some properties of the delta function that arise from the sifting property. As two simple examples,

$$\int_{-\infty}^{\infty} f(x)\delta(x - x_0)\, dx = \int_{-\infty}^{\infty} f(x + x_0)\delta(x)\, dx = f(x_0) \tag{2.1.6}$$

and

$$\int_{-\infty}^{\infty} f(x)\delta(ax)\, dx = \frac{1}{|a|} \int_{-\infty}^{\infty} f(x/a)\delta(x)\, dx$$

$$= \frac{1}{|a|} f(0) \quad \Rightarrow$$

$$\delta(ax) = \frac{1}{|a|} \delta(x). \tag{2.1.7}$$

Similarly, let us suppose that $g(x)$ is a function for which $g(x_0) = 0$ but $g'(x_0) \neq 0$. Then, on an interval in x containing only this zero of $g(x)$,

$$\delta(g(x)) = |g'(x_0)|^{-1}\delta(x - x_0). \tag{2.1.8}$$

INTEGRALS AND DERIVATIVES OF THE DELTA FUNCTION

We consider now the function

$$H(x) = \int_{-\infty}^{x} \delta(x')\,dx'. \tag{2.1.9}$$

We can verify by using the delta sequence (2.1.1) that

$$H(x) = \begin{cases} 0, & x < 0, \\ 1, & x > 0. \end{cases} \tag{2.1.10}$$

The fact that the function is not defined at one point will cause no difficulty. We shall return to this later. This function is known as the *Heaviside function*. Since the Heaviside function can be identified as an ordinary function, there is no need to discuss further integrals of the delta function as distributions. On the other hand, we introduce the derivative $\delta'(x)$, which again must be understood by its action on test functions:

$$\int_{-\infty}^{\infty} f(x)\delta'(x)\,dx = -\int_{-\infty}^{\infty} f'(x)\delta(x)\,dx = -f'(0). \tag{2.1.11}$$

Thus, we interpret $\delta'(x)$ as that distribution for which the sifting property is that it produces $-f'(0)$. Clearly, this process can be repeated to interpret the higher derivatives of the delta function.

PRODUCTS OF DELTA FUNCTIONS

The product $\delta(x)\delta(x)$ makes no sense at all, since the sifting property would require that the delta function itself be evaluated at a point. That is, the delta function is itself outside the class of test functions for which it itself makes sense. On the other hand, it is not unreasonable to define a delta function in two variables. Thus, we introduce the distribution $\delta(x)\delta(y)$ in the (x, y) plane as that distribution for which

$$\int_{-\infty}^{\infty} f(x, y)\delta(x)\delta(y)\,dx\,dy = f(0, 0). \tag{2.1.12}$$

Again, we think of this result as being applied on sufficiently differentiable test functions that vanish outside a finite region. Clearly, this idea can be extended to higher dimensions. Furthermore, the derivatives and integrals of these multidimensional delta functions can be defined by extending the ideas of differentiation and integration from one dimension. We remark that at times the two-dimensional delta function will be written as $\delta(x, y)$ or $\delta(\mathbf{x})$, the latter used equally well for delta functions of higher dimension.

Now we consider the distribution $\delta(g_1(x, y))\delta(g_2(x, y))$. Here we assume that the level curves $g_1(x, y) = 0$ and $g_2(x, y) = 0$ intersect nontangentially at a point (x_0, y_0). More precisely, we assume that the cross product of the gradients has nonzero magnitude, that is, we introduce the Jacobian

$$G = \det \begin{bmatrix} \dfrac{\partial g_1(x, y)}{\partial x} & \dfrac{\partial g_1(x, y)}{\partial y} \\[2ex] \dfrac{\partial g_2(x, y)}{\partial x} & \dfrac{\partial g_2(x, y)}{\partial y} \end{bmatrix} \tag{2.1.13}$$

and require that

$$J = G(x_0, y_0) \neq 0. \tag{2.1.14}$$

Let us consider now the integral

$$I = \int_D f(x, y)\delta(g_1(x, y), g_2(x, y)) \, dx \, dy. \tag{2.1.15}$$

Here D is some domain containing (x_0, y_0) but no other zero of both g_1 and g_2. We further assume that G is not zero over the domain D. If it were, we could "shrink" the domain without affecting the integral at all to a domain on which this were true. Now, we introduce g_1 and g_2 as new variables of integration. Thus,

$$I = \int_{D'} \frac{f(x(g_1, g_2), y(g_1, g_2))}{|G|} \delta(g_1)\delta(g_2) \, dg_1 \, dg_2. \tag{2.1.16}$$

Here D' is the image of D under the transformation from (x, y) to (g_1, g_2). Since (x_0, y_0) was in D, $(0, 0)$ must be in D'. The integral has now been reduced to the form (2.1.12). Thus,

$$I = f(x_0, y_0)/|J|. \tag{2.1.17}$$

This result can be extended to higher dimensions.

Let us now consider the distribution $\delta(x)\delta(y)$ in cylindrical coordinates. Thus, we write the identity

$$\int_D \delta(x)\delta(y) \, dx \, dy = \int_{D'} \delta(x)\delta(y)\rho \, d\rho \, d\phi = 1, \tag{2.1.18}$$

with D denoting the entire (x, y) plane and ρ, ϕ the polar radius and angle, respectively. We seek the representation of the product of delta functions in terms of distributions in ρ and ϕ.

First, we observe that the support of such a distribution—the domain in which the distribution is nonzero—must be the origin itself. Thus, only the *form* of the distribution with support at the origin is at issue. We will only consider the class of functions that are single valued at the origin. Therefore, there will be no angular dependence of the distribution; integration with respect to ϕ only introduces a multiplier of 2π, and we must have

$$\delta(x)\delta(y) = \frac{a(\rho)}{2\pi}\delta(\rho) \qquad (2.1.19)$$

such that

$$\int_0^\infty a(\rho)\delta(\rho)\rho\, d\rho = 1. \qquad (2.1.20)$$

Integration of the delta function with one endpoint being its support point is not well defined. Returning to the definition of the delta function, we remark that the sequence of test functions that are used to obtain the limiting integral (2.1.5) could be chosen with any fraction of its weight being on the positive half interval and the complementary weight on the negative half interval. Thus, we must make a convention consistent with the fact that we will consider test function that are nonzero only for ρ positive. Hence, a sequence $\{s_n(\rho)\}$ used to define $\delta(\rho)$ must also be nonzero only for positive values of ρ. That is,

$$\int_0^\infty \delta(\rho)\, d\rho = 1. \qquad (2.1.21)$$

This, in turn, leads us to conclude that

$$a(\rho) = 1/\rho \qquad (2.1.22)$$

and

$$\delta(x)\delta(y) = \delta(\rho)/2\pi\rho. \qquad (2.1.23)$$

Similarly,

$$\delta(x)\delta(y)\delta(z) = \delta(\rho)\delta(z)/2\pi\rho = \delta(r)/4\pi r^2, \qquad (2.1.24)$$

with r the spherical radius. These results play an important role in determining Green's functions in polar coordinates in that they provide the proper "weight" to the delta function in the radial variable.

Exercises

2.1 Suppose that $f(x)$ is a differentiable function everywhere on the interval $[a, b]$ except at x_0 in the interval, where it is discontinuous. Define the jump in $f(x)$ at that point to be α. Define $g(x)$ by

$$g(x) = \begin{cases} f(x), & a \le x < x_0, \\ f(x) - \alpha, & x_0 \le x \le b. \end{cases}$$

(a) Show that

$$f(x) = g(x) + \alpha H(x - x_0).$$

(b) Show that

$$f'(x) = g'(x) + \alpha\delta(x - x_0).$$

(c) Suppose that y is a solution of the ordinary differential equation

$$y'' + a(x)y' + b(x)y = \delta(x - x_0).$$

If $a(x)$ and $b(x)$ are continuous, examine the possibility that $y(x)$ is discontinuous at x_0, and explain why this cannot occur. Then conclude that this equation is equivalent to the homogeneous equation plus the conditions that $y(x)$ be continuous at x_0 but that y' be discontinuous with

$$y'(x_0 +) - y'(x_0 -) = 1.$$

2.2 Suppose that a surface S is parameterized by two variables s_1, s_2. Introduce a third coordinate σ, which is an oriented distance normal to S (i.e., a "signed" distance, positive on one side of S and negative on the other side). Define the *singular function* $\gamma(\mathbf{x})$ of the surface S to be

$$\gamma(\mathbf{x}) = \delta(\sigma).$$

Show that for any "test function" $f(\mathbf{x})$,

$$\int_{-\infty}^{\infty} f(\mathbf{x})\gamma(\mathbf{x})\,dV = \int_{S} f(\mathbf{x})\,dS.$$

That is, the singular function provides a means of writing a surface integral as a volume integral.

2.3 Suppose that the point (x_0, y_0), different from the origin, has the polar coordinates (ρ_0, ϕ_0). Set

$$\int_{D} \delta(x - x_0)\delta(y - y_0)\,dx\,dy = \int_{D'} A(\rho, \phi)\delta(\rho - \rho_0)\delta(\phi - \phi_0)\rho\,d\rho\,d\phi = 1.$$

Conclude that $A(\rho_0, \phi_0) = 1/\rho_0$ and hence that

$$\delta(x - x_0)\delta(y - y_0) = (1/\rho)\delta(\rho - \rho_0)\delta(\phi - \phi_0).$$

2.4 Define

$$s_n(x) = (\sin nx)/\pi x, \qquad n = 1, 2, \ldots.$$

(a)   Show that

$$\int_{-\infty}^{\infty} s_n(x)\, dx = 1$$

for all n.

(b)   Use integration by parts to obtain an absolutely convergent integral, and estimate that integral to show that

$$\left| \int_{x_0}^{\infty} s_n(x)\, dx \right| \le \frac{2}{n\pi x_0}, \qquad x_0 > 0.$$

(c)   Use (b) to conclude that for any $x_0 > 0$,

$$\lim_{n \to \infty} \int_{-x_0}^{x_0} s_n(x)\, dx = 1.$$

(d)   Show that $s_n(x)$ does not have a limit as $n \to \infty$ at $x = \pi/2$.

2.2   FOURIER TRANSFORMS

We begin our discussion of Fourier transforms from the analysis of complex Fourier series. Thus, let us suppose that $U(x)$ is defined on the interval $(-L, L)$. The complex Fourier series representation for $U(x)$ is

$$U(x) = \sum_{n=-\infty}^{+\infty} u_n e^{in\pi x/L}. \tag{2.2.1}$$

Here the coeficients n_n are defined by

$$u_n = \frac{1}{2L} \int_{-L}^{L} U(\xi) e^{-in\pi\xi/L}\, d\xi. \tag{2.2.2}$$

If U^2 has a finite integral on $(-L, L)$, the sum of the squares of the coefficients is finite as well, and the series converges to the function U in a "square integral sense." That is,

$$\lim_{n \to \infty} \int_{-L}^{L} \left| U(\xi) - \sum_{n=-N}^{N} u_n e^{in\pi\xi/L} \right|^2 d\xi = 0. \tag{2.2.3}$$

If $U(x)$ is "piecewise smooth"; that is, if it has a continuous derivative at all but a finite number of points on $[-L, L]$, then (i) the series converges to the function at every point of continuity and (ii) the series converges to the mean of the left and right limits $\frac{1}{2}[f(x + 0) + f(x - 0)]$ at each point of discontinuity.

The series representation extends $U(x)$ as a periodic function of period $2L$. Alternatively, we can take the point of view that the series provides a representation for periodic functions of period $2L$. We propose now to extend the period $2L$, ultimately to define a representation for functions whose period is the whole line, that is, to functions that are not periodic at all. We introduce a continuous variable k, discretized by setting

$$k_n = n\pi/L, \qquad \Delta k = \pi/L. \tag{2.2.4}$$

Then we rewrite (2.2.1) and (2.2.2) as

$$U(x) = \frac{1}{2\pi} \sum_{n=-\infty}^{\infty} \Delta k \, e^{ik_n x} \int_{-L}^{L} U(\xi) e^{-ik_n \xi} \, d\xi. \tag{2.2.5}$$

We can now contemplate taking the limit $L \to \infty$. Heuristically, in this limit the sum would appear to be the discretization of an integral with respect to a continuous variable k on the whole line. Furthermore, the interval of integration on the explicit integral here also becomes the entire line in the limit. Thus, formally at least, if we define the Fourier transform $u(k)$ of the function $U(x)$ by

$$u(k) = \int_{-\infty}^{\infty} U(\xi) e^{-ik\xi} \, d\xi, \tag{2.2.6}$$

then

$$U(x) = \frac{1}{2\pi} \int_{-\infty}^{\infty} u(k) e^{ikx} \, dk. \tag{2.2.7}$$

The functions $U(x)$ and $u(k)$ are called a *Fourier transform pair*; $u(k)$ is the *forward transform* of $U(x)$, and $U(x)$ is the *inverse* transform of $u(k)$. Notice that the two transforms differ only in a sign in the exponent and in the multiplier. In fact, if $u(k)$ were redefined with a multiplier $\{2\pi\}^{-1/2}$, then the same factor would appear in the inverse transform, and only the change in sign in the exponent would distinguish the two. Thus, we could have as easily defined the *forward transform* (2.2.6) with a plus sign in the exponent and the *inverse transform* in (2.2.7) with a minus sign in the exponent. We shall use these transform pairs for spatial variables and the pair with the opposite sign for temporal (time) transforms. The reason for this will be explained when we discuss applications of multifold transforms.

One class of functions for which (2.2.6) and (2.2.7) hold is the class of square integrable functions. Furthermore,

$$\int_{-\infty}^{\infty} |U(x)|^2 \, dx = \frac{1}{2\pi} \int_{-\infty}^{\infty} |u(k)|^2 \, dk. \tag{2.2.8}$$

A second class of functions are those for which $|U(x)|$ has a bounded integral. In this case, $u(k)$ is bounded and (Riemann–Lebesgue lemma) $\lim_{|k| \to \infty} |u(k)| = 0$. This still does not guarantee that the transform can be inverted for this class of functions. However, it is sufficient to assume, in addition, that the function $U(x)$ is of *bounded variation*, that is, that the sum of increments of the function plus the sum of decrements of the function on the entire line is bounded. In this case,

$$\frac{U(x + 0) + U(x - 0)}{2} = \frac{1}{2\pi} \int_{-\infty}^{\infty} u(k) e^{ikx} \, dk. \tag{2.2.9}$$

The set of numbers $\{u_n\}$ in the discrete case or the function $u(k)$ in the continuous case is called the *spectrum* of the function $U(x)$. For the discrete case, the coefficients u_n are the amplitudes of each discrete *wave number*, or inverse wavelength, $2n\pi/L$. For the continuous case, $u(k) \, dk$ is an average amplitude for the packet of continuous wave numbers in an interval dk containing k. Thus, $u(k)$ itself is a *spectral density*, that is, spectrum per unit length in k, for the function $U(x)$.

An alternative representation uses in place of the wave number k the *spatial frequency*

$$k = 2\pi f_x. \tag{2.2.10}$$

Direct substitution in (2.2.6) and (2.2.9) yields

$$u(k) \equiv \tilde{u}(f_x) = \int_{-\infty}^{\infty} U(x) e^{-2\pi i f_x x} \, dx,$$

$$\frac{U(x + 0) + U(x - 0)}{2} = \int_{-\infty}^{\infty} \tilde{u}(f_x) e^{2\pi i f_x x} \, df. \tag{2.2.11}$$

Note that with this definition the square integrals of U and u are equal. This form of the Fourier transform is particularly useful for numerical computation. However, in analytical work, it requires writing many more factors of 2π and so is used less often.

Another version of the basic Fourier inversion theorem is particularly useful when dealing with solutions of initial value problems in which the function of interest is zero up to some finite time. Thus, we will state this result as the transform of a function of the variable t with transform variable ω. Let us define the *half transforms*

$$u_+(\omega) = \int_0^\infty U(t) e^{i\omega t} \, dt, \qquad u_-(\omega) = \int_{-\infty}^0 U(t) e^{i\omega t} \, dt. \tag{2.2.12}$$

Now the Fourier transform of $U(t)$ might not exist because of the behavior of the function at infinity. However, let us suppose that the exponential decay provided by Im ω being sufficiently large and positive in the first integral (say, Im $\omega > a$) and sufficiently large and negative in the second integral (say, Im $\omega < b$) that the function $U(t)e^{-\text{Im}\,\omega t}$, defined on one of these half intervals, does satisfy one of the possible sets of conditions for Fourier inversion to be valid. Then with $\omega = \mu + i\eta$,

$$\frac{1}{2\pi} \int_{-\infty}^{\infty} u_+(\mu + i\eta)e^{-i\mu t}\,d\mu = \begin{cases} U(t)e^{-\eta t}, & t > 0, \\ 0, & t < 0, \end{cases} \quad \eta > a, \quad (2.2.13)$$

or

$$\frac{1}{2\pi} \int_{-\infty}^{\infty} u_+(\mu + i\eta)e^{-i(\mu + i\eta)t}\,d\mu = \begin{cases} U(t), & t > 0, \\ 0, & t < 0, \end{cases} \quad \eta > a. \quad (2.2.14)$$

A similar result holds for u_-, namely,

$$\frac{1}{2\pi} \int_{-\infty}^{\infty} u_-(\mu + i\eta)e^{-i(\mu + i\eta)t}\,d\mu = \begin{cases} 0, & t > 0, \\ U(t), & t < 0, \end{cases} \quad \eta < b. \quad (2.2.15)$$

These two half-transform inversion formulas can be viewed as contour integrals in the complex ω plane. The right sides, $U(t)$, should be viewed in the context of what has been discussed. That is, equality here could be, for example, in the square integral sense, analogous to Eq. (2.2.3) for the discrete case or in a mean value sense, as in (2.2.9), for example. If $b > a$, then the two choices of η could be the same, the results could be added, and the ordinary transform would be obtained, however, for real μ but not necessarily for real $\omega = \mu + i\eta$.

Given any of a number of conditions on $U(t)$, u_+ and u_- are analytic in their respective half planes of validity. Indeed, formally, differentiation with respect to ω amounts to multiplication by it under the integral sign. That is,

$$\frac{du_+}{d\omega} = \int_0^{\infty} itU(t)e^{i\omega t}\,dt, \quad (2.2.16)$$

with a similar result for u_-. This result is valid under many circumstances. A simple constraint would be U continuous and bounded or alternatively, U of bounded variation. Unfortunately, the latter choice would preclude functions that might grow no worse than a linear exponential. Thus, we take our functions to be of bounded variation on any finite interval and growing no worse than a linear exponential. Now with these transforms analytic in their respective half planes, all of the theory of contour integration for analytic functions of a complex variable can be brought to bear to calculate the inverse transform. Furthermore, for these functions. the contour of integration in (2.2.14) need only be chosen to be *above all singularities* of u_+ and in (2.2.15) *below all singularities* of u_-. When $U(t)$ is a solution of an initial value problem, we define U to be zero for $t < 0$. Consequently, for such functions,

$$u_-(\omega) \equiv 0,$$

and we have only (2.2.14) with the integral interpreted as a contour integral in the complex ω plane and with η chosen above the singularities of u_+.

The dual result to (2.2.16) is not quite the same. That is,

$$\int_0^\infty U'(t)e^{i\omega t}\, dt = -U(0+) - i\omega \int_0^\infty U(t)e^{i\omega t}\, dt$$

$$= -U(0+) - i\omega u_+(\omega). \tag{2.2.17}$$

Thus, the Fourier transform of the derivative has multiplier $-i\omega$ on the transform of the function itself and an additive term $-U(0+)$. This result is valid for any class of functions for which the integration by parts operation used here is justified. These results are completely equivalent to results for the one-sided Laplace transform with $i\omega = -s$; see Exercise 3.2.

One should note that the signs of the multipliers $i\omega$ in (2.2.17) and *it* in (2.2.16) depend on the definition of the transform. The signs here would be reversed for the choice of signs used in the spatial transforms (2.2.6) and (2.2.7). Furthermore, for those transforms, there are no "endpoint contributions" from integration by parts. Thus, we have the transform pairs

$$u'(k) \longrightarrow -ixU(x); \qquad U'(x) \longrightarrow +iku(k). \tag{2.2.18}$$

We close this section with an identity known as the *convolution theorem*. We define the convolution of the functions $U(x)$ and $V(x)$ by

$$W(x) = U(x) * V(x) = \int_{-\infty}^\infty V(\xi)U(x - \xi)\, d\xi \tag{2.2.19}$$

and calculate

$$w(k) = \int_{-\infty}^\infty W(x)e^{-ikx}\, dx$$

$$= \int_{-\infty}^\infty dx \int_{-\infty}^\infty d\xi\, V(\xi)U(x - \xi)e^{-ikx}$$

$$= \int_{-\infty}^\infty d\xi \int_{-\infty}^\infty dx\, V(\xi)e^{-ik\xi}U(x - \xi)e^{-ik(x - \xi)}$$

$$= \int_{-\infty}^\infty d\xi\, V(\xi)e^{-ik\xi} \int_{-\infty}^\infty dx\, U(x)e^{-ikx}$$

$$= u(k)v(k). \tag{2.2.20}$$

Here we have moved the differentials close to the integral signs to emphasize the order of integration. The interchange of order in the third line is valid for test functions and then for functions that can be attained as a limit of test functions. The fourth line is achieved by "shifting" the variable of integration x by ξ. Thus, the Fourier transform of the convolution of two functions is the product of their two transforms.

Exercises

2.5   (a)   Show that

$$U(ax - b) \longrightarrow \frac{1}{|a|} u\left[\frac{k}{a}\right] e^{-ikb/a},$$

with the transform defined by (2.2.6).

(b)   Verify that if $W(x) = aU(x) + bV(x)$, then

$$w(k) = au(k) + bv(k).$$

(c)   If $F(x)$ has Fourier transform $ikg(k)$, then

$$\int_0^x F(x') \, dx'$$

and $G(x)$ differ at most by a constant.

2.6   (a)   Suppose that the function $U(x)$ in (2.2.6) is a test function from the class of infinitely differentiable functions that vanish outside a finite interval. Show that if $U(x)$ is even (odd), then $u(k)$ is also even (odd).

(b)   Suppose that $U_n(x)$ is a sequence of test functions as in (a). Suppose that the limit $u(k)$ of the sequence $u_n(k)$ exists. Show that if every function in the sequence $U_n(x)$ is even (odd), then $u(k)$ is even (odd).

2.7   Define the *box* function $B(x, L)$ by

$$B(x, L) = H(x + L/2) - H(x - L/2),$$

with H defined by (2.1.10). That is, B is the function that is equal to unity on the interval $(-L/2, L/2)$.

(a)   Show that

$$\tilde{b}(f_x, L) = (\sin f_x \pi L)/\pi f_x.$$

(b)   Check by direct calculation that with the possible exceptions of the points $\pm L/2$, $B(x, L)$ is the inverse transform of $\tilde{b}(x, L)$.

(c)   Use Exercise 2.4 to conclude that

$$\delta(x) \longrightarrow 1.$$

That is, the Dirac delta function and the function that is identically equal to one are a Fourier transform pair.

2.8 Define

$$T(x, L) = B(x, L) * B(x, L).$$

(a) Show that the graph of T is an isosceles triangle centered at the origin with base length $3L$ and height L.

(b) Use the convolution theorem to conclude that

$$t(x, L) = \left[\frac{2 \sin(kL/2)}{k} \right]^2.$$

2.9 (a) Show from (2.2.19) that convolution is commutative. That is,

$$V * U = U * V.$$

(b) Show from (2.2.20) that convolution is commutative.

2.3 FOURIER TRANSFORMS OF DISTRIBUTIONS

We have already seen in Exercise 2.7 that the Dirac delta function and the function that is identically equal to one are a Fourier transform pair. That is,

$$\int_{-\infty}^{\infty} \delta(x) e^{-ikx} \, dx = 1, \tag{2.3.1}$$

and, perhaps more surprisingly,

$$\frac{1}{2\pi} \int_{-\infty}^{\infty} 1 e^{ikx} \, dk = \delta(x). \tag{2.3.2}$$

Both of these results follow from the exercises.

Alternatively, from (2.2.6) and (2.2.7), we can write for $U(x)$ at a point at which the function is continuous that

$$U(x) = \frac{1}{2\pi} \int_{-\infty}^{\infty} dk \int_{-\infty}^{\infty} d\xi \, U(\xi) e^{ik(x - \xi)}. \tag{2.3.3}$$

The iterated integration here merely performs the sifting operation of the Dirac delta function. Thus, if we rewrite the right side as an integral of U with some other "function," that function must indeed be the Dirac delta function. Formally, then, by interchanging orders of integration, we conclude that

$$\delta(x - \xi) = \frac{1}{2\pi} \int_{-\infty}^{\infty} e^{ik(x - \xi)} \, dk. \tag{2.3.4}$$

In order to justify that interchange, it would be necessary first to truncate the integral in k, and then we shall have returned to Exercise 2.7.

Let us now consider the function

$$U(x) = \text{sign}(x) = \begin{cases} -1, & x < 0, \\ 0, & x = 0, \\ +1, & x > 0. \end{cases} \tag{2.3.5}$$

For this function,

$$U'(x) = 2\delta(x), \tag{2.3.6}$$

and therefore by using (2.2.18),

$$U'(x) \longleftrightarrow iku(k) = 2. \tag{2.3.7}$$

We cannot find the Fourier transform of $U(x)$ by simply dividing by ik. The difficulty is that there is a whole class of functions that have the same derivative $U'(x)$ but differ from $U(x)$ only by an arbitrary constant. Division here produces one of these but not necessarily the function defined by (2.2.6). However, if the functions differ by an arbitrary constant, their Fourier transforms differ at worst by the Fourier transform of a constant. Since we know the Fourier transform of the constant one, we know all of these transforms. Thus, we conclude that

$$u(k) = (2/ik) + C\delta(k), \tag{2.3.8}$$

with C an arbitrary constant to be determined. To find C, we note that $U(x)$ is *odd* and, therefore, so is $u(k)$ by Exercise 2.6. However, by specializing (2.1.7) to the case $a = -1$, we see that the delta function is *even*. Consequently, $C = 0$ and

$$\text{sign}(x) \longleftrightarrow 2/ik \tag{2.3.9}$$

after all!

We have now created a Fourier transform that as an ordinary function is singular at the origin. We remark, however, that this transform was not deduced from operations on ordinary functions but by imposing a property about the Fourier transform of ordinary functions (2.2.18) on the transform of a distribution (2.3.7). We show later in Exercise 2.10 how (2.3.9) can be deduced as the limit of a sequence of transforms. The limiting process leads us to the conclusion that we should interpret $1/ik$ as a *distribution* whose integral with test functions should be interpreted as a *Cauchy principal value*. That is,

$$\fint \frac{u(k)}{ik}\, dk \equiv \lim_{\varepsilon \to 0} \int_{-\infty}^{-\varepsilon} + \int_{\varepsilon}^{\infty} \frac{u(k)}{ik}\, dk. \tag{2.3.10}$$

We remark that except for the value at the origin,

$$H(x) = \tfrac{1}{2}\{\text{sign}(x) + 1\}, \tag{2.3.11}$$

from which it follows that

$$h(k) = (1/ik) + \pi\delta(k). \tag{2.3.12}$$

We see here that the Fourier inversion of $h(k)$ produces a function that has the value one-half at $x = 0$, consistent with (2.2.9). Often, $H(x)$ is defined this way so that it agrees with the transform of $h(k)$.

Exercise

2.10   Consider the sequence of functions defined by

$$S_n(x) = \begin{cases} -e^{x/n}, & x < 0, \\ 0, & x = 0, \\ e^{-x/n}, & x > 0. \end{cases}$$

(a)   Use the fact that

$$1 - e^{-x/n} = \int_0^{x/n} e^{-\xi}\, d\xi$$

to estimate the difference between S_n and $\text{sign}(x)$ to conclude that

$$\lim_{n \to \infty} |S_n(x) - 1| = 0$$

for each fixed $x > 0$ and

$$\lim_{n \to \infty} |S_n(x) + 1| = 0$$

for each fixed $x < 0$.

(b)   Let $s_n(k)$ be the Fourier transform of $S_n(x)$ defined by (2.2.6). Show that

$$s_n(k) = s_n^+(k) + s_n^-(k), \qquad s_n^\pm(k) = [i(k \mp i/n)]^{-1}.$$

(c)   Introduce a class of test functions that are analytic for complex k in some strip containing the real k axis and that decays to zero at least to some algebraic order in k as $|k| \to \infty$ in that strip. For $u(k)$ such a test function, consider the integrals

$$I_n = \int_{-\infty}^{\infty} s_n(k) u(k)\, dk.$$

Rewrite this integral as the sum

$$I_n = I_n^+ + I_n^-, \qquad I_n^\pm = \int_{-\infty}^{\infty} s_n^\pm(k) u(k)\, dk.$$

Show that

$$\lim_{n \to \infty} I_n^\pm = \int \frac{u(k)}{ik}\, dk \mp \pi u(0),$$

and by summing conclude that the Fourier transform of sign(x) is a distribution that when integrated with test functions, produces twice the Cauchy principal value.

2.4 MULTIDIMENSIONAL FOURIER TRANSFORMS

Let us now define the multifold Fourier transform

$$u(\mathbf{k}) = \int_{-\infty}^{\infty} U(\mathbf{x}) e^{-i\mathbf{k}\cdot\mathbf{x}}\, dx_1 \cdots dx_m,$$

$$\mathbf{k} = (k_1 \cdots k_m), \quad \mathbf{x} = (x_1 \cdots x_m). \qquad (2.4.1)$$

The inversion formula associated with this transform is

$$U(\mathbf{x}) = \frac{1}{[2\pi]^m} \int_{-\infty}^{\infty} u(\mathbf{k}) e^{i\mathbf{k}\cdot\mathbf{x}}\, dk_1 \cdots dk_m. \qquad (2.4.2)$$

As mentioned earlier, we will use transforms of opposite sign in space and time. Thus, for example, given a function $U(\mathbf{x}, t)$ that is identically equal to zero for t negative, we define

$$u(\mathbf{k}, \omega) = \int_0^{\infty} dt \int_{-\infty}^{\infty} dx_1 \cdots dx_m\, U(\mathbf{x}, t) e^{i(\omega t - \mathbf{k}\cdot\mathbf{x})}. \qquad (2.4.3)$$

The inverse transform in this case is

$$U(\mathbf{x}, t) = \frac{1}{[2\pi]^{m+1}} \int_C d\omega \int_{-\infty}^{\infty} dk_1 \cdots dk_m\, u(\mathbf{k}, \omega) e^{i(\mathbf{k}\cdot\mathbf{x} - \omega t)}. \qquad (2.4.4)$$

Here C is a contour in the complex ω plane, passing above all singularities of the integrand and along which Re ω ranges from $-\infty$ to $+\infty$. When we can justify moving the contour down to the real axis, the integrand has the form of a plane wave whose phase is constant when

$$\mathbf{k}\cdot\mathbf{x} - \omega t = \text{const.}$$

These planes can be seen to propagate in the direction of $\mathbf{k}$ with velocity vector $\omega \mathbf{k}/\mathbf{k} \cdot \mathbf{k}$. Had we not used transforms of opposite sign in space and time, the velocity vector would have carried a (nuisance) minus sign, which would have persisted throughout our computations.

As in the one-dimensional case, we conclude that the multidimensional Dirac delta function and unity are a Fourier transform pair; in particular,

$$\delta(\mathbf{x} - \boldsymbol{\xi}) \equiv \delta(x_1 - \xi_1) \cdots \delta(x_m - \xi_m)$$

$$= \frac{1}{[2\pi]^m} \int_{-\infty}^{\infty} e^{\pm i\mathbf{k}\cdot(\mathbf{x}-\boldsymbol{\xi})} \, dk_1 \cdots dk_m. \qquad (2.4.5)$$

We shall discuss now an example of a Fourier inversion that will be of use in the study of the wave equation. We consider the following function in three dimensions ($m = 3$):

$$U(\mathbf{x}, t) = \frac{c^2}{[2\pi]^4} \int_C d\omega \int_{-\infty}^{\infty} dk_1 \, dk_2 \, dk_3 \frac{e^{i(\mathbf{k} \cdot \mathbf{x} - \omega t)}}{c^2 k^2 - \omega^2}, \qquad k^2 = \mathbf{k} \cdot \mathbf{k}. \quad (2.4.6)$$

Here the contour C must pass above the poles of the integrand at $\omega = \pm ck$.

Let us consider the ω integration first and view it as the limit of finite integrals on which $|\omega| \le R$. For $t < 0$, we close this path of integration with a semicircle of radius R in the upper half ω plane. The integral around this closed contour—C extended by a semicircle—is zero because the integrand is analytic in the interior of the domain of integration. We leave it to the exercises to verify that the integral on the semicircle decays to zero as the radius, $R \to \infty$. Thus,

$$U(\mathbf{x}, t) = 0, \qquad t < 0. \qquad (2.4.7)$$

Now let us suppose that $t > 0$. In this case, we can close the contour of integration in the *lower* half plane and determine the integral by summing residues. The result is

$$U(\mathbf{x}, t) = \frac{-ic}{2[2\pi]^3} \int_{-\infty}^{\infty} e^{i\mathbf{k}\cdot\mathbf{x}} \left[e^{ickt} - e^{-ickt} \right] k^{-1} \, dk_1 \, dk_2 \, dk_3. \quad (2.4.8)$$

We introduce spherical polar coordinates r, k, θ, ϕ, with the angle θ measured from the vector $\mathbf{x}$. Thus,

$$U(\mathbf{x}, t) = \frac{-ic}{2[2\pi]^3} \int_0^{\infty} k \, dk \int_{\Omega} \sin\theta \, d\theta \, d\phi \, e^{ikr\cos\theta} \left[e^{ickt} - e^{-ickt} \right]. \quad (2.4.9)$$

Clearly, the integral in k makes no sense as an ordinary integral. However, in light of our definition of the delta function as a Fourier integral, we reserve judgment on this integral until the angular integration has been carried out. The symbol Ω indicates integration over the entire 4π sr (steradians) of solid angle in three-dimensional space.

Integration in ϕ produces a multiplier of 2π; integration in θ then yields

$$U(\mathbf{x}, t) = \frac{-c}{2r(2\pi)^2} \int_0^\infty dk \left[e^{ikr} - e^{-ikr}\right]\left[e^{ickt} - e^{-ickt}\right]. \qquad (2.4.10)$$

In the pair of integrals containing the exponent $-ikr$, we replace k by $-k$ and obtain in place of (2.4.10)

$$U(\mathbf{x}, t) = \frac{c}{2r(2\pi)^2} \int_{-\infty}^\infty dk \left[e^{ik(r-ct)} - e^{ik(r+ct)}\right]. \qquad (2.4.11)$$

We now use the result (2.3.2) to conclude that

$$U(\mathbf{x}, t) = \frac{c}{4\pi r} \left[\delta(r - ct) - \delta(r + ct)\right]. \qquad (2.4.12)$$

In this expression, both t and r are positive. Consequently, the function $r + ct$ is never zero in the range of these variables, and we can replace the second distribution by zero. In the first distribution we use the result (2.1.7) to rewrite U as

$$U(\mathbf{x}, t) = \delta(t - r/c)/4\pi r. \qquad (2.4.13)$$

In this example, we have exploited the complex Fourier transform and our interpretation of a divergent integral as a distribution to obtain a meaningful (in the sense of distributions) expression for a multifold Fourier transform. This is not atypical. The function U is the response to a point source at $(\mathbf{0}, 0)$ for the acoustic wave equation. It is an idealized model of the propagation of sound from a handclap. To the extent that speech can be viewed as a series of such *pulses* of sound, it models the propagation of an elemental component of speech. We see then that each such pulse of sound passes a point in space in an extremely short interval of time. Thus, we are able to hear the elements of speech separately and synthesize them into words that carry meaning.

Exercise 2.13 carries out the same analysis in two dimensions. Here it is seen that an algebraic (square root) singularity propagates with speed c but is followed by a "tail" that decays to zero at each spatial point at a rate of only $1/t$.

Pity the poor resident of flatland! Each sound pulse impacts on the listener but leaves a trailing residue to obscure the sounds that follow as

```
HHHHHHHHHHHHHHHHHHHHHHHHHHHHHHHHHHHHHHHHHHHHHHHHHHHHHH . . .
OOOOOOOOOOOOOOOOOOOOOOOOOOOOOOOOOOOOOOOOOOOOOOOOOOOOOO . . .
 WWWWWWWWWWWWWWWWWWWWWWWWWWWWWWWWWWWWWWWWWWWWWWWWWWWWWW. . .

HHHHHHHHHHHHHHHHHHHHHHHHHHHHHHHHHHHHHHHHHHHHHHHHHHHHHHHHH. . .
AAAAAAAAAAAAAAAAAAAAAAAAAAAAAAAAAAAAAAAAAAAAAAAAAAAAAAAAA. . .
RRRRRRRRRRRRRRRRRRRRRRRRRRRRRRRRRRRRRRRRRRRRRRRRRRRRRRRRR. . .
DDDDDDDDDDDDDDDDDDDDDDDDDDDDDDDDDDDDDDDDDDDDDDDDDDDDDDDDD. . .

 TTTTTTTTTTTTTTTTTTTTTTTTTTTTTTTTTTTTTTTTTTTTTTTTTTTTTTTT . . .
OOOOOOOOOOOOOOOOOOOOOOOOOOOOOOOOOOOOOOOOOOOOOOOOOOOOOOOOOO. . .

 HHHHHHHHHHHHHHHHHHHHHHHHHHHHHHHHHHHHHHHHHHHHHHHHHHHHHHHHHH. . .
 EEEEEEEEEEEEEEEEEEEEEEEEEEEEEEEEEEEEEEEEEEEEEEEEEEEEEEEEEEEE. . .
 AAAAAAAAAAAAAAAAAAAAAAAAAAAAAAAAAAAAAAAAAAAAAAAAAAAAAAAAAA. . .
 RRRRRRRRRRRRRRRRRRRRRRRRRRRRRRRRRRRRRRRRRRRRRRRRRRRRRRRRRR. . .
```

when all he/she wants to receive is the leftmost diagonal; but at each instant of time, the flatlander is receiving the column of this matrix.

Exercises

2.11 Let

$$I = \int_C e^{iz}\, dz,$$

where C is the semicircle on which

$$R = |z| = \text{const}, \qquad 0 < \theta = \arg z < \pi.$$

(a) Express the integral in polar coordinates, and show that on C

$$|e^{iz}| \le e^{-2R\theta/\pi}, \qquad 0 \le \theta \le \pi/2;$$

$$|e^{iz}| \le e^{-2R(\pi-\theta)/\pi}, \qquad \pi/2 \le \theta \le \pi.$$

(b) Use the estimate in (a) to show that (Jordan's lemma)

$$|I| \le 2\pi.$$

(c) Suppose that $f(z)$ is analytic in the upper half z plane and

$$|f(z)| \to 0 \qquad \text{as} \quad |z| \to \infty$$

in that half plane. Let

$$I = \int_C f(z)e^{iz}\, dz,$$

with C as defined in (a). Show that

$$\lim_{R \to \infty} I = 0.$$

2.12   Suppose that $m = 2$ and that U in (2.4.1) is a function of $r = \{\mathbf{x} \cdot \mathbf{x}\}^{1/2}$ only.

(a)   Set $U = V(r)$, and show that its Fourier transform is a function of k only and is given by

$$v(k) = 2\pi \int_0^\infty V(r) J_0(kr) r\, dr,$$

with J_0 the zero-order Bessel function of the first kind.

(b)   Apply (2.4.2) to show that

$$V(r) = \frac{1}{2\pi} \int_0^\infty v(k) J_0(kr) k\, dk.$$

(c)   Define

$$W(r) = 2\pi V(r) r^{1/2} \qquad \text{and} \qquad w(k) = v(k) k^{1/2},$$

and deduce the formulas for the direct and inverse *Hankel transform*

$$w(k) = \int_0^\infty W(r) J_0(kr) \{kr\}^{1/2}\, dr, \qquad W(r) = \int_0^\infty w(k) J_0(kr) \{kr\}^{1/2}\, dk.$$

(d)   Deduce that

$$\frac{\delta(r - r')}{r} = \int_0^\infty J_0(kr) J_0(kr') k\, dk, \qquad r, r' \neq 0.$$

2.13   The purpose of this exercise is to carry out the analysis of the two-dimensional analog of (2.4.6). Define

$$U(\mathbf{x}, t) = \frac{c^2}{[2\pi]^3} \int_C d\omega \int_{-\infty}^\infty dk_1\, dk_2\, \frac{e^{\cdot(\mathbf{k} \cdot \mathbf{x} - \omega t)}}{c^2 k^2 - \omega^2}.$$

(a)   Carry out the ω integration as in the discussion of (2.4.6) to obtain

$$\frac{-ic}{2(2\pi)^2} \int_0^{2\pi} d\theta \int_0^\infty dk \left[e^{ik(r \cos \theta + ct)} - e^{ik(\cdot \cos \theta - ct)} \right].$$

(b)   Use the discussion of the Fourier transform of the Heaviside function $H(x)$ in Section 2.3 to conclude that

$$U(\mathbf{x}, t) = \frac{c}{2(2\pi)^2} \int_0^{2\pi} d\theta \left[\frac{1}{r \cos \theta + ct} - \frac{1}{r \cos \theta - ct} \right].$$

Note that it is necessary to verify that the sum of integrals with Dirac delta functions in the integrand is zero.

(c)   For $ct < r$, rewrite the principal value integral as a contour integral passing above the poles in the complex θ plane plus *half residues* that sum to zero.

(d)   Still for $ct < r$, introduce $z = e^{i\theta}$, and interpret the integrals of (c) as contour integrals in the complex z plane. Show that the integrals are zero.

(e)   Now take $ct > r$ and use the same transformation as that in (d) to write U as a contour integral. Show that there are two poles interior to the contour of integration given by

$$z_{\pm} = \mp \frac{ct - \{c^2 t^2 - r^2\}^{1/2}}{r}.$$

(f)   *Finally*, show that

$$U(\mathbf{x}, t) = \frac{H(ct - r)}{2\pi \{t^2 - r^2/c^2\}^{1/2}}.$$

2.5   ASYMPTOTIC EXPANSIONS

An asymptotic expansion of a function is an approximation of a function in the neighborhood of a point. However, unlike the Taylor series for an analytic function, an asymptotic expansion need not converge to the function at any point at all! Nonetheless, asymptotic expansions are quite adequate for approximating functions in a sense to be defined. (Extensive discussions of error analysis can be found in Olver [1974].) Furthermore, as will be seen in the exercises, the statement of Taylor's theorem for the approximation of a function by a power series with remainder is, in fact, a statement about an asymptotic expansion. Indeed, Taylor's theorem does not guarantee the convergence of the series to the function; when the series converges, the function is *analytic*. Thus, the reader has already encountered asymptotic expansions early in the study of calculus.

Asymptotic expansions are usually of greatest value in approximating functions in the neighborhood of an *essential singularity* in the complex plane of the independent variable. In our implementation, the essential singularity will be the *point at infinity*. Thus, we will develop all of our machinery for deriving asymptotic expansions about the point at infinity, although in the exercises, we will also discuss expansions about the origin, which for this discussion will be equivalent to any finite point.

We remark that the reader may well have already been introduced to asymptotic expansions in another context without having had the fact pointed out. For example, Stirling's formula for the approximation to $n!$ for large values of n is the leading term of an asymptotic expansion. Also, the representation of the Bessel functions as a linear combination of sines and cosines, each multiplied by a series in inverse powers of the independent variable, is an asymptotic expansion. Finally, the representation of a radiation pattern as e^{ikr}/kr multiplied by an amplitude function is really the leading term of an asymptotic series.

The reader might well wonder at this point what the independent variable is that is "in the neighborhood of infinity" in wave problems in order to make these expansions useful. These approximations will be valid "in the high-frequency" limit. Then the question arises of what high frequency is. One answer to this question is that the frequencies of interest should be associated with wave periods that are "small" compared to the "natural" physical time scales of the problem at hand. Thus, if T is a time scale and f a frequency in hertz, then

$$2\pi f T \gg 1.$$

In practice, the left side being at least equal to 3 seems to suffice for asymptotic expansions to be sufficiently accurate for interpretational purposes. Often at this value, the actual numerical accuracy is well within the bounds of interest as well. We remark that these condition can be reinterpreted as a condition on wavelengths when we have a "rule" for relating frequency and wave number (a dispersion relation). In this case, the requirement would be that the wavelengths of interest should be sufficiently small compared to the natural length scales of the problem at hand.

In the next few chapters, the functions whose asymptotic expansions we seek will be represented by Fourier integrals. Thus, after a preliminary discussion of asymptotic expansions in general, we shall focus our attention in subsequent sections of this chapter on the asymptotic expansion of Fourier integrals.

ORDER ESTIMATES

We begin our development of asymptotic expansions with the concept of O estimates (read *large OH*) and o estimates (read *small oh*). Given two functions $f(\lambda)$ and $g(\lambda)$, we shall say that $f(\lambda)$ *is large OH of* $g(\lambda)$ in some neighborhood of infinity if

$$|f(\lambda)| \le k|g(\lambda)|, \qquad \lambda \ge \lambda_0. \tag{2.5.1}$$

for some positive constant k and some positive λ_0. We write then

$$f(\lambda) = O(g(\lambda)), \qquad \lambda \to \infty. \tag{2.5.2}$$

We shall say that $f(\lambda)$ *is small oh of* $g(\lambda)$ if the constant k can be taken to be arbitrarily small. That is, if

$$\lim_{\lambda \to \infty} \frac{f(\lambda)}{g(\lambda)} = 0, \tag{2.5.3}$$

we write

$$f(\lambda) = o(g(\lambda)), \qquad \lambda \to \infty. \tag{2.5.4}$$

Both of these types of estimates are called *order estimates*. In particular, (2.5.2) is to read as $f(\lambda)$ is of the order of $g(\lambda)$. Simple examples of O and o are

$$e^{-\lambda} = o(\lambda^{-n}), \qquad \lambda \to \infty, \quad \text{any} \quad n,$$

and

$$e^{-1/\lambda} = O(1), \qquad \lambda \to \infty.$$

Qualitatively, the first example is a statement of the fact that $e^{-\lambda}$ vanishes faster than any power of λ as λ approaches infinity; the second example states that $e^{-1/\lambda}$ behaves like 1 as λ approaches infinity. This latter estimate is *sharp* in that any positive power of λ would provide a grosser estimate than $O(1)$ of the behavior of the function—$e^{-1/\lambda} = o(\lambda^n)$, $\lambda \to \infty$, $n > 0$—and any negative power would not provide an estimate—$e^{-1/\lambda} \neq O(\lambda^n)$, $\lambda \to \infty$, $n < 0$.

We remark that the first example is not *symmetric* while the second example is; that is,

$$\lambda^{-n} \neq o(e^{-\lambda}), \qquad \text{any} \quad n; \qquad 1 = O(e^{-1/\lambda}), \qquad \lambda \to \infty.$$

Another nonsymmetric example is

$$\sin \lambda = O(1), \qquad \lambda \to \infty.$$

On the one hand, the right side provides a useful bound on the function $\sin \lambda$; however, it fails to characterize the *oscillatory* behavior of this function. Thus, the right side provides a *gauge* of the gross features of the sine function, perhaps useful in error estimates but not sufficient to be a simplification that retains useful qualitative information.

ASYMPTOTIC POWER SERIES

An extremely important set of functions to be compared to one another are the powers of λ themselves; thus,

$$\lambda^{-(n+k)} = o(\lambda^{-n}), \qquad \lambda \to \infty, \quad \text{any} \quad k > 0, \quad \text{any} \, n. \tag{2.5.5}$$

That is, the more negative the power of λ, the more rapid the decay to zero as λ approaches infinity. This motivates our first definition of an asymptotic expansion, namely, *an asymptotic power series*. We will say that $f(\lambda)$ has an asymptotic power series to order N at infinity if

$$f(\lambda) - \sum_{n=1}^{N-1} a_n \lambda^{-n} = O(\lambda^{-N})$$

or
(2.5.6)

$$f(\lambda) - \sum_{n=0}^{N-1} a_n \lambda^{-n} = o(\lambda^{-(N-1)})$$

and write

$$f(\lambda) \sim \sum_{n=0}^{N-1} a_n \lambda^{-n}. \tag{2.5.7}$$

Here the coefficients a_n are constants. We read the symbol $\sim$ as is *asymptotically equal to* or, more succinctly, is asymptotic to. If (2.5.6) is true for all N, then we can replace $N - 1$ in the upper limit, here by ∞. However, this does not imply convergence of the series but only that (2.5.6) is true for all N.

Simple examples of asymptotic expansions at infinity can be obtained by considering Taylor series about the origin and replacing λ by $1/\lambda$. A richer example is provided by the function

$$I(\lambda) = \lambda e^\lambda \int_\lambda^\infty \frac{e^{-t}}{t} \, dt. \tag{2.5.8}$$

This function is related to the *exponential integral* but has been scaled so as to produce a power series as λ approaches infinity. Integration by parts—integrating the exponential and differentiating the power of t—yields the result

$$I(\lambda) - \sum_{n=0}^{N-1} \frac{(-1)^n n!}{\lambda^n} = E(\lambda; N)$$

$$= (-1)^N N! \lambda e^\lambda \int_\lambda^\infty \frac{e^{-t}}{t^{N+1}} \, dt, \qquad N = 1, 2, \ldots. \tag{2.5.9}$$

We claim that, in fact, $E(\lambda; N)$ is $O(\lambda^{-N})$. To verify this, estimate the algebraic term in the integrand by its maximum value, namely, λ^{-N-1}, and integrate the exponential. The result is

$$|E(\lambda; N)| \le N! \lambda^{-N} = O(\lambda^{-N}). \tag{2.5.10}$$

For fixed λ, this bound on the error can be seen to decrease for increasing N so long as $N \le \lambda$ but to increase for increasing N above this limit. Thus, the approximation would seem to be optimal for λ and N approximately equal. Furthermore, from the exact expression for the error, it can be seen that at fixed λ the error is never zero and, in fact, increases beyond all bounds as N approaches infinity.

Therefore, it is prudent to ask how good the approximation (2.5.9) really is. Table 2.1, taken from Bleistein and Handelsman [1975], demonstrates the accuracy of this expansion. In Table 2.1, S_N is the sum to N. It can be seen that S_1 is the best approximation at $\lambda = 1$, S_2 the best approximation at $\lambda = 2$, and so on.

Since (2.5.10) is true for each choice of N, we can write

$$I(\lambda) \sim \sum_0^\infty \frac{(-1)^n n!}{\lambda^n}. \tag{2.5.11}$$

It can be seen that this series does not converge for any finite value of λ. Nonetheless, the sum has meaning as an asymptotic power series in the sense defined by (2.5.6) and confirmed for this example by (2.5.9) and (2.5.10).

ASYMPTOTIC SEQUENCES

We shall find in the applications in subsequent chapters that asymptotic power series will not suffice to characterize the approximate representations we derive. Thus, we introduce the concept of an *asymptotic sequence* of functions. Let us suppose that the sequence $\{\phi_n(\lambda)\}$ is such that

$$\phi_{n+1} = o(\phi_n(\lambda)), \qquad \lambda \to \infty, \quad n = 1, 2, \dots. \tag{2.5.12}$$

Such a sequence of functions will be called an *asymptotic sequence*. We shall say that $f(\lambda)$ has an asymptotic expansion with respect to the sequence $\{\phi_n(\lambda)\}$ to order N if

$$f(\lambda) - \sum_{n=0}^{N-1} a_n \phi_n(\lambda) = O(\phi_N(\lambda)), \qquad \lambda \to \infty,$$

or

$$f(\lambda) - \sum_{n=0}^{N-1} a_n \phi_n(\lambda) = o(\phi_{(N-1)}(\lambda)), \qquad \lambda \to \infty,$$

$$\tag{2.5.13}$$

and we write

$$f(\lambda) \sim \sum_{n=0}^{N-1} a_n \phi_n(\lambda). \tag{2.5.14}$$

Again, if (2.5.13) holds for all N, we can replace N by ∞. This is understood only to mean that (2.5.13) is true for all N and implies nothing about the convergence of the series.

We remark that the negative integer powers that arose in the asymptotic

Table 2.1 [a]

x	$I(x)$	S_1	S_2	S_3	S_4	S_5	S_6	S_7	S_8	S_9	S_{10}
1	0.59634	1	0	2.0000	−4.0000						
2	0.72266	1	0.5000	1.0000	0.2500	1.7500					
3	0.78625	1	0.667	0.8999	0.6667	0.9626	0.4688				
5	0.85212	1	0.8000	0.8800	0.8352	0.8736	0.8352	0.8820			
10	0.91563	1	0.9000	0.9200	0.9140	0.9164	0.9152	0.91592	0.91542	0.91581	0.91544
100	0.99019	1	0.9900	0.9902	0.99019						

[a] From Bleistein and Handelsman [1975b].

power series above are an asymptotic sequence. Thus, the fundamental asymptotic expansion is subsumed under this broader class of asymptotic expansions.

Another example of elements of an asymptotic sequence is

$$\phi_n(\lambda) = e^{ia\lambda}/\lambda^n, \qquad n = 1, 2, \dots \tag{2.5.15}$$

The reader should verify that these functions satisfy (2.5.12). A sequence of exactly this form arises in analyzing the integral

$$I(\lambda) = \int_1^\infty \frac{e^{i\lambda t}}{1 + t}\, dt. \tag{2.5.16}$$

Integration by parts here—integrating the oscillatory exponential—yields

$$I(\lambda) = -\sum_{n=0}^{N-1} \frac{n!\, e^{i\lambda}}{(2i\lambda)^{n+1}} + \frac{(N)!}{(i\lambda)^N} \int_1^\infty \frac{e^{i\lambda t}}{(1+t)^{N+1}}\, dt. \tag{2.5.17}$$

We leave it as an exercise to confirm that this result is an asymptotic expansion with respect to the sequence defined by (2.5.15).

Other examples of elements of asymptotic sequences are

$$\lambda^{-n+a}, \qquad \text{any} \quad \alpha; \qquad \lambda^{-n/3} e^{-\lambda^2}; \qquad J_n(1/\lambda). \tag{2.5.18}$$

The function J_n is the Bessel function of the first kind of order n.

AUXILIARY SEQUENCES

One further extension of the basic concept of an asymptotic expansion will prove useful, namely, the concept of an *auxiliary sequence*. To motivate this extension, let us suppose that by some formal method we have obtained a series representation of $f(\lambda)$ of the form

$$f(\lambda) \approx \sum_{n=0}^\infty \frac{n!\cos n\lambda}{\lambda^n}.$$

The occurrence of the inverse powers of λ here would suggest that this is an asymptotic expansion. However, if we set

$$\phi_n(\lambda) = (\cos n\lambda)/\lambda^n, \qquad n = 1, 2, \ldots,$$

then we cannot verify (2.5.12). The reason is that for each n, the right side of (2.5.12) would have zeros at finite λ values arbitrarily large, that is, "arbitrarily close to infinity" at points at which the left side is not zero; thus, the requirement (2.5.3) cannot be satisfied for this sequence. On the other hand, it would be reasonable to expect that along with the formal result here one might have derived the following result for finite sums:

$$f(\lambda) - \sum_{n=0}^{N-1} \frac{n! \cos n\lambda}{\lambda^n} = O(\lambda^{-N}), \qquad N = 1, 2, \ldots.$$

Thus, the error terms at each step are bounded by an asymptotic sequence even though the λ dependence of each term in the series is not an element of an asymptotic sequence. It would be reasonable to expect that an approximation such as this would be as accurate as an asymptotic power series. This suggests the following extension. Suppose that $\{\phi_n(\lambda)\}$ is an asymptotic sequence as above. Suppose further that there is *another* sequence $\{\xi_n(\lambda)\}$ such that

$$f(\lambda) - \sum_{n=0}^{N-1} a_n \xi_n(\lambda) = O(\phi_N(\lambda)), \qquad \lambda \to \infty,$$

or $\qquad\qquad\qquad\qquad\qquad\qquad\qquad\qquad\qquad\qquad\qquad\qquad$ (2.5.19)

$$f(\lambda) - \sum_{n=0}^{N-1} a_n \xi_n(\lambda) = o(\phi_{(N-1)}(\lambda)), \qquad \lambda \to \infty.$$

Then the series here is an asymptotic expansion *with respect to the auxiliary sequence* $\{\phi_n(\lambda)\}$ to order N, and we write

$$f(\lambda) \sim \sum_{n=0}^{N-1} a_n \xi_n(\lambda). \qquad\qquad (2.5.20)$$

It is to be understood in context that this is an asymptotic expansion with respect to an auxiliary sequence. Again, if (2.5.19) is true for all N, then we replace $N - 1$ by ∞ in (2.5.20), and this is to be understood to mean only that (2.5.19) is true for all N.

Exercises

2.14   (a)   With the definitions of O and o estimates at infinity as a model, define O and o estimates at a finite point, say, $x = 0$.

(b)   In the definition of an asymptotic power series at infinity, set $\lambda = 1/x$ and $f(1/x) = g(x)$ and deduce a definition of an asymptotic powers series near $x = 0+$. Note that the O estimate of the remainder produced by direct substitution should be consistent with the definition stated in (a).

(c)   State Taylor's theorem with remainder, and verify that it is of the form of the definition in (b) except that now the coefficients are of a prescribed form.

2.15   (a)   Use the Riemann–Lebesgue lemma to show that the remainder in (2.5.17) is $o(\lambda^{-N})$.

(b)   Show that· the remainder in (2.5.17) is actually $O(\lambda^{-N-1})$. [*Hint*: Integrate by parts one more time.]

2.16   Show that the following are asymptotic sequences in the indicated limits.

(a) $\{\lambda^{-\alpha_n}\}$, $\operatorname{Re}\alpha_n > \operatorname{Re}\alpha_{n-1}$, $n = 1, 2, \ldots$, $\lambda \to \infty$.

(b) $\{(x - x_0)^{\alpha_n}\}$, α_n as in (a), $x \to x_0$.

(c) $\{[g(\lambda)]^n\}$, $\lambda \to \infty$, $\lim_{\lambda \to \infty} g(\lambda) = 0$, $g(\lambda) \neq 0$, $\lambda_0 \leq \lambda < \infty$.

2.6   ASYMPTOTIC EXPANSIONS OF FOURIER INTEGRALS WITH MONOTONIC PHASE

Our objective is to develop methods for obtaining the asymptotic expansion or, more realistically, one or two terms of the expansion of a function defined by an integral of the form

$$I(\lambda) = \int_a^b f(t)e^{i\lambda\phi(t)}\, dt, \tag{2.6.1}$$

with f and ϕ real. We shall begin more modestly and consider the integral

$$I(\lambda) = \int_a^b f(t)e^{i\lambda t}\, dt. \tag{2.6.2}$$

We have already encountered one example such as this [(2.5.16)] and were able to obtain an expansion by the straightforward technique of *integration by parts*. Qualitatively, the method "works" because each integration of the exponential produces a power of λ in the denominator. Therefore, under the assumption that f has sufficient derivatives to allow the indicated operations to be carried out, we integrate by parts to find that

$$I(\lambda) = \sum_{n=0}^{N-1} \frac{(-1)^n}{(i\lambda)^{n+1}} \left[f^{(n)}(b)e^{i\lambda b} - f^{(n)}(a)e^{i\lambda a} \right] + \frac{(-1)^N}{(i\lambda)^N} \int_a^b f^{(N)}(t)e^{i\lambda t}\, dt. \tag{2.6.3}$$

If f has $(N + 1)$ integrable derivatives, we can integrate by parts one more time and conclude that the second line here is $O(\lambda^{-N-1})$. If only the Nth derivative is absolutely integrable, then by the Riemann–Lebesgue lemma, $I(\lambda) = o(1)$ and the remainder is $o(\lambda^{-N})$. In either case, the series in the first line provides an asymptotic expansion to order N of the integral $I(\lambda)$.

As a first extension of this result, let us consider (2.6.1) under the assumption that

$$\phi'(t) \neq 0, \qquad a \leq t \leq b. \tag{2.6.4}$$

We rewrite (2.6.1) as

$$I(\lambda) = \int_a^b \frac{f(t)}{\phi'(t)} \phi'(t) e^{i\lambda\phi(t)} \, dt. \tag{2.6.5}$$

Now integration by parts can be carried out once to yield

$$I(\lambda) = \frac{f(t)}{i\lambda\phi'(t)} e^{i\lambda\phi(t)} \Big|_a^b - \frac{1}{i\lambda} \int_a^b \frac{d}{dt}\left[\frac{f(t)}{\phi'(t)}\right] e^{i\lambda\phi(t)} \, dt. \tag{2.6.6}$$

Repeating this process yields

$$I(\lambda) = \sum_{n=0}^{N-1} \frac{(-1)^n e^{i\lambda\phi(t)}}{(i\lambda)^{n+1}} \left[\frac{1}{\phi'(t)}\frac{d}{dt}\right]^n \left[\frac{f(t)}{\phi'(t)}\right]\Big|_a^b$$
$$+ \frac{(-1)^N}{(i\lambda)^N} \int_a^b e^{i\lambda\phi(t)} \frac{d}{dt}\left[\frac{1}{\phi'(t)}\frac{d}{dt}\right]^{N-1} \left[\frac{f(t)}{\phi'(t)}\right] dt. \tag{2.6.7}$$

As in the discussion of (2.6.3), if another integration by parts could be carried out, we would estimate that the second line here is $O(\lambda^{-N-1})$; if the integrand were only absolutely integrable, by transforming to $\phi(t)$ as a new variable of integration and applying the Riemann–Lebesgue lemma, we would find the second line to be $o(\lambda^{-N})$. In either case, the first line is an asymptotic expansion to order N.

We shall make some qualitative remarks about the result (2.6.7). First, it is apparent that this process can be carried out so long as

(i) $\phi'(t)$ is nonzero on the closed interval of integration and
(ii) all of the derivatives of $f(t)$ and $\phi(t)$ exist up to the order required.

Therefore, we see that there are certain *important points* insofar as the asymptotic expansion of the integral is concerned. These important points are called *critical points*; they are exactly the points at which the preceding list fails. That is, the critical points are points at which

(i)   points at which $\phi'(t) = 0$,
(ii) points at which some derivative of f or ϕ fails to be continuous, and
(iii) endpoints of integration.

We can think of the endpoints of integration as a special case of case (ii), in which the function f itself fails to be continuous in the sense that it "jumps" from zero to its interior value. Indeed, if f were zero at one of its endpoints, we would see that the series associated with that endpoint would be of lower order than if the value of f there were nonzero. In fact, if f and all of its derivatives vanished at each of the endpoints, then in (2.6.7) $I(\lambda) = 0$ to all algebraic orders in λ.

Another observation from (2.6.7) is that the result can be viewed as a sum of contributions from each of the critical points, that is, from each endpoint of integration. Indeed, if f and all of its derivatives vanished only at $t = b$, we would obtain an asymptotic expansion that depended only on the local properties of f near $t = a$. Conversely, for a function that with all of its derivatives vanished at a, we would obtain an expansion that depended only on the values of the function f at $t = b$.

It would be desirable when an integral has many critical points (more than one) to devise a mechanism for isolating those critical points. Then for the purposes of derivation and calculation, we could think of *the contribution from* each critical point separately. A mechanism for doing this has been devised. It is called the *van der Corput neutralizer*. This is a function that isolates a critical point without itself introducing new contributions to the asymptotic expansion of an integral.

The simplest type of neutralizer is a function $q = q(t, \alpha_1, \alpha_2)$ that depends on two parameters α_1 and α_2, with $\alpha_1 < \alpha_2$, and is infinitely differentiable for all t. Furthermore,

$$q(t, \alpha_1, \alpha_2) = \begin{cases} 0, & t \le \alpha_1, \\ 1, & t \ge \alpha_2. \end{cases} \tag{2.6.8}$$

In Exercise 2.19, an explicit van der Corput neutralizer is developed.

Let us suppose now that we are given an integral with many critical points. We first decompose the integral into a sum of integrals in which each of the latter has only two critical points at its left and right endpoints. In particular, let us suppose that the integral (2.6.1) is just such an integral with endpoints a and b and any of the critical points cited earlier at these endpoints. We then set

$$I(\lambda) = I_a(\lambda) + I_b(\lambda), \tag{2.6.9}$$

where

$$I_a(\lambda) = \int_a^b [1 - q(t, \alpha_1, \alpha_2)] f(t) e^{i\lambda\phi(t)} \, dt,$$

$$I_b(\lambda) = \int_a^b q(t, \alpha_1, \alpha_2) f(t) e^{i\lambda\phi(t)} \, dt, \tag{2.6.10}$$

and $a < \alpha_1 < \alpha_2 < b$. The integral I_a has a critical point at $t = a$ only, while I_b has a critical point at $t = b$ only. Furthermore, since $1 - q \equiv 1$ at $t = a$, with all of its derivatives equal to zero there, we can expect that the asymptotic expansion of I_a will depend only on the local properties of f and ϕ near $t = a$. In particular, let us suppose that we introduce the neutralizer function in the case in which (2.6.4) is true near $t = a$. Then choosing α_2 so that this is true everywhere on $[a, b]$ where $1 - q$ is nonzero, we find that

$$I_a(\lambda) = \sum_{n=0}^{\infty} \frac{(-1)^{n+1}}{(i\lambda)^{n+1}} \left[\frac{1}{\phi'(t)} \frac{d}{dt} \right]^n \left[\frac{f(t)}{\phi'(t)} \right] e^{i\lambda\phi(t)} \Bigg|_{t=a} . \qquad (2.6.11)$$

We remark that a similar result holds for I_b, the only differences being that a is replaced by b and that the power of -1 is reduced by one.

Thus, viewing our integration by parts result in the larger context, we can now think of these differences as characterizing the distinctions between contributions at left ($t = a$) and right ($t = b$) endpoints, contributions in which both f and ϕ are differentiable to all orders and ϕ' is nonzero at the endpoint.

Exercises

2.17 Calculate the asymptotic expansion as $\lambda \to \infty$ of

$$\int_1^2 e^{-t^2} e^{i\lambda t^2} \, dt.$$

2.18 Rederive the result (2.6.7) in the following manner: Introduce a new variable of integration $u = \phi(t)$ in (2.6.1) and thereby express that integral as one of the form (2.6.2). Express your answer in the form (2.6.3) in the new variable u, and reintroduce t to obtain (2.6.7).

2.19 Let

$$p(t) = \begin{cases} 0, & t \le 0, \\ e^{-1/t}, & t \ge 0. \end{cases}$$

Show that one choice of q is

$$q(t, \alpha_1, \alpha_2) = \frac{p(t - \alpha_1)}{p(t - \alpha_1) + p(\alpha_2 - t)}.$$

2.20 Identify the critical points of the following integrals.

(a) $\displaystyle\int_0^{2\pi} \frac{\exp(i\lambda \cos t)}{\{1 - t^2\}^{1/3}} \, dt.$

(b) $\displaystyle\int_0^{2\pi} \frac{\exp\{i\lambda[x \cos \theta + y \sin \theta - ct \, \text{sign}(\sin \theta)]\}}{\{1 - \cos^2 \theta\}^{1/2}} \, d\theta.$

2.21   Suppose that i is replaced by $-i$ in (2.6.2). Derive the asymptotic expansion of this integral as a special case of (2.6.7).

2.7   THE METHOD OF STATIONARY PHASE

We are now prepared to consider a critical point at which ϕ' vanishes. Such a point is called a *stationary point*. If the second derivative ϕ'' does not vanish there, the stationary point is called *simple* or of order one. Higher-order stationary points are classified by the last vanishing derivative. Thus, at a stationary point of order two, both the first and second derivatives vanish, and so on.

Let us consider, therefore, the integral

$$I_a(\lambda) = \int_a^b f(t)e^{i\lambda\phi(t)}\,dt. \tag{2.7.1}$$

Here by the notation I_a, we mean to imply that f vanishes "infinitely smoothly" away from some right neighborhood about the left endpoint a. That is, we assume the integrand has been multiplied by a *neutralizer* even though we do not explicitly write it down. The left endpoint $t = a$ is assumed to be a *simple stationary point*, so that

$$\phi'(a) = 0, \qquad \phi''(a) \neq 0. \tag{2.7.2}$$

Because of the neutralizer, there are no other critical points on the domain of integration.

Let us introduce a new variable of integration u by the equation

$$u^2 = \mu\{\phi(t) - \phi(a)\}, \qquad \mu = \operatorname{sign}\,\phi''(a).$$

Here multiplication by μ ensures that u^2 is positive everywhere of interest on the integral of integration (why?); then we set

$$u = \sqrt{|\phi(t) - \phi(a)|}. \tag{2.7.3}$$

We have chosen the positive square root here so that u increases from zero as t increases from a. In fact, near $t = a$,

$$u = \sqrt{\tfrac{1}{2}|\phi''(a)|}\,(t - a) + O((t - a)^2). \tag{2.7.4}$$

This transformation can be inverted to yield

$$t = a + \sqrt{2/|\phi''(a)|}\,u + O(u^2). \tag{2.7.5}$$

In terms of the new variable of integration I_a can be written as

$$I_a = e^{i\lambda\phi(a)} \int_0^{|\phi(\alpha_2) - \phi(a)|} F(u)e^{i\lambda\mu u^2}\,du. \tag{2.7.6}$$

Here the function $F(u)$ is defined by

$$F(u) = f(t(u)) \frac{dt}{du}, \tag{2.7.7}$$

and α_2 is the right endpoint of the support of $f(t)$; α_2 plays the same role here as it did in the discussion of the neutralizer in the preceding section. We remark that since $f(t)$ vanishes infinitely smoothly at α_2, so must $F(u)$ at $\phi(\alpha_2)$. That is, the only critical point is still the left endpoint.

Remarkably, the asymptotic expansion of this integral can be calculated by integration by parts. The method of derivation was formulated by Erdelyi [1955, 1956]. To carry out the integration, introduce the iterated integrals

$$k^{(-n-1)}(u; \lambda) = \frac{(-1)^{n+1}}{n!} \int_u^{\infty e^{i\mu\pi/4}} (\sigma - u)^n e^{i\lambda\mu\sigma^2} \, d\sigma, \qquad n = 0, 1, \ldots. \tag{2.7.8}$$

Here the path of integration is a ray in the complex σ plane starting from the endpoint u and ending at infinity in a direction in which the exponential decays to zero. We leave it to the exercises to verify that these functions are iterated integrals of the exponential appearing in (2.7.6). Furthermore, we can show that

$$k^{(-n-1)}(0; \lambda) = \frac{(-1)^{n+1}}{n!} \frac{1}{2} \Gamma\left[\frac{n+1}{2}\right] \left[\frac{e^{i\mu\pi/2}}{\lambda}\right]^{(n+1)/2}. \tag{2.7.9}$$

Here the gamma function

$$\Gamma(z) = \int_0^\infty t^{z-1} e^{-t} \, dt, \qquad \mathrm{Re} \, z > 0; \qquad \Gamma(n+1) = n!. \tag{2.7.10}$$

We are now prepared to calculate the integral (2.7.6) by integration by parts. The result is

$$I_a(\lambda) = e^{i\lambda\phi(a)} \left[\sum_{n=0}^{N-1} \frac{F^{(n)}(0)}{n!} \Gamma\left[\frac{n+1}{2}\right] \left[\frac{e^{i\mu\pi/4}}{\lambda}\right]^{(n+1)/2} \right.$$
$$\left. + \int_0^{|\phi(b)-\phi(a)|} k^{-N}(u; \lambda) F^{(N)}(u) e^{i\lambda\mu u^2} \, du \right]. \tag{2.7.11}$$

Another integration by parts here allows us to conclude that the error is $O(\lambda^{-(N+1)/2})$. Therefore, the first line provides an asymptotic expansion with respect to the auxiliary sequence $\{\lambda^{-(n+1)/2}\}$.

This asymptotic expansion has the disadvantage that it is expressed in terms of the auxiliary function F, which arose through a transformation of coordinates. We see, however, that the result does not depend on *global* properties of F but only on local properties, namely, F and its derivatives, near the critical point $u = 0$. This, in turn, depends on the power series for f and ϕ near $t = a$. It is straightforward but tedious to carry out the power series expansions necessary to obtain all $F^{(n)}(0)$. We first write F in terms of f and ϕ, expand those functions in powers of $(t - a)$, expand this variable in terms of u from (2.7.3), and substitute into the expression for F. This yields the coefficient of each power of u in the series expansion of F, thereby yielding the coefficients in (2.7.11).

According to Bleistein and Handelsman [2], this process has been carried out sufficiently far to yield three terms of the asymptotic expansion in terms of the original functions. For our purposes, two terms will suffice. The result is

$$
I_a(\lambda) \sim \frac{1}{2} e^{\{i\lambda\phi(a) + i\mu\pi/4\}} \left\{ f(a) \left[\frac{2\pi}{\lambda|\phi''(a)|} \right]^{1/2} \right.
$$
$$
\left. + \frac{2}{\lambda|\phi''(a)|} \left[f'(a) - \frac{\phi'''f(a)}{3\phi''(a)} \right] e^{i\mu\pi/4} \right\}. \tag{2.7.12}
$$

Here we have used the facts that $\Gamma(\tfrac{1}{2}) = \sqrt{\pi}$, $\Gamma(1) = 1$.

If the stationary point of interest were at the *right* endpoint of integration $t = b$, then the derivation would change as described briefly below. First, replace (2.7.1) by

$$
I_b(\lambda) = \int_a^b f(t) e^{i\lambda\phi(t)} \, dt. \tag{2.7.13}
$$

Here it is understood that the only critical point is at $t = b$, where

$$
\phi'(b) = 0, \qquad \phi''(b) \neq 0. \tag{2.7.14}
$$

We again introduce a change of variable of integration

$$
u^2 = \mu\{\phi(t) - \phi(b)\}; \qquad \mu = \text{sign } \phi''(b). \tag{2.7.15}
$$

However, we now choose the *negative* square root in solving for u,

$$
u = -|\phi(t) - \phi(b)|^{1/2}, \tag{2.7.16}
$$

so that u *increases* from zero as t *decreases* from b.

In the result [(2.7.11)], we need only replace a by b and change the sign of the second line in order to obtain the asymptotic expansion of $I_b(\lambda)$. The result is

$$
I_b(\lambda) \sim \frac{1}{2} e^{\{i\lambda\phi(b) + i\mu\pi/4\}} \left\{ f(b) \left[\frac{2\pi}{\lambda|\phi''(b)|} \right]^{1/2} \right.
$$
$$
\left. - \frac{2}{\lambda|\phi''(b)|} \left[f'(b) - \frac{\phi'''f(b)}{3\phi''(b)} \right] e^{i\mu\pi/4} \right\}. \tag{2.7.17}
$$

In fact, if the entire series for I_a and I_b is generated, the signs of the latter will alternate compared with the signs of the former. Consequently, for an *interior* simple stationary point, the first, third, etc., terms will add, while the second, fourth, etc., terms will cancel, leaving a series in powers $\{\lambda^{-1/2}, \lambda^{-3/2}, \ldots\}$. That is, only half-integer powers will appear in this combined series, while all powers of the form $\{\lambda^{-n/2}\}$ will appear in either of the former series.

Thus, if $t = c$ is an interior simple stationary point of $I(\lambda)$, then to the same order of accuracy as the preceding expansions,

$$I_c(\lambda) \sim e^{\{i\lambda\phi(c) + i\mu\pi/4\}} f(c) \left[\frac{2\pi}{\lambda |\phi''(c)|} \right]^{1/2}. \qquad (2.7.18)$$

This last result is often referred to as *the stationary phase formula*. The reader is cautioned, however, that this phenomenon of left and right-endpoint terms adding is not generally true for higher-order stationary points.

As a simple application of (2.7.18), we consider the integral representation of the Bessel function of the first kind of integer order n:

$$J_n(\lambda) = \frac{1}{\pi} \int_0^\pi \cos(nt - \lambda \sin t) \, dt = \frac{1}{2\pi} \sum_\pm \int_0^\pi e^{\pm int} e^{\mp i\lambda \sin t} \, dt. \qquad (2.7.19)$$

This is a sum of two integrals for which $\phi = \mp \sin t$. Both of these phases have a simple interior stationary point at $t = \pi/2$, with $\phi''(\pi/2) = \pm 1$. Applying (2.7.18) to this example yields

$$J_n(\lambda) \sim \frac{1}{\sqrt{2\pi\lambda}} \sum_\pm e^{\{\pm i[\lambda - n\pi/2 - \pi/4]\}} \sim \left[\frac{2}{\pi\lambda} \right]^{1/2} \cos[\lambda - n\pi/2 - \pi/4]. \qquad (2.7.20)$$

Actually, this is the leading order expansion for noninteger values of n as well.

The method used here to derive the stationary phase formulas can be generalized to treat a stationary point of arbitrary (not necessarily integer) order and also to integrable algebraic behavior in $f(t)$. Furthermore, multiples of $\log t$ can be introduced as well. The method requires an appropriate transformation to a new variable u, definition of the iterated integrals completely analogous to our functions $k^{(-n)}$, and expression of the resulting integrals from $u = 0$ in terms of gamma functions. We shall state some results along these lines below.

Let us suppose that near $t = a+$

$$\phi(t) - \phi(a) = \phi_a(t - a)^\alpha + o((t - a)^\alpha), \qquad \alpha > 0,$$
$$f(t) = f_a(t - a)^{(\gamma - 1)} + o((t - a)^{\gamma - 1}), \qquad \gamma > 0. \qquad (2.7.21)$$

Then

$$I_a(\lambda) \sim \frac{f_a \Gamma [\gamma/\alpha]}{\alpha(\lambda |\phi_a|)^{\gamma/\alpha}} e^{i\lambda\phi(a) + i\mu\pi\gamma/2\alpha}, \qquad \mu = \operatorname{sign} \phi_a. \qquad (2.7.22)$$

Similarly, let us suppose that near $t = b-$

$$\phi(t) - \phi(b) = \phi_b(b - t)^\beta + o((b - t)^\beta), \qquad \beta > 0,$$
$$f(t) = f_b(b - t)^{\delta - 1} + o((b - t)^{\delta - 1}), \qquad \delta > 0. \qquad (2.7.23)$$

Then

$$I_b(\lambda) \sim \frac{f_b \Gamma [\delta/\beta]}{\beta(\lambda |\phi_b|)^{\delta/\beta}} e^{i\lambda\phi(b) + i\mu\pi\delta/2\beta}, \qquad \mu = \operatorname{sign} \phi_b. \qquad (2.7.24)$$

In both of these results, the asymptotic expansion of the integral depends only on asymptotic properties of the amplitude and phase near the endpoint of interest. The only *global* dependence in these results is manifested in the gamma function with complex multiplier, which really arises from the feature that the function with the large parameter in it (and consequent *rapid variation*) is the Fourier kernel. Another feature of interest is the dependence of the asymptotic order of the results (2.7.22) and (2.7.24) on the order of vanishing of the phase and amplitude at the critical point. Let us consider I_a (or I_b), which is $O(\lambda^{-\gamma/\alpha})$ when $f = O((t - a)^{\gamma - 1})$ and $\phi = O((t - a)^\alpha)$. We see that this power of λ *becomes more negative* (I_a becomes asymptotically smaller) when the order of vanishing of the amplitude f increases, but this power of λ *becomes less negative* (I_a becomes asymptotically larger) when the order of vanishing of the phase ϕ (order of stationarity) increases.

Exercises

2.22 Consider the sequence of functions $\{k^{(-n-1)}\}$ defined by (2.7.8).
  (a) Show that

$$\frac{dk^{(-1)}(u; \lambda)}{du} = e^{i\lambda\mu u^2}.$$

  (b) Show that

$$\frac{dk^{(-n-1)}(u; \lambda)}{du} = k^{(-n)}(u; \lambda), \qquad n = 1, 2, \ldots.$$

  (c) For $u = 0$ in (2.7.8), introduce the variable of integration

$$\sigma = \lambda^{-1/2}\tau e^{i\mu\pi/4},$$

and verify (2.7.9) by using the definition (2.7.10).

2.23 Write the explicit results for (2.7.22) and (2.7.24) for the cases listed below.

(a) $I_a + I_b$, $\alpha = \beta = \gamma = \delta = 1$. Compare with the leading term obtained by integration by parts. (They should agree!)

(b) I_a, $\alpha = 3$, $\gamma = 1$. Express the coefficients in terms of derivatives of f and ϕ.

(c) I_b, $\beta = 3$, $\delta = 1$. Express the result as in (b).

(d) I_a, $\alpha = 1$, $\gamma = \frac{1}{2}$.

(e) I_b, $\beta = 1$, $\delta = \frac{1}{2}$.

(f) I_a, $\alpha = 2$, $\gamma = \frac{3}{2}$.

2.24 Calculate the leading order contribution from each critical point to the asymptotic expansion of the following integrals.

(a) $I(\lambda) = \int_0^\pi e^{i\lambda(t + \cos t)} \, dt$.

(b) $I(\lambda) = \int_0^1 \{1 - t^2\}^{-1/2} e^{i\lambda t^3} \, dt$.

(c) Both of these examples have a second-order stationary point. Consider the introduction of a change of variable of integration such as (2.7.3), and develop the asymptotic expansion along the lines of the discussion in this section sufficiently far to estimate that the error in each expansion here is $O(\lambda^{-2/3})$.

2.25 (a) From the definition of the Bessel function [(2.7.19)], write a representation for $J_\lambda(\lambda r)$ valid for λ, an integer. Show that for $r > 1$,

$$J_\lambda(\lambda r) \sim \left[\frac{\pi}{2\lambda}\right]^{1/2} (r^2 - 1)^{-1/4} \cos\{\lambda[\sqrt{r^2 - 1} - \cos^{-1}(r^{-1})] - \pi/4\}.$$

(b) For $r = 1$, show that

$$J_\lambda(\lambda) \sim \Gamma(\tfrac{1}{3})/2^{2/3}\pi 3^{1/6}\lambda^{1/3}.$$

2.8 MULTIDIMENSIONAL FOURIER INTEGRALS

We consider integrals of the form

$$I(\lambda) = \int_D f(\mathbf{x})e^{i\lambda\phi(\mathbf{x})} \, d\mathbf{x}, \qquad \mathbf{x} = (x_1, x_2, \ldots, x_m). \tag{2.8.1}$$

Here ϕ is real, D is a bounded connected domain in Euclidean m-dimensional space, and $d\mathbf{x}$ denotes the volume element in m-space.

Our treatment of one-dimensional integrals in the previous two sections might lead us to expect that the asymptotic expansion of this integral will depend on the nature of the integrand in the neighborhood of certain possible *critical points. These are*

(i) points in D at which $\nabla\phi = \mathbf{0}$,
(ii) points at which f or ϕ fails to be infinitely differentiable, and
(iii) all points on ∂D, the boundary of D.

Points at which $\nabla\phi$ vanishes are *stationary points,* and points on ∂D correspond to the endpoints of integration in the one-dimensional integrals. Of course, the critical nature of these points can only be established by determining associated contributions to the asymptotic expansion of $I(\lambda)$. We will assume for the moment that f and ϕ are infinitely differentiable on $\bar{D}$, D, and its boundary ∂D and that $\nabla\phi \neq \mathbf{0}$ on that domain. Then if our list is correct, only the critical points of type (iii) are possible.

In order to check this, we will calculate an alternative representation of $I(\lambda)$ by integration by parts, which in higher dimensions means *by the divergence theorem.* In order to use the divergence theorem, we need the identity

$$fe^{i\lambda\phi} = \frac{1}{i\lambda}\nabla\cdot\left[\frac{f\,\nabla\phi}{|\nabla\phi|^2}e^{i\lambda\phi}\right] - \frac{1}{i\lambda}\nabla\cdot\left[\frac{f\,\nabla\phi}{|\nabla\phi|^2}\right]e^{i\lambda\phi}. \tag{2.8.2}$$

We now substitute the right side here into (2.8.1) and apply the divergence theorem to the first term to obtain

$$I(\lambda) = \frac{1}{i\lambda}\int_{\partial D}\frac{f\hat{\boldsymbol{\eta}}\cdot\nabla\phi}{|\nabla\phi|^2}e^{i\lambda\psi}\,d\boldsymbol{\sigma} - \frac{1}{i\lambda}\int_D\nabla\cdot\left[\frac{f\,\nabla\phi}{|\nabla\phi|^2}\right]e^{i\lambda\phi}\,d\mathbf{x}, \tag{2.8.3}$$

$$\boldsymbol{\sigma} = (\sigma_1, \sigma_2, \ldots, \sigma_{m-1}), \quad \psi = \psi(\boldsymbol{\sigma}) = \phi(\mathbf{x}(\boldsymbol{\sigma})).$$

Here $d\boldsymbol{\sigma}$ is the "surface area element" on the $(m - 1)$-dimensional boundary ∂D and $\hat{\boldsymbol{\eta}}$ the unit outward normal to that boundary. We see that the first term is an integral over the boundary while the second term is an integral *of the same form as I itself* except that it is multiplied by $(i\lambda)^{-1}$. It is reasonable to expect that this second term is asymptotically of lower order than I and, therefore, that the leading term must arise from the first integral, that is, from the *boundary.*

Admittedly, we know no more about the asymptotic expansion of $(m - 1)$-dimensional integrals (except when $m = 2$) than we do about m-dimensional integrals. However, that will be remedied later. For now, let us proceed to apply this process recursively to obtain

$$I(\lambda) = \sum_{n=0}^{N-1}\frac{(-1)^n}{(i\lambda)^{n+1}}\int_{\partial D}g_{n+1}(\boldsymbol{\sigma})e^{i\lambda\psi(\boldsymbol{\sigma})}\,d\boldsymbol{\sigma} + \frac{(-1)^N}{(i\lambda)^N}\int_D f_N(\mathbf{x})e^{i\lambda\phi}\,d\mathbf{x};$$

$$\mathbf{H}_n = \frac{f_n\nabla\phi}{|\nabla\phi|^2}; \quad f_{n+1} = \nabla\cdot\mathbf{H}_n; \tag{2.8.4}$$

$$g_{n+1} = \hat{\boldsymbol{\eta}}\cdot\mathbf{H}_n; \quad n = 1, 2, \ldots; \quad f_0 = f.$$

We have replaced $I(\lambda)$ by a sum of integrals over the boundary plus a volume integral like I itself multiplied by an arbitrarily large negative power of λ. Thus, we conclude that the asymptotic expansion in this case is dominated by the boundary integrals. Furthermore, if $f(\mathbf{x})$ and all of its derivatives vanished on the boundary, then, in fact, all of the boundary integrals are zero and

$$I(\lambda) = O(\lambda^{-N}), \qquad \text{any} \quad N.$$

We can use this last observation to isolate the critical points, much as we did for one-dimensional integrals. Let us suppose that we have identified a number of critical points in the domain D. Introduce a multidimensional van der Corput neutralizer function to isolate each of those critical points. This can be accomplished by introducing either a neutralizer function of radial distance for each critical point or a product of one-dimensional neutralizers in each independent variable for each critical point. Subtracting the sum of these neutralizers from unity yields a function that isolates the boundary as the only set of critical points. Then, of course, we would have to apply our methods for multidimensional integrals to this $(m - 1)$-dimensional integral. Thus, we can now think of our integral I in (2.8.1) as having only one critical point for the purposes of deriving asymptotic expansions. Then, a sum of contributions from the critical points of interest will yield an asymptotic expansion.

Actually, our goal will be much more modest. We shall derive the asymptotic expansion of $I(\lambda)$ for the case in which the integrand has only one singularity, a *simple stationary point*. A stationary point $\mathbf{x}_0$ will be called simple if

$$\nabla \phi(\mathbf{x}_0) = \mathbf{0},$$

$$\det A \neq 0, \qquad A_{jk} = \left[\frac{\partial^2 \phi(\mathbf{x}_0)}{\partial x_j \, \partial x_k}\right], \quad j, k = 1, 2, \ldots, m, \tag{2.8.5}$$

where A is called the Hessian matrix for ϕ. We will now assume that I as defined by (2.8.1) has only this critical point in the domain of integration D and vanishes "infinitely smoothly" at the boundary. Our goal is to obtain the leading term of the asymptotic expansion of $I(\lambda)$.

The properties of the matrix A will play a crucial role in this development. To begin, let us denote the positive eigenvalues of A by $\lambda_1, \lambda_2, \ldots, \lambda_r, r \leq m$, and the negative eignevalues of A by $\lambda_{r+1}, \lambda_{r+2}, \ldots, \lambda_m$. The signature of A, to be denoted by sgn A, is the difference in the number of positive and negative eigenvalues given by

$$\text{sgn } A = 2r - m. \tag{2.8.6}$$

Because A is a *symmetric* matrix, there exists an *orthogonal* matrix Q,[†] which "diagonalizes" the matrix A; that is,

$$Q^T A Q = \text{diag}[\lambda_1, \ldots, \lambda_m].$$

Here diag denotes a matrix with only diagonal elements as listed inside the brackets.

We think of the vector $\mathbf{x} - \mathbf{x}_0$ as a *horizontal* array and introduce a new set of variables of integration by the matrix equation

$$(\mathbf{x} - \mathbf{x}_0)^T = QRz^T,$$

$$R = \text{diag}[\lambda_1^{-1/2}, \ldots, \lambda_r^{-1/2}, |\lambda_{r+1}|^{-1/2}, \ldots, |\lambda_m|^{-1/2}]. \quad (2.8.7)$$

We remark that

$$\phi(\mathbf{x}) - \phi(\mathbf{x}_0) \approx \tfrac{1}{2}(\mathbf{x} - \mathbf{x}_0)A(\mathbf{x} - \mathbf{x}_0)^T. \quad (2.8.8)$$

After the transformation (2.8.7),

$$\theta(\mathbf{z}) = \phi(\mathbf{x}(\mathbf{z})) - \phi(\mathbf{x}_0) \approx \frac{1}{2} \mathbf{z} R Q^T A Q R \mathbf{z}^T \approx \frac{1}{2}\left[\sum_{j=1}^{r} z_j^2 - \sum_{j=r+1}^{m} z_j^2 \right]. \quad (2.8.9)$$

Remarkably, this approximate result can be made exact. That is, there exists a change of variables that in some neighborhood of the stationary point, now $\mathbf{z} = \mathbf{0}$, is (i) infinitely differentiable, (ii) one-to-one, and (iii) transforms the phase exactly into a sum of signed squares;[‡] thus,

$$\xi_j = h_j(\mathbf{z}) = z_j + o(|\mathbf{z}|), \qquad j = 1, \ldots, m. \quad (2.8.10)$$

Furthermore,

$$\theta(\mathbf{z}(\boldsymbol{\xi})) = \phi(\mathbf{x}(\mathbf{z}(\boldsymbol{\xi}))) - \phi(\mathbf{x}_0) = \frac{1}{2}\left[\sum_{j=1}^{r} \xi_j^2 - \sum_{j=r+1}^{m} \xi_j^2 \right]. \quad (2.8.11)$$

The transformation from $\mathbf{x}$ to $\boldsymbol{\xi}$ yields a new representation of the integral (2.8.1):

$$I(\lambda) = e^{i\lambda\phi(\mathbf{x}_0)} \int_{\tilde{D}} F_0 \tilde{n}(\boldsymbol{\xi}) e^{i\lambda \boldsymbol{\rho} \cdot \boldsymbol{\xi}/2} \, d\boldsymbol{\xi}. \quad (2.8.12)$$

Here we have explicitly introduced the neutralizer function denoted by $\tilde{n}(\boldsymbol{\xi})$. This will play a crucial role in the analysis that follows. Also, in this equation, $\tilde{D}$ is the image of D under the change of variable of integration

$$\boldsymbol{\rho} = (\xi_1, \ldots, \xi_r, -\xi_{r+1}, \ldots, -\xi_m), \qquad F_0(\boldsymbol{\xi})\tilde{n}(\boldsymbol{\xi}) = f(\mathbf{x}(\boldsymbol{\xi}))J(\boldsymbol{\xi}), \quad (2.8.13)$$

[†] A real matrix Q is orthogonal if its transpose is its inverse: $QQ^T = I$; the superscript T denotes transpose.

[‡] Proof by J. Milnor.

and J is the Jacobian of the change of variables

$$J = \left| \det \left[\frac{\partial x_j}{\partial \xi_k} \right] \right|. \tag{2.8.14}$$

We remark that J can be calculated as a product of Jacobians of the transformations from $\mathbf{x}$ to $\mathbf{z}$ and from $\mathbf{z}$ to $\boldsymbol{\xi}$. At the stationary point, from (2.8.10) the latter transformation has Jacobian equal to unity, while from (2.8.7) the former transformation has Jacobian equal to $|\det A|^{-1/2}$. Thus,

$$J(\mathbf{0}) = |\det A|^{-1/2}. \tag{2.8.15}$$

We now set

$$F_0(\boldsymbol{\xi}) = F_0(\mathbf{0}) + \boldsymbol{\rho} \cdot \mathbf{H}_0. \tag{2.8.16}$$

The vector function $\mathbf{H}_0$ is not uniquely determined. However, one choice of this vector is

$$H_{01} = \frac{1}{\xi_1} [F_0(\xi_1, \xi_2, \ldots, \xi_m) - F_0(0, \xi_2, \ldots, \xi_m)],$$

$$H_{02} = \frac{1}{\xi_2} [F_0(0, \xi_2, \xi_3, \ldots, \xi_m) - F_0(0, 0, \xi_3, \ldots, \xi_m)],$$

$$\vdots$$

$$H_{0r} = \frac{1}{\xi_r} [F_0(0, \ldots, \xi_r, \xi_{r+1}, \ldots, \xi_m) - F_0(0, \ldots, 0, \xi_{r+1}, \ldots, \xi_m)],$$

$$H_{0r+1} = -\frac{1}{\xi_{r+1}} [F_0(0, \ldots, \xi_{r+1}, \xi_{r+2}, \ldots, \xi_m) - F_0(0, \ldots, 0, \xi_{r+2}, \ldots, \xi_m)],$$

$$\vdots$$

$$H_{0m} = -\frac{1}{\xi_m} [F_0(0, \ldots, 0, \xi_m) - F_0(0, \ldots, 0, 0)].$$

The ambiguity in $\mathbf{H}$ will not affect the asymptotic expansion.

We substitute (2.8.16) into (2.8.12) and write this result as

$$I(\lambda) = e^{i\lambda\phi(\mathbf{x}_0)} [I_0(\lambda) + I_1(\lambda)], \tag{2.8.17}$$

where

$$I_0(\lambda) = F_0(\mathbf{0}) \int_{\tilde{D}} \tilde{n}(\boldsymbol{\xi}) e^{i\lambda\boldsymbol{\rho}\cdot\boldsymbol{\xi}/2} \, d\boldsymbol{\xi} \tag{2.8.18}$$

and

$$I_1(\lambda) = \int_{\bar{D}} (\boldsymbol{\rho} \cdot \mathbf{H}_0) \tilde{n}(\boldsymbol{\xi}) e^{i\lambda \boldsymbol{\rho} \cdot \boldsymbol{\xi}/2} \, d\boldsymbol{\xi}. \tag{2.8.19}$$

We shall deal with I_1 first. We remark that the factor $\boldsymbol{\rho}$ in the amplitude is proportional to the gradient of the phase. Thus, this integral can be calculated by the divergence theorem. That is, we write

$$\boldsymbol{\rho} \cdot \mathbf{H}_0 \tilde{n}(\boldsymbol{\xi}) e^{i\lambda \boldsymbol{\rho} \cdot \boldsymbol{\xi}} = \frac{1}{i\lambda} \left[\mathbf{V} \cdot [\mathbf{H}_0 \tilde{n}(\boldsymbol{\xi}) e^{i\lambda \boldsymbol{\rho} \cdot \boldsymbol{\xi}}] - \mathbf{V} \cdot [\mathbf{H}_0 \tilde{n}(\boldsymbol{\xi})] e^{i\lambda \boldsymbol{\rho} \cdot \boldsymbol{\xi}} \right]$$

and apply the divergence theorem to the integral of the first term on the right side of this equation. The result is

$$I_1(\lambda) = -\frac{1}{i\lambda} \int_{\bar{D}} [\tilde{n} \, \mathbf{V}_\xi \cdot \mathbf{H}_0 + \mathbf{H}_0 \cdot \mathbf{V}_\xi \tilde{n}] e^{i\lambda \boldsymbol{\rho} \cdot \boldsymbol{\xi}/2} \, d\boldsymbol{\xi}. \tag{2.8.20}$$

There are no boundary terms here because the neutralizer and its derivatives vanish on the boundary. This integral is of the same form as $I(\lambda)$ itself except for the multiplier $1/i\lambda$. Thus, we anticipate that this integral is of lower order asymptotically than $I(\lambda)$ and that the leading order expansion must come from the first term. Furthermore, the second term of the integrand in $I_1(\lambda)$ has no critical points at all because the gradient of the neutralizer is identically zero at the critical point and is infinitely differentiable. The first term in $I_1(\lambda)$ is an integrand exactly like the integrand in $I(\lambda)$, and we could contemplate continuing the expansion process that led to (2.8.17) recursively, continuing from this integral. This process is discussed by Bleistein and Handelsman [1975a,b].

Let us now turn to $I_0(\lambda)$. We contemplate the effect of replacing the neutralizer $\tilde{n}$ by a product of one-dimensional neutralizers along the coordinate axes. The difference of these two functions is an infinitely differentiable function that is identically zero at the critical point. The integral of this difference has no critical points at all and as has been shown, is asymptotically zero to all orders in λ. Therefore, we can write

$$I_0 \sim \int_{-\infty}^{\infty} \prod_{j=1}^{m} \tilde{n}_j e^{i\lambda \boldsymbol{\rho} \cdot \boldsymbol{\xi}/2} \, d\boldsymbol{\xi} \sim \prod_{j=1}^{m} \int_{-\infty}^{\infty} \tilde{n}_j e^{i\lambda \rho_j \xi_j/2} \, d\xi_j. \tag{2.8.21}$$

Each integral in this product can by approximated asymptotically by our one-dimensional stationary phase formula for an interior stationary point [(2.7.19)]. The result is

$$I_0(\lambda) \sim [2\pi/\lambda]^{m/2} e^{i \operatorname{sgn} A \pi/4}. \tag{2.8.22}$$

Here one factor of $[2\pi/\lambda]^{1/2}$ arises from each one-dimensional integral. For each $\rho_j = \xi_j$, we obtain a factor $e^{i\pi/4}$; for each $\rho_j = -\xi_j$, we obtain a factor

$e^{-i\pi/4}$. The multiplier sgn A counts those occurrences of positive and negative signs. In order to determine the leading term of the asymptotic expansion of $I(\lambda)$, it is only necessary to determine the amplitude F_0. This is done by using (2.8.13) and (2.8.15). Finally, we collect these results and substitute into (2.8.18) to obtain

$$I(\lambda) \sim I_0(\lambda)e^{i\lambda\phi(\mathbf{x}_0)} \sim \left[\frac{2\pi}{\lambda}\right]^{m/2} \frac{f(\mathbf{x}_0)}{\sqrt{|\det A|}} e^{\{i\lambda\phi(\mathbf{x}_0) + i\,\mathrm{sgn}\,A\pi/4\}}. \quad (2.8.23)$$

This is the multidimensional stationary phase formula.

Notice that the one-dimensional stationary phase formula [(2.7.18)] agrees with this result for $m = 1$. The fact that the leading terms in (2.7.12) and (2.7.13) are half of the result (2.7.18) has its analogy in m dimensions. In particular, if ϕ has a simple stationary point at a "smooth" point on the boundary of D, then the contribution from that point is indeed half the result (2.8.23). For a stationary point at a "corner" on the boundary, only a few results are known; this is still an open area for research.

Exercises

2.26 Suppose in (2.8.1) that ϕ has no stationary points in $\bar{D}$ so that (2.8.3) is true. Show that the stationary points of ϕ are those points on the boundary where $\nabla\phi$ is normal to the boundary. Thus, conclude that even a *linear* phase const $\cdot\, x$ can have boundary stationary points in an integral in two or more dimensions.

2.27 In (2.8.1), let $m = 3$ and view the integral after diagonalization of the exponent as an *iterated* integral in three variables. Using your knowledge of the asymptotic order of one-dimensional integrals, estimate the asymptotic order of $I(\lambda)$ for each of the following cases.

(a) ϕ has a simple stationary point, and f and ϕ are "smooth."

(b) As in (a) except that one of the eigenvalues of the Hessian matrix is zero and the *third* directional derivative in the direction of that eigenvector is nonzero.

(c) As in (a) except that the amplitude has a simple zero in one of the principal directions.

(d) Phase as in (b) but amplitude as in (c).

2.28 In (2.8.1), suppose that

$$f(\mathbf{x}) = \sin x_1,$$

$$\phi(\mathbf{x}) = \sin x_1 \cos x_2 \sin \alpha \cos \beta + \sin x_1 \sin x_2 \sin \alpha \sin \beta + \cos x_1 \cos \alpha,$$

and D is the domain $0 \le x_1 \le \pi$, $0 \le x_2 \le 2\pi$. Find the asymptotic of I to order λ^{-1}.

References

Bleistein, N., and Handelsman, R. A. [1975a]. Multidimensional stationary phase, an alternative derivation, *SIAM J. Math. Anal.* **6,** 480–487.

Bleistein, N., and Handelsman, R. A. [1975b]. "Asymptotic Expansions of Integrals." Holt, Rinehard and Winston, New York. [Also to be published by Dover, New York.]

Bracewell, R. N. [1978]. "The Fourier Transform and Its Applications," 2nd ed. McGraw-Hill, New York.

van der Corput, J. G. [1948]. On the method of critical points. *Nederl. Akad. Wetensch. Proc.* **51,** 650–658.

Erdélyi, A. [1955]. Asymptotic representations of Fourier integrals and the method of stationary phase, *SIAM J. Appl. Math.* **3,** 17–27.

Erdélyi, A. [1956]. "Asymptotic Expansions." Dover, New York.

Jones, D. S. [1966]. "Generalized Functions." McGraw-Hill, London.

Jones, D. S., and Kline, M. [1958]. Asymptotic expansion of multiple integrals and the method of stationary phase, *J. Math. Phys.* **37,** 1–28.

Lewis, R. M. [1964]. Asymptotic methods for the solution of dispersive hyperbolic equations, *in* "Asymptotic Solutions of Differential Equations and Their Applications" (C. Wilcox, ed.), pp. 53–108. Wiley, New York.

Milnor, J. [1963]. "Morse Theory." Princeton Univ. Press, Princeton, New Jersey.

Olver, F. W. J. [1974]. "Asymptotics and Special Functions." Academic Press, New York.

Sneddon, I. H. [1972]. "The Use of Integral Transforms." McGraw-Hill, New York.

Stakgold, I. [1967]. "Boundary Value Problems of Mathematical Physics," Vol. 1. Macmillan, New York.

Titchmarsh, E. C. [1948]. "Introduction to the Theory of Fourier Integrals," 2nd ed. Clarendon Press, Oxford.

Tolstov, G. P. [1962]. "Fourier Series" (R. A. Silverman, transl.). Dover, New York.

3 SECOND-ORDER PARTIAL DIFFERENTIAL EQUATIONS

Before beginning our discussions of specific second-order partial differential equations, we shall make some general remarks about such equations in this chapter. By *second order* we mean an equation that includes derivatives up to and including second-order partial derivatives of the unknown function and none of higher order. If the equation is written in the form

$$F(x_1, \ldots, x_m, U_{x_1 x_1}, \ldots, U) = 0,$$

then the partial derivative of F with respect to one of its second derivative arguments should be nonzero. We shall not be dealing with such a general equation but with "much simpler" linear second-order equations.

We shall be concerned with analytical solution techniques, both exact and approximate. The kind of solution we mean here is the solution to a *problem* that consists of both a partial differential equation for an unknown function in some prescribed domain of the independent variables and a set of data on the boundary[†] of the domain. These data will consist of functional relationships among the unknown and its directional derivatives directed out of the boundary.

Such a problem for an unknown function arises from an attempt to use mathematics to model some physical or other observable phenomenon. Thus, a priori there is no reason to expect that this problem will have a solution at all. In fact, there are three properties of interest, namely,

  (i)   the existence of a solution,
  (ii) the uniqueness of the solution,
  (iii) the continuous dependence of the solution on the data.

[†] In this general context, initial data can be viewed as a form of boundary data on the boundary $t = 0$.

When a particular type of problem—equation plus data—is known to exhibit all three of these properties, it is said to be *well posed* in the sense of Hadamard.

There is an interesting result as regards the existence and uniqueness of a solution to a particular class of problems. It deals with the most natural extension of the initial value problem for ordinary differential equations to partial differential equations.

Suppose that U satisfies a partial differential equation of order n in m independent variables. Suppose that one particular derivative of order n with respect to one variable is singled out so that the form of the equation is $\partial^n U/\partial x_1^n$ equals a function G of the independent variables, U, and all other derivatives to order n. We suppose further that "initial data," U, and its derivatives to order $n-1$ with respect to x_1 are given at $x_1 = 0$. Choose a fixed point on this initial manifold by fixing the values of the other independent variables, and consider finding a solution about this point by formal power series methods. If in some neighborhood of this initial point, all of the power series deduced from these initial data converge and the power series expansion of G converges in all its variables, then the power series solution will also converge in some neighborhood of this point to provide a unique solution of the initial value problem.

The preceding is a paraphrase of the *Cauchy–Kowaleski theorem*. More generally, the $(m-1)$-dimensional initial manifold could be any analytic surface in m-dimensional space, with the function and $(n-1)$ directional derivatives out of the manifold being prescribed. Also, the power series need not be about the origin in all variables; straightforward translation of the dependent and independent variables overcomes this constraint. It is only additionally necessary now that the partial differential equation can be solved for the nth derivative off the surface.

This generalization of the initial value problem is called the *Cauchy problem*. The data, U, and $(n-1)$ directional derivatives out of the surface are called *Cauchy data*. Thus, in Chapter 1, we studied the Cauchy problem for first-order equations. We did not solve that problem by power series. However, in order to solve on the manifold (curve, surface, etc.) where U was given, it was necessary that we be able to solve for the first directional derivative out of that manifold. In this case, existence and uniqueness in some neighborhood of the initial manifold were confirmed. Although we did not speak to the question of continuous dependence on the initial data, that is true as well.

While the Cauchy–Kowaleski theorem does guarantee existence and uniqueness, it does not speak to the issue of continuous dependence on the data. In fact, as we shall see from subsequent examples, it cannot, because the initial value problem does not always satisfy this criterion; that is, for some problems, the initial value problem is *ill posed*.

3.1 PROTOTYPE SECOND-ORDER EQUATIONS

There are three second-order partial differential equations of interest to us here. One of these is the *wave equation*

$$\Delta U - \frac{1}{c^2} U_{tt} = F(\mathbf{x}, t). \tag{3.1.1}$$

Here

$$\Delta = \nabla^2 = \frac{\partial^2}{\partial x_1^2} + \cdots + \frac{\partial^2}{\partial x_m^2} \tag{3.1.2}$$

is the *Laplace operator*. In one spatial dimension, this equation models the "vibrating string," with U being the small-amplitude transverse motions of the string under tension and applied transverse force F. The equation also models the dynamics of a vibrating wire, such as a guitar strong, when the amplitude of the transverse motions is sufficiently small to make the *stiffness* of the wire unimportant to the model. An adequately precise meaning of *small* requires a discussion of the assumptions made to deduce this simple equation from the governing physical laws of the dynamical system at hand. This discussion is outside the scope of this book. In two spatial dimensions, the wave-equation models the dynamics of small-amplitude vibrations of a stretched membrane or certain small-amplitude waves on the surface of the water. In three dimensions, it models the propagation of acoustic waves and special cases of elastic and electromagnetic waves.

The second prototype equation is the *Helmholtz* or *reduced wave equation*

$$\Delta U + (\omega^2/c^2)U = F(\mathbf{x}). \tag{3.1.3}$$

This equation arises by applying the Fourier transform in t to the wave equation. The case $\omega = 0$ is of interest in its own right; it is called the *potential equation* or *Laplace equation*:

$$\Delta U = F(\mathbf{x}). \tag{3.1.4}$$

The function U could be an electrostatic potential (voltage), with its gradient being a force per unit charge. Alternatively, U might be a fluid potential with its gradient a flow velocity vector. Note from the second example that the lack of time dependence does not imply a lack of motion, but only an absence of variation of that motion with time. This equation can model wave motion through the time dependence of boundary data, such as when it models fluid flow in the interior of the ocean with a time-dependent boundary condition at the surface. (See Whitham [1974, Section 13.3].)

The third prototype equation is the *heat* or *diffusion equation*

$$\Delta U - K U_t = F(\mathbf{x}, t). \tag{3.1.5}$$

As its name implies, this equation models the flow of heat and other *diffusive* processes. One discussion of this equation will be limited to comparisons with the wave equation and reduced wave equation for qualitative purposes only.

We remark that the second-order operators appearing in (3.1.4) are extremely special cases of the most general second-order operator

$$LU = \sum_{i,j=1}^{n} a_{ij} \frac{\partial^2 u}{\partial x_i \, \partial x_j}.$$

In the preceding prototypes, the coefficients a_{ij} are ± 1 or 0. Thus, we might contemplate the process of "diagonalizing" the general second-order operator by a transformation of coordinates to yield an operator with coefficients ± 1 or 0, as in our prototypes. When the coefficients a_{ij} are constants, a linear transformation of type (2.8.7) will achieve this end. When the coefficients are variable, it is possible to achieve the same diagonalization globally only in two dimensions. In higher dimensions, we can diagonalize the operator only locally in the neighborhood of a point. (For a more complete discussion, see Garabedian [1964].)

In either case, whether the diagonalization is achieved locally or globally, the process suggests a means of classifying second-order operators by means of the properties of the eigenvalues of the matrix of coefficients. Thus, if all of the eigenvalues are nonzero and of one sign, the second-order operator can be reduced to the Laplace operator in all independent variables. This type of equation is called *elliptic*. If all eigenvalues are nonzero and exactly one of them is of opposite sign from the others, the second-order operator can be diagonalized to the wave operator in all variables. This type of equation is called *hyperbolic*. Again, with all eigenvalues nonzero, if there are at least two of each sign, the equation is called *ultrahyperbolic*. Finally, if at least one eigenvalue is zero, the equation is called *parabolic*. Note that the heat equation is a special case of the latter in which all of the nonzero eigenvalues are of the same sign and exactly one eigenvalue is equal to zero but the first derivative in the same "direction," time, is nonzero.

The question arises as to what type of problems we should consider for each of the prototype equations. Surely, one guide to the answer would be to consider a phenomenon that each equation models and to examine the sort of information that "naturally" arises from the modeled problem. Returning to the wave equation in one space dimension, we can think of the guitar string as a model. Here we would expect to know how each end of the string is held as well as how the string is "plucked." The former constitutes one piece of boundary data at each end of the string, that is, at each boundary point. The latter takes the form of both source information and initial data. That is, the following actions are possible:

(i)   an initial displacement is imparted to the string,
(ii) an initial velocity is imparted to the string, and
(iii) a force is continuously applied to the string.

Thus, we are led to an *initial boundary-value problem* for the displacement of the string, which involves the wave equation in one space dimension, Cauchy data at the initial time, and a boundary condition at each point of the boundary. Clearly, analogous problems can be devised in higher dimensions. The extent of well-posedness for these problems has been extensively studied and will be discussed in subsequent chapters.

As we have already mentioned, the Helmholtz equation (3.1.3) arises as the Fourier transform of the wave equation. Thus, we might expect that an appropriate problem for this equation can be deduced from the problems for the wave equation. One might expect here to consider problems in which the initial data for the wave equation are exchanged for additional source terms, while boundary data for the wave equation would lead to (Fourier time-transformed) boundary data for the Helmholtz equation. Hence, an appropriate problem to consider would be one in which a source term is prescribed along with one boundary condition at each point of the boundary.

Similarly, a problem in which boundary data and source information are prescribed would be expected for Laplace's equation.

Finally, for the heat equation, we note that only a first derivative in t appears. Thus, for this equation, it is appropriate to impose one initial condition and one boundary condition at each boundary point.

Implicit in our discussion has been the assumption that the domain of interest in time is semi-infinite, extending from zero to infinity. It is possible that the spatial domain of interest is semi-infinite or infinite as well. In this case, an appropriate "condition at infinity" will replace the boundary condition. This will be clarified in context.

3.2 SOME SIMPLE EXAMPLES

Let us consider the following problem for U:

$$\frac{\partial^2 U}{\partial x^2} - \frac{1}{c^2}\frac{\partial^2 U}{\partial t^2} = 0, \qquad -\infty < x < \infty,$$

$$U(x, 0) = f(x), \qquad U_t(x, 0) = 0. \tag{3.2.1}$$

Here c is a constant. By direct substitution, we can verify that the solution to this problem for U is

$$U(x, t) = \tfrac{1}{2}[f(x - ct) + f(x + ct)] \tag{3.2.2}$$

so long as the function f has two derivatives with respect to its argument.

Although this problem is extremely simple, some interesting observations of a general nature are exhibited by the solution. First, we note that small changes in f produce small changes in the solution U. We shall later prove that the solution is unique, as well, for data taken from a large class of functions. That is, the solution to this problem is *well posed*. Second, suppose that f were discontinuous at a point x_0. Then, so long as the "rays" $x \pm ct = x_0$ are avoided, the solution (3.2.2) remains valid. Of course, the limits as $x \pm ct$ approaches x_0 from above or below are not the same but exhibit half the discontinuity at the initial point $(x_0, 0)$. That is, the discontinuity in the data propagates along these rays, or *characteristics*, of the second-order equation. We shall have much more to say about the characteristics of the wave equation in the next chapter.

To recapitulate: We have demonstrated (i) existence, (ii) claimed uniqueness, and (iii) indicated continuous dependence of the solution on the data. Furthermore, we have seen, at least formally, that this equation admits solutions with discontinuous data and the discontinuities of the data propagate on trajectories in space–time, which we shall again call characteristics.

In Eq. (3.2.1), there is really no distinction between space and time except the writer's arbitrary choice of labeling one independent variable by x and the other by t. Thus, we could as easily consider an "initial value problem" in x by imposing data at $x = 0$ for U and U_x. Indeed, we could just consider the problem above with x and t interchanged and thereby demonstrate the existence and continuous dependence of the solution on the data as well as the propagation of the discontinuities on characteristics.

This interchangeability of x and t does not carry over into higher spatial dimensions. In particular, let us now consider the problem

$$\frac{\partial^2 U}{\partial x^2} + \frac{\partial^2 U}{\partial y^2} - \frac{\partial^2 U}{\partial t^2} = 0;$$

$$U(0, y, t) = \frac{1}{n^2} \sin ny \sin t, \tag{3.2.3}$$

$$U_x(0, y, t) = \frac{\sqrt{n^2 - 1}}{n^2} \sin ny \sin t, \qquad n > 1.$$

The solution to this problem is

$$U(x, y, t) = \frac{1}{n^2} \sin ny \sin t \exp[x\{n^2 - 1\}^{1/2}]. \qquad (3.2.4)$$

It can be seen here that as $n \to \infty$, the data approach zero while the solution increases beyond all bounds for any $x > 0$. Indeed, even for finite $n > 1$, the solution increases beyond all bounds as $x \to \infty$. Most often, the phenomenological process being modeled will preclude such solutions. When such solutions are not precluded by the model, the researcher would be well advised to reexamine his/her assumptions leading to the mathematical model.

This example is a slight variant on an example suggested by Hadamard for Laplace's equation. This form has relevance to exploration geophysics. Let us suppose that an acoustic source is set off at or near the surface of the earth. A signal propagates into the earth and reflects energy from *inhomogeneities* in the earth back up to the surface. The objective is to map those inhomogeneities from observations of the upward propagating wave. In the simplest model, this reduces to a Cauchy problem of the type (3.2.4) to *downward continue* the upward propagating wave back to the anomalies that produced it. From this example, it can be seen that this problem admits exponentially growing solutions. Whether or not they occur depends on the sign of $n^2 - 1$, that is, on the relative oscillation rates in the transverse dimension and time of the observed data. Even when one can guarantee on physical grounds that the actual wave field does not have such exponential modes, it may be that they arise through the presence of *noise* in the observed data. Thus, ill-posedness in the sense of Hadamard is an extremely practical and meaningful concept.

We turn now to two problems for Laplace's equation in two dimensions. First, we consider the problem

$$\Delta U = 0, \qquad r = \sqrt{x^2 + y^2} < 1; \qquad U = f(\theta), \qquad r = 1. \quad (3.2.5)$$

Here θ is the polar angle measured from the x axis. We obtain the solution by Fourier series, assuming that

$$U(r, \theta) = \frac{a_0(r)}{2} + \sum_{n=1}^{\infty} a_n(r) \cos n\theta + b_n(r) \sin n\theta. \qquad (3.2.6)$$

Recalling that in polar coordinates Laplace's equation has the form

$$\frac{1}{r} \frac{\partial}{\partial r}\left(r \frac{\partial U}{\partial r}\right) + \frac{1}{r^2} \frac{\partial^2 U}{\partial \theta^2} = 0, \qquad (3.2.7)$$

we obtain the following equations for the coefficients of the series:

$$\frac{d^2 u}{dr^2} + \frac{1}{r} \frac{du}{dr} - \frac{n^2}{r^2} u = 0, \qquad u = a_n, b_n, \qquad n = 0, 1, \ldots. \qquad (3.2.8)$$

Modulo a constant scale factor, these equations have the solutions

$$a_0 = 1, \log r; \qquad a_n, b_n = r^{\pm n}, \qquad n = 1, 2, \ldots. \qquad (3.2.9)$$

In order that U and its derivatives remain bounded in the domain of interest, we choose from among these solutions only the nonnegative powers of r. Thus, the solution U is of the form

$$U(r, \theta) = \frac{c_0}{2} + \sum_{n=1}^{\infty} r^n [c_n \cos n\theta + d_n \sin n\theta], \qquad (3.2.10)$$

with constants c_n and d_n to be determined. We now impose the boundary condition in (3.2.5) and find that these constants are the Fourier coefficients of the function f:

$$c_n = \frac{1}{2\pi} \int_0^{2\pi} f(\phi) \cos n\phi \, d\phi,$$

$$\qquad (3.2.11)$$

$$d_n = \frac{1}{2\pi} \int_0^{2\pi} f(\phi) \sin n\phi \, d\phi, \qquad n = 0, 1, \ldots.$$

As with the wave equation, we have demonstrated existence by exhibiting a solution. Again, we do not address the question of uniqueness (which holds). As for continuous dependence on the data, that holds but is not so transparent from the Fourier representation of the solution. The reader should be cautioned, however, that this need not be the case for all elliptic partial differential equations or for all types of boundary conditions. In particular, the Helmholtz equation on a bounded domain admits eigensolutions for particular choices of ω. See Exercise 3.1.

We present now the more traditional Hadamard example. Thus, we consider the problem

$$\Delta U = 0, \qquad x > 0; \qquad U(0, y) = 0, \qquad U_x(0, y) = n^{-1} \sin ny. \quad (3.2.12)$$

The solution to this problem is

$$U = n^{-2} \sinh nx \sin ny. \qquad (3.2.13)$$

Again, it is seen that the data approach zero with increasing n but that the solution increases beyond all bounds for increasing n and $x \neq 0$. Furthermore, for any n, the solution increases beyond all bounds for $|x|$ approaching infinity.

Finally, we turn to problems for the heat equation. Let us consider first the initial value problem

$$\frac{\partial^2 U}{\partial x^2} - \frac{\partial U}{\partial t} = 0, \qquad U(x, 0) = f(x). \qquad (3.2.14)$$

We leave it to the exercises to show that the solution to this problem is

$$U(x, t) = \frac{1}{2\sqrt{\pi t}} \int_{-\infty}^{\infty} f(\xi) \exp(-|x - \xi|^2/4t) \, d\xi. \tag{3.2.15}$$

It is straightforward to check by formal differentiation under the integral sign that U does indeed satisfy the differential equation. It is less apparent that U satisfies the initial condition. In fact, it does so only in the limit as t approaches $0+$; that is, the limiting process here defines a Dirac delta function.

The solution exhibits continuous dependence on the data. Indeed, this problem is well posed.

The ill-posed problem for Laplace's equation [(3.2.12)] also provides an example for the heat equation in two spatial dimensions and time.

We have demonstrated now both well-posed and ill-posed problems for the three major classifications of second-order equations. These issues will remain of concern to us in the chapters that follow.

Exercises

3.1 Consider the Helmholtz equation in the form

$$\Delta U + \lambda^2 U = 0.$$

(a) Let the domain of interest be the square $|x| \le \pi, |y| \le \pi$. Verify that the functions

$$U_{nm}(x, y) = \sin nx \sin my, \qquad n, m = 1, 2, \ldots,$$

satisfy this equation so long as

$$\lambda = \lambda_{nm} = \sqrt{n^2 + m^2}.$$

Furthermore, each U_{nm} is zero everywhere on the boundary.

(b) Derive the analogous result for the interior of the unit circle.

(c) What are the implications of this exercise as regards the uniqueness of solutions of the Helmholtz equation on a bounded domain?

3.2 Using the results (2.2.12) and (2.2.14) for the one-sided Fourier transform, introduce $s, \omega = is$, and deduce the Laplace transform formulas

$$u_+(is) = u(s) = \int_0^{\infty} U(t)e^{-st} \, dt, \qquad \mathrm{Re}\, s > \eta,$$

and

$$\frac{1}{2\pi i} \int_{c-i\infty}^{c+i\infty} u(s)e^{st} \, ds = \begin{cases} U(t), & t > 0, \quad c > \eta, \\ 0, & t < 0, \quad c > \eta. \end{cases}$$

3.3   Solve the problem (3.2.14) as follows:

(a)   Introduce the Laplace transform of $U(x, t)$ as defined in Exercise 3.2. Call it $u(x, s)$. By applying this transform to (3.2.14), obtain the problem

$$u''(x, s) - su(x, s) = -f(x).$$

(b)   Derive the solution to this problem, analytic in the right half s plane,

$$u(x, s) = \frac{1}{2\sqrt{s}} \int_{-\infty}^{\infty} f(\xi) \exp(-\sqrt{s}\,|x - \xi|)\, d\xi,$$

where $\sqrt{s}$ is defined by $-\pi < \arg s \le \pi$.

(c)   Conclude that

$$U(x, t) = \frac{1}{4\pi i} \int_{c-i\infty}^{c+i\infty} ds \frac{e^{st}}{\sqrt{s}} \int_{-\infty}^{\infty} d\xi\, f(\xi) \exp(-\sqrt{s}\,|x - \xi|), \qquad c > 0.$$

(d)   Interchange the order of integration and verify that the integral over s is zero for $t < 0$, except possibly when $x = \xi$.

(e)   For $t > 0$, except possibly at $x = \xi$, justify deforming the contour in s onto a "keyhole" contour around the negative s axis, i.e., about the line $\arg s = \pm \pi$.

(f)   Introduce $\sqrt{s} = \eta$ and obtain

$$U(x, t) = \frac{1}{2\pi i} \int_{-\infty}^{\infty} d\xi\, f(\xi) \int_{c-i\infty}^{c+i\infty} d\eta \exp(-\eta\,|x - \xi| + \eta^2 t),$$

where again $c > 0$.

(g)   Finally, set $\eta = i\sigma$ and calculate the integral in η to obtain (3.2.15).

References

Courant, R., and Hilbert, D. [1964]. "Methods of Mathematical Physics," Vol. 2, Partial Differential Equations. Wiley, New York.

Garabedian, P. R. [1964]. "Partial Differential Equations." Wiley, New York.

John, F. [1982]. "Partial Differential Equations," 4th ed. Springer-Verlag, New York.

Weinberger, H. F. [1965]. "A First Course in Partial Differential Equations." Blaisdell, New York.

Whitham, G. B. [1974]. "Linear and Nonlinear Waves." Wiley, New York.

4 THE WAVE EQUATION IN ONE SPACE DIMENSION

This chapter will be devoted to the study of the wave equation in one space dimension. In Chapter 5, the wave equation in two and three dimensions will be discussed. As noted in Chapter 3, in one space dimension, there is really no distinction between space and time as far as the equation itself is concerned, while in higher dimensions, there is one second derivative whose coefficient is opposite in sign to the coefficients of all the other second derivatives. A number of the consequences of this observation will be demonstrated in this chapter. We begin with a discussion of the wave equation in one space dimension and then discuss the equation in higher dimensions.

4.1 CHARACTERISTICS FOR THE WAVE EQUATION IN ONE SPACE DIMENSION

Let us consider the wave equation

$$U_{xx} - c^{-2}U_{tt} = F(x, t, U, U_x, U_t). \qquad (4.1.1)$$

We contemplate the possibility of introducing new variables in this equation:

$$\xi = \phi(x, t), \qquad \eta = \psi(x, t) \qquad (4.1.2)$$

to be chosen so that the equation in the new variables ξ, η does not contain a second derivative with respect to at least one of these new variables. Thus, we set

$$U(x, t) = V(\xi, \eta) \qquad (4.1.3)$$

and note that the second derivatives with respect to x and t can be rewritten as second derivatives with respect to ξ and η to yield the following form of the wave operator:

$$U_{xx} - c^{-2}U_{tt} = V_{\xi\xi}[\phi_x^2 - c^{-2}\phi_t^2] + V_{\eta\eta}[\psi_x^2 - c^{-2}\psi_t^2]$$

$$+ 2V_{\xi\eta}[\phi_x\psi_x - c^{-2}\phi_t\psi_t] + \cdots. \qquad (4.1.4)$$

Here the ellipses $(\cdots)$ denote terms involving first derivatives with respect to ξ and η.

Setting the coefficient of $V_{\xi\xi}$ or $V_{\eta\eta}$ equal to zero leads to the requirement that ϕ or ψ be a solution of the *same* quadratic first-order partial differential equation. We can set *both* of these coefficients equal to zero by choosing ϕ and ψ to satisfy the two distinct *linear* first-order equations

$$\phi_x + c^{-1}\phi_t = 0, \qquad \psi_x - c^{-1}\psi_t = 0. \qquad (4.1.5)$$

Then (4.1.1) has the form

$$V_{\xi\eta} = F_1(\xi, \eta, V, V_\xi, V_\eta). \qquad (4.1.6)$$

Thus, we see that (4.1.1) can be satisfied by a function that has a discontinuous first directional derivative along one of the space–time directions defined by the normals to the level curves of the solutions to (4.1.5). That is, the level curves themselves are boundaries across which the solution may have a discontinuous first derivative. As in the one-dimensional theory of Chapter 1, these curves are called *characteristics*.

Associated with the solutions of (4.1.5) are characteristics for those two first-order partial differential equations. These are called *bicharacteristics* of the second-order equation (4.1.1). With t as an independent parameter on the bicharacteristics, we find that the characteristic equations for (4.1.5) take the form

$$\frac{dx}{dt} = c, \quad \frac{d\phi}{dt} = 0 \quad \text{and} \quad \frac{dx}{dt} = -c, \quad \frac{d\psi}{dt} = 0. \qquad (4.1.7)$$

We see here that the bicharacteristics are just the curves along which ϕ or ψ is constant; that is, in this one-space-dimensional problem, they are the characteristics themselves. Furthermore, if a characteristic is viewed as a point in space moving at a prescribed speed as time progresses, then from (4.1.7), that speed has magnitude c, and we call this the *characteristic speed*. As in the first-order theory of Chapter 1, we see that the discontinuities propagate along the characteristics; that is, second derivatives with respect to ξ and η may be singular without affecting the solvability of (4.1.4).

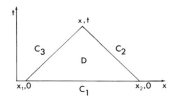

Fig. 4.1. The domain of integration for the integral in (4.1.8).

We shall now develop an integral equation for U from which we can develop some insight into the nature of the solution. Toward this end, we consider the integral

$$I = \int_D \left[U_{xx} - \frac{1}{c^2} U_{tt} \right] dx' \, dt', \qquad (4.1.8)$$

where the domain of integration D is as shown in Fig. 4.1. The curves C_2 and C_3 are the characteristics through the point (x, t). They are depicted as straight lines, as they would be if c were constant. However, we will proceed for a few lines with variable c, then specialize below to constant c, and continue the discussion of the case of variable c in Exercise 4.2.

We apply the divergence theorem to rewrite (4.1.8) as

$$I = \int_{\partial D} \hat{\mathbf{n}} \cdot \left[U_x, \ -\frac{1}{c^2} U_t \right] d\sigma. \qquad (4.1.9)$$

Here ∂D is the boundary of the domain D, $\hat{\mathbf{n}}$ the unit outward normal to ∂D, and σ an arc-length variable along ∂D directed counterclockwise. We remark that if x and t are given parametrically in terms of σ, then the unit normal is given by

$$\hat{\mathbf{n}} = [(dt/d\sigma), \ -(dx/d\sigma)]. \qquad (4.1.10)$$

On C_1, σ is just x and $\hat{\mathbf{n}} = (0, -1)$. On C_2 and C_3, σ is an arc-length parameter on the characteristics of (4.1.5) for ψ and ϕ, respectively. Thus, on C_2 and C_3, respectively,

$$\frac{dx}{d\sigma} = \frac{-c}{\sqrt{1 + c^2}}, \quad \frac{dt}{d\sigma} = \frac{1}{\sqrt{1 + c^2}}, \quad \hat{\mathbf{n}} = \left[\frac{1}{\sqrt{1 + c^2}}, \frac{c}{\sqrt{1 + c^2}} \right],$$

$$\frac{dx}{d\sigma} = \frac{-c}{\sqrt{1 + c^2}}, \quad \frac{dt}{d\sigma} = \frac{-1}{\sqrt{1 + c^2}}, \quad \hat{\mathbf{n}} = \left[\frac{-1}{\sqrt{1 + c^2}}, \frac{c}{\sqrt{1 + c^2}} \right]. \qquad (4.1.11)$$

Now we can rewrite the integrand in (4.1.9) as follows: On C_1,

$$\hat{\mathbf{n}} \cdot \left[U_x, \ -\frac{1}{c^2} U_t \right] = -\frac{\partial U}{\partial t}; \qquad (4.1.12)$$

on C_2,

$$\hat{n} \cdot \left[U_x, -\frac{1}{c^2} U_t \right] = \frac{1}{\sqrt{1+c^2}} \left[U_x - \frac{1}{c} U_t \right]$$

$$= \frac{1}{c} \left[-\frac{dx}{d\sigma} U_x - \frac{dt}{d\sigma} U_t \right] = -\frac{1}{c} \frac{dU}{d\sigma}; \quad (4.1.13)$$

and on C_3,

$$\hat{n} \cdot \left[U_x, -\frac{1}{c^2} U_t \right] = \frac{1}{c} \frac{dU}{d\sigma}. \quad (4.1.14)$$

We now use these results in (4.1.9) to conclude that

$$I = \frac{1}{c^2} \int_{x_1}^{x_2} U_t(x', 0) \, dx' - \int_{C_2} \frac{1}{c} \frac{dU}{d\sigma} \, d\sigma + \int_{C_3} \frac{1}{c} \frac{dU}{d\sigma} \, d\sigma. \quad (4.1.15)$$

We specialize to the case of constant c. Now the integrals over C_2 and C_3 can be carried out explicitly. We use these results and the partial differential equation (4.1.1) for U to conclude that

$$U(x, t) = \frac{U(x - ct, 0) + U(x + ct, 0)}{2} + \frac{1}{2c} \int_{x-ct}^{x-ct} U_t(x', 0) \, dx'$$

$$- \frac{c}{2} \int_D F(x', t', U, U_x, U_t) \, dx' \, dt'. \quad (4.1.16)$$

Here we have used the fact that for constant c, the values of x_1 and x_2 are explicitly known in terms of (x, t). This formula expresses $U(x, t)$ in terms of its values and the values of its first derivatives in between and on the characteristics that pass through (x, t).

Let us suppose that we are given a Cauchy problem with data on $t = 0$, that is, an initial value problem with U and U_t given at $t = 0$. In the simplest case, where $F = 0$, the first line is a solution, the d'Alembert solution. When F is a function of x and t only, (4.1.16) is still a solution now to the inhomogeneous problem with known source. When F depends on any or all of its last three arguments, (4.1.16) is an integral equation for U. One approach to solving this integral equation is the method of successive approximations, that is, iteration. Under relatively "mild" conditions on F, one can prove existence, uniqueness, and continuous dependence of the solution on the data. This remains true even if the Cauchy data are given on some more general curve in (x, t). In the next chapter, we shall discuss the constraints on such more general initial curves in the context of problems in two and three dimensions. Those results are easily specialized to one space dimension. (Also see Garabedian [1964].)

If the data and F are "smooth," then so is the solution. Now let us suppose that $x_3 < x_1$ is a point of discontinuity for the initial data for U. The solution formula (4.1.16) remains valid because this discontinuity does not affect this formula at all. Indeed, we can take a limit from the right; that is, we can decrease x so long as in doing so we keep C_3 to the right of x_3. We can further take this characteristic right up to x_3 so long as we interpret the data as a right limit and the solution likewise. Similarly, we can start from the left side and take left limits so long as we interpret the data and the solution likewise. This is simply another demonstration that the discontinuity of the data propagates along the characteristics. However, here we have allowed a discontinuity in the function U itself by the expedient of interpreting all processes as one-sided limits along the characteristic emanating from the discontinuity.

Note that there is also a characteristic of type ψ through the point x_3. Correspondingly, the discontinuity in the data propagates in the opposite direction along this characteristic.

It can also be seen from (4.1.16) that data outside of the domain D do not affect the solution at (x, t). That is, data cannot propagate faster than the characteristic speed c. This property is called *causality*, and the domain D itself is called the *domain of dependence* of the point (x, t). On the other hand, let us consider the region above any point (forward in time) bounded left and right by the two characteristics through that point. This is the region in which the solution is affected by the data at the given point. This region is called the *range of influence of that point*.

Exercises

4.1 (a) Suppose that $F = 0$ in (4.1.1) and c is constant. By direct integration in (4.1.6), conclude that

$$V(\xi, \eta) = f(\xi) + g(\eta), \qquad U(x, t) = f(x - ct) + g(x - ct),$$

with f and g arbitrary functions of ξ and η having one derivative.

(b) Suppose now that these initial data are given

$$U(x, 0) = U_0(x), \qquad U_t(x, 0) = U_1(x).$$

Express f and g in terms of U_0 and U_1 for this problem.

4.2 (a) Suppose that AB is a characteristic curve of type ψ, that is, of the same type as C_2 as defined by (4.1.11) directed from $t' = 0$ to $t' = t$. Suppose that σ is the directed arc length from A to B and $\hat{\mathbf{n}}$ is a unit normal directed to the right of the oriented characteristic. Show that

$$\int_{AB} \hat{n} \cdot \left[U_x, -\frac{1}{c^2} U_t \right] d\sigma = -\frac{1}{c} U(x, t) \Big|_A^B - \int_{AB} \frac{U(x', t') \, dc(x')}{c^2 \, dx'} \, dx'.$$

Here t' is a function of x' as defined by the characteristic differential equation (4.1.11).

(b)   Repeat the process for the case in which AB is a characteristic of type ϕ as defined by (4.1.11) for C_3, and show that the result here is just the negative of the result in (a). In this case, the characteristic is directed from time $t' = t$ to $t' = 0$.

4.3   Suppose that in (4.1.1) c is a function of x. Use the results of the preceding problem to show that in this case the formula (4.1.16) is replaced by

$$U(x, t) = \frac{c(x)}{2} \left[\frac{U(x_1, 0)}{c(x_1)} + \frac{U(x_2, 0)}{c(x_2)} + \int_{x_1}^{x_2} \frac{U_t(x', 0)}{c^2(x')} \, dx' \right.$$
$$\left. - \int_D F \, dx' \, dt' + \int_{C_3} - \int_{C_2} \frac{U \, dc(x')}{c^2 \, dx'} \, dx' \right].$$

What are the implications of this result as regards the domain of dependence of the point (x, t)?

4.4   (a)   Assume that $F = 0$ in (4.1.1). Use the solution formula (4.1.16) to solve the two initial value problems

$$U(x, 0) = U_0(x), \qquad U_t(x, 0) = \pm c U_0'(x).$$

(b)   Specialize the solution of (a) to the case $U_0(x) = \delta(x - x_3)$.

(c)   Apply the solution formula to the problem in which

$$U(x, 0) = 0, \qquad U_t(x, 0) = c^2 \delta(x - x_3), \qquad F = 0.$$

(d)   Solve the problem in which

$$U(x, 0) = U_t(x, 0) = 0, \qquad F(x, t) = -\delta(x - x_3)\delta(t - 0+).$$

4.2   THE INITIAL BOUNDARY VALUE PROBLEM

We consider now the following problem for U with constant c:

$$U_{xx} - c^{-2} U_{tt} = F(x, t), \qquad -L < x < L, \quad t > 0, \quad c = \text{const};$$
$$U(x, 0) = f(x), \qquad U_t(x, 0) = g(x); \qquad\qquad (4.2.1)$$
$$U(\pm L, t) = h_\pm(t), \qquad h_\pm(0) = f(\pm L), \qquad dh_\pm(0)/dt = g(\pm L).$$

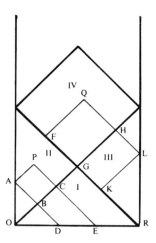

Fig. 4.2. The (x, t) domain for this problem.

In Fig. 4.2, we show the domain of interest for this problem, decomposed into subdomains I–IV, whose boundaries are characteristics. We shall demonstrate how the solution is obtained on each of these subdomains by an extension of the method used in the preceding section. In this section, we shall use the single letters of the figure to denote points in the domain. In the next section, we shall return to use of the explicit (x, t) values. The method described here will explicitly demonstrate the domains of dependence for points in the separate regions and will also introduce the solution to problems other than the Cauchy problem for U. In each case, we shall consider the integral of $U_{xx} - c^{-2}U_{tt}$ over a prescribed domain and show by the divergence theorem that this integral is given in terms of the value of U at the point of interest and in terms of other points at which the solution is known. Thus, by integrating the source F over the same domain, we obtain a simple linear algebraic equation for U at the point of interest and hence determine U.

We have already determined the solution (4.1.16) in region I in this manner. Therefore, we shall consider the solution at a point (x, t) in region II. By integrating over the diamond-shaped region $ABCP$, we find that

$$\int_{ABCP} \left[U_{xx} - \frac{1}{c^2} U_{tt} \right] dx' \, dt' = -\frac{1}{c}[U(P) - U(C)]$$

$$+ \frac{1}{c}[U(A) - U(P)] - \frac{1}{c}[U(B) - U(A)] + \frac{1}{c}[U(C) - U(B)]. \quad (4.2.2)$$

Because this sum is equal to the integral of the source term over the same domain $ABCP$, we can solve for $U(P)$. The result is

$$U(P) = U(A) + U(C) - U(B) - \frac{c}{2} \int_{ABCP} F(x', t')\, dx'\, dt'. \quad (4.2.3)$$

Here the solution is expressed in terms of the values of U on the left boundary and on the characteristic through BC. The problem for U in which the data are given on a characteristic and a noncharacteristic is called a *Goursat problem*. Thus, (4.2.3) is a solution to a Goursat problem for U.

By integrating over the domain $BDEC$, we can express the difference $U(B) - U(C)$ in terms of initial values as

$$U(C) - U(B) = \frac{1}{2}\left[U(E) - U(D)\right] + \frac{1}{2c} \int_{DE} U_t(x', 0)\, dx'$$

$$- \frac{c}{2} \int_{BDEC} F(x', t')\, dx'\, dt'. \quad (4.2.4)$$

Using this result in (4.2.3) yields

$$U(P) = U(A) + \frac{1}{2}\left[U(E) - U(D)\right] + \frac{1}{2c} \int_{DE} U_t(x', 0)\, dx'$$

$$- \frac{c}{2} \int_{ADEP} F(x', t')\, dx'\, dt'. \quad (4.2.5)$$

The solution formula (4.2.5) expresses $U(P)$ in terms of the initial and boundary values and the source and demonstrates the domain of dependence for this point. We remark that this formula could also be obtained as follows. Extend the initial data and source as *odd functions* about the left endpoint. Remove the boundary and use the method of the preceding section to solve for $U(P)$. Add the effect due to the nonzero boundary data. We leave the verification as an exercise.

In contrast, when U_x is prescribed on the boundary instead of U, we would extend the data as an *even function* about the left endpoint. The solution to this problem is outlined in Exercise 4.7.

For the special case in which the only nonzero datum is a Dirac delta for U itself located at D, we see that half of this delta function propagates to the left at characteristic speed and reflects at the boundary *but also changes sign* and then propagates to the right from the boundary in region II, again at characteristic speed, but now as the negative of its value in region I.

This result can be seen experimentally as follows. Fix the end of a length of rope or wire (3–5 m long) to a wall or pole. At the other end, snap the rope smartly in the vertical direction to create a ripplelike disturbance, which will propagate along the rope. When it reflects from the fixed end, it will exhibit the sign reversal demonstrated by the solution formula.

In region III, the solution is determined in exactly the same way as was used here to obtain the solution in region II except that now the domains of integration touch the right boundary. We leave this calculation to the exercises.

Let us now consider the point Q, and integrate over the domain $FGHQ$:

$$\int_{FGHQ} \left[U_{xx} - \frac{1}{c^2} U_{tt} \right] dx'\, dt' = -\frac{1}{c} [U(Q) - U(H)]$$

$$+ \frac{1}{c}[U(H) - U(G)] - \frac{1}{c}[U(G) - U(F)] + \frac{1}{c}[U(F) - U(Q)]. \quad (4.2.6)$$

From this result it follows that

$$U(Q) = U(F) + U(H) - U(G) - \frac{c}{2} \int_{FGHQ} F(x', t')\, dx'\, dt'. \quad (4.2.7)$$

The solution at Q has now been expressed in terms of function values on the two characteristics below Q. The problem in which boundary data are given on two characteristics is called the *characteristic Goursat problem*. Thus, (4.2.7) is a solution formula for the characteristic Goursat problem. The solution at G is determined by the method introduced in the preceding section, which provides the solution for points in region I. The point F is on the boundary of region II, and, hence, the solution at this point is obtained by the method described above for points in region II. The solution at H is determined in a completely analogous manner applied to region III.

Exercises

4.5 Let S be a point in region III of Fig. 4.2. Develop solution formulas analogous to (4.2.3) and (4.2.5) for $U(S)$.

4.6 Derive the solution formula (4.2.5) in the following manner: Extend the initial data and source as odd functions about the left endpoint. Solve the initial value problem (disregarding the boundary) by the method of Section 4.1. Use the method of Section 4.2 to solve the boundary value problem with zero source and initial data. Add the solutions.

4.7 (a) Suppose that on the left boundary U_x is given instead of U. Integrate $U_{xx} - c^{-2}U_{tt}$ over the domain AOD to determine $U(A)$ in terms of the source in AOD and the data on OA and OD. The result is

$$U(P) = \frac{1}{2}[U(E) + U(D)] + \frac{1}{c}\left[\int_{OD} + \frac{1}{2}\int_{DE}\right]U_t(x', 0)\, dx'$$

$$+ c\int_{OA} U_x(0, t')\, dt' - c\left[\frac{1}{2}\int_{ADEP} + \int_{AOD}\right]F(x', t')\, dx'\, dt'.$$

(b)   Obtain the solution at P in the following manner: Extend the initial data and source as *even functions* about the left endpoint. Solve for $U(P)$ by the method of Section 4.1. Now solve the problem for the prescribed boundary data with zero source and initial data by the method of (a). Add the two solutions.

4.3   THE INITIAL BOUNDARY VALUE PROBLEM CONTINUED

We consider again the problem defined by (4.2.1). Now we shall use Fourier transform to derive the solution. Thus, we introduce the one-sided Fourier transform (2.1.12)

$$u(x, \omega) = \int_0^\infty U(x, t)e^{i\omega t}\, dt. \tag{4.3.1}$$

By applying this transform to the problem (4.2.1), we obtain the following problem for u:

$$\frac{d^2 u}{dx^2} + \frac{\omega^2}{c^2}u = G(x, \omega), \qquad -L < x < L;$$

$$G(x, \omega) = \tilde{F} - \frac{g}{c^2} + i\omega\frac{f}{c^2}; \tag{4.3.2}$$

$$u(\pm L, \omega) = \tilde{h}_\pm(\omega).$$

In this equation, we have used the tilde to denote the Fourier transform of the data. We remark that we expect that the solution will be bounded and, therefore, we expect that the Fourier transform (4.3.1) is defined for Im $\omega > 0$.

In order to solve for u, we introduce two solutions of the homogeneous differential equation in (4.3.2):

$$u_1(x, \omega) = \sin[\omega(x + L)/c], \qquad u_2(x, \omega) = \sin[\omega(x - L)/c]. \tag{4.3.3}$$

The function u_1 is zero at the left endpoint of the interval $[-L, L]$ whereas it is nonzero at the right endpoint; the function u_2 is zero at the right endpoint of the interval $[-L, L]$ but nonzero at the left endpoint. We remark further that the *Wronskian*

$$W(\omega) = u_1 \frac{du_2}{dx} - u_2 \frac{du_2}{dx} = \frac{\omega}{c} \sin \frac{2\omega L}{c} \tag{4.3.4}$$

of the two solutions is nonzero for ω in the upper half plane. Thus, these solutions are *linearly independent* in the upper half ω plane.

We solve the problem (4.3.2) by the method of *variation of parameters* [Ince, 1956]. The result is

$$u(x, \omega) = \frac{1}{W} \left\{ u_1(x, \omega) \left[\frac{\omega}{c} \tilde{h}_+(\omega) + \int_x^L G(x', \omega) u_2(x', \omega)\, dx' \right] \right.$$
$$\left. + u_2(x, \omega) \left[-\frac{\omega}{c} \tilde{h}_-(\omega) + \int_{-L}^x G(x', \omega) u_1(x', \omega)\, dx' \right] \right\}. \tag{4.3.5}$$

This leads to the following solution formula for U:

$$U(x, t) = \frac{c}{2\pi} \int \frac{d\omega\, e^{-i\omega t}}{\omega \sin(2\omega L/c)}$$
$$\cdot \left\{ \sin\left(\frac{\omega(x+L)}{c}\right) \left[\frac{\omega}{c} \tilde{h}_+(\omega) + \int_x^L G(x', \omega) \sin\left(\frac{\omega(x'-L)}{c}\right) dx' \right] \right.$$
$$\left. + \sin\left(\frac{\omega(x-L)}{c}\right) \left[-\frac{\omega}{c} \tilde{h}_-(\omega) + \int_{-L}^x G(x', \omega) \sin\left(\frac{\omega(x'+L)}{c}\right) dx' \right] \right\}.$$
$$\tag{4.3.6}$$

Insight into this representation can be gained by considering special cases. Let us suppose that

$$F = g = h_\pm = 0, \tag{4.3.7}$$

so that only f, the initial value of U itself, is nonzero. We use (4.3.2) to express the integrand directly in terms of f, and we also rewrite the integrand in a somewhat contracted form as

$$U(x, t) = \frac{i}{2\pi c} \int \frac{d\omega\, e^{-i\omega t}}{\sin(2\omega L/c)}$$
$$\cdot \int_{-L}^L f(x') \sin\left[\frac{\omega(x_< + L)}{c} \right] \sin\left[\frac{\omega(x_> - L)}{c} \right] dx'. \tag{4.3.8}$$

In this equation,

$$x_< = \min(x, x'), \qquad x_> = \max(x, x'). \tag{4.3.9}$$

The convergence of the ω integral here is not guaranteed. To ensure convergence, we require that the x' integral produce a locally integrable function of ω that vanishes as $|\mathrm{Re}\,\omega| \to \infty$. The latter requirement is guaranteed by the *Riemann–Lebesgue lemma* so long as $|f(x)|$ is integrable. Actually, this assures integrability in ω as well, but we will assume the stronger condition that f is piecewise differentiable. While this will not affect the exponential behavior in ω of the x' integral, it will introduce algebraic decay $1/\omega$ as can be confirmed by integration by parts.

The objective now is to calculate the ω integral by using complex function theory and, more precisely, by "closing" the contour of integration in an appropriate half plane and summing residues when appropriate. Thus, we view the integral as the limit of finite integrals on which $|\mathrm{Re}\,\omega| < R$. First, we consider closing the contour of integration with a semicircle of radius R in the upper half plane. When $R \to \infty$, the integral on the semicircular contour will approach zero so long as the coefficient of $i\omega$ in the exponent is positive. By writing the sine functions as complex exponentials, we can verify that the worst case to be considered is

$$-t + \frac{2L}{c} - \frac{L + x_<}{c} + \frac{x_> - L}{c} = -t + \frac{x_> - x_<}{c} > 0. \quad (4.3.10)$$

This is always true for t negative. Thus, the integral on the semicircle approaches zero with increasing R for t negative. Because the closed contour contains no singularities in its interior, we conclude that the representation (4.3.8) for U is zero for t negative.

For t positive, (4.3.10) is not true for all choices of x and x', and we consider replacing the semicircle in the upper half plane with one in the lower half plane. The integral on this contour will approach zero with increasing radius whenever the coefficient of $i\omega$ in the exponent is negative. Worst-case analysis in this case leads to the requirement that

$$-t - \frac{2L}{c} + \frac{L + x_<}{c} - \frac{x_> - L}{c} = -t + \frac{x_< - x_>}{c} < 0. \quad (4.3.11)$$

This will always be true for t positive.

Thus, for t positive, we must examine the singularities of the integrand below the path of integration in (4.3.8). These are (i) a removable singularity at $\omega = 0$ and (ii) simple poles at the zeros of the sine function in the denominator in which $2\omega L/c\pi$ is an integer. The denominator arose from the Wronskian of the two solutions u_1 and u_2. Thus, when the denominator is zero, these two solutions must be *linearly dependent*. Indeed, let us define

$$\omega_n = n\pi c/2L, \qquad k_n = \omega_n/c, \qquad n = \pm 1, \pm 2, \ldots. \quad (4.3.12)$$

Then it follows from (4.3.3) that

$$u_1(x, \omega_n) = \sin[k_n x + (n\pi/2)] = (-1)^n u_2(x, \omega_n). \qquad (4.3.13)$$

These functions are *eigenfunctions* of the differential operator d^2u/dx^2, and the sequence $-(\omega_n/c)^2$ are *eigenvalues* of the same operator. That is, these functions satisfy the homogeneous equation in the form

$$d^2u/dx^2 = -(\omega_n^2/c^2)u,$$

subject to homogeneous boundary conditions.

It can now be seen that the semicircular contour with which we close the given path of integration must not pass through these poles on the real ω axis. Thus, choose the radius of the semicircle to be a half-integer multiple of $\pi c/2L$ and let that integer increase beyond all bounds. The residue sum then becomes a series solution for U. After some manipulation, which we leave to the exercises, the following result is obtained:

$$U(x, t) = \sum_{n=1}^{\infty} \cos \omega_n t \sin\left[k_n x + \frac{n\pi}{2}\right] \frac{1}{L} \int_{-L}^{L} f(x') \sin\left[k_n x' + \frac{n\pi}{2}\right] dx'. \qquad (4.3.14)$$

Evaluation of this result at $t = 0$ yields

$$U(x, 0) = \sum_{n=1}^{\infty} \sin\left[k_n x + \frac{n\pi}{2}\right] \frac{1}{L} \int_{-L}^{L} f(x') \sin\left[k_n x' + \frac{n\pi}{2}\right] dx',$$

$$U_t(x, 0) = 0. \qquad (4.3.15)$$

In the first line of this equation, the right side is to be recognized as the Fourier series for the function $f(x)$ on the interval $(-L, L)$ continued to the interval $(-3L, -L)$ as an odd function of x about $x = -L$. As noted in the preceding section, this is exactly the extension of the initial data that is required in order to solve a problem in which the boundary datum at $x = -L$ is the value of U itself. Thus, the series solution satisfies the initial conditions; each term in the series satisfies the boundary condition and because of (4.3.12), each term satisfies the homogeneous wave equation as well.

Digression (Eigenfunction Expansions)   It has been seen here that the function f had a series expansion in terms of the eigenfunctions of the differential equation with prescribed boundary conditions. When a class of functions have series representations in terms of a sequence of functions, the sequence is called *complete*. Completeness of eigenfunctions of a differential operator is a more general property. Indeed, consider the eigenvalue problem on $(-L, L)$:

$$-\frac{d}{dx}\left[p\frac{du}{dx}\right] + qu = \lambda u, \qquad \alpha u(-L) + \beta u'(-L) = 0, \qquad \gamma u(L) + \delta u'(L) = 0,$$

with α, β, γ, and δ real; p differentiable; and q continuous. Then the eigenvalues are discrete, and the eigenfunctions are complete and orthogonal. More precisely, suppose that we denote the sequence of eigenfunctions by $\{v_n\}$. Then, with appropriate scaling,

$$\int_{-L}^{L} v_n(x')v_m(x')\,dx' = \delta_{nm} = \begin{cases} 1, & n = m, \\ 0, & n \neq m; \end{cases}$$

that is, the set of $\{v_n\}$ is *orthonormal*. Furthermore, square integrable functions have series representations in terms of the eigenfunctions in the sense that

$$\lim_{N \to \infty} \int_{-L}^{L} \left[f - \sum_{n=0}^{N} f_n v_n \right]^2 dx = 0.$$

Here

$$f_n = \int_{-L}^{L} f(x')v_n(x')\,dx'$$

are the Fourier coefficients of the function f with respect to the sequence of eigenfunctions. (For further discussion, see Titchmarsh [1962] and Coddington and Levinson [1955].)

We return now to the representation (4.3.8) and consider another method of analyzing this result. We begin by setting

$$\frac{1}{\sin(2\omega L/c)} = \frac{-2ie^{2i\omega L/c}}{1 - e^{4i\omega L/c}} = -2ie^{2i\omega L/c} \sum_{n=0}^{\infty} e^{4in\omega L/c}. \tag{4.3.16}$$

This expansion converges absolutely and uniformly in $\mathrm{Re}\,\omega$ for $\mathrm{Im}\,\omega$ bounded from below by a positive number, which is true on the contour of integration.

This result is substituted into (4.3.8), and the sine functions are rewritten as complex exponentials to yield the representation

$$U(x, t) = \sum_{n=0}^{\infty} U_n(x, t);$$

$$\tag{4.3.17}$$

$$U_n(x, t) = \frac{1}{4\pi c} \int d\omega \exp\left(\frac{-i\omega t_n + 2i\omega L}{c} \right)$$

$$\cdot \int_{-L}^{L} \left[\exp\left(\frac{i\omega[|x - x'| - 2L]}{c} \right) + \exp\left(\frac{-i\omega[|x - x'| - 2L]}{c} \right) \right.$$

$$\left. - \exp\left(\frac{i\omega[x + x']}{c} \right) - \exp\left(\frac{-i\omega[x + x']}{c} \right) \right] f(x')\,dx'.$$

Here

$$t_n = t - 4nL/c, \qquad n = 0, 1, 2, \ldots. \tag{4.3.18}$$

We remark that U_n is exactly like U_0 with $t = t_0$ replaced by t_n. Furthermore, the analysis that allowed us to conclude that U was zero for t negative allows us here to conclude that the same is true for U_n when t_n is negative. Thus, let us consider only U_0.

We can perform the ω integration first if we view the result as a distribution. Indeed, each of the four integrals to be calculated can be recognized as a variant on the Fourier representation on the Dirac delta function (2.3.2). Thus, the result of calculating the integral in ω is

$$U_0(x, t) = \frac{1}{2c} \int_{-L}^{L} f(x') \, dx' \left[\delta\left(t - \frac{|x - x'|}{c} \right) + \delta\left(t - \frac{[2L - |x - x'|]}{c} \right) \right.$$
$$\left. - \delta\left(t - \frac{[2L + x + x']}{c} \right) - \delta\left(t - \frac{[2L - x - x']}{c} \right) \right].$$

$$\tag{4.3.19}$$

The delta functions "act" or *have their support* at the zeros of their arguments, which are given by

$$x'_1 = x - ct, \qquad x'_2 = x + ct, \qquad x'_3 = x + ct - 4L,$$
$$x'_4 = x - ct + 4L, \quad x'_5 = -x + ct - 2L, \quad x'_6 = -x - ct + 2L. \tag{4.3.20}$$

In Fig. 4.3, Fig. 4.2 is repeated and extended. The boundaries between the subregions and the equations for these boundaries are indicated. Because the x' integration in (4.3.19) is over the interval $(-L, L)$, the integration will yield a nonzero contribution from a delta function only for its zero, as given by (4.3.20), in this interval. Thus, we obtain contributions to the integral in (4.3.19) only in the following regions:

$$x'_1, \quad \text{I, III}; \qquad x'_2, \quad \text{I, II}; \qquad x'_3, \quad \text{VI, VII};$$
$$x'_4, \quad \text{V, VII}; \qquad x'_5, \quad \text{II, IV, VI}; \qquad x'_6, \quad \text{III, IV, V.} \tag{4.3.21}$$

These results can be confirmed by checking that in the indicated ranges, the zero of the delta function lies between $-L$ and L. For (x, t) in the unshown triangular regions above V and VI, the roots x'_3 and x'_4 still lie in $(-L, L)$. But for this exception, none of these delta functions "acts" in any but the indicated regions. Thus, for example, in region I, the solution is

$$U(x, t) = \tfrac{1}{2}[f(x + ct) + f(x - ct)]. \tag{4.3.22}$$

Note here that the factor of $1/c$ in (4.3.19) is used in the evaluation of the delta function in accordance with (2.1.7). This solution is exactly the contribution

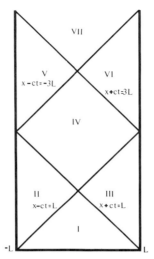

Fig. 4.3. The regions in which the delta functions of (4.3.19) have support.

from initial data in the solution formula (4.1.16). Clearly, evaluation of (4.3.19) with the guidance of (4.3.20) and (4.3.21) will yield the contribution from initial data in each of the regions and agree with the results of Section 4.2.

Let us now consider the case in which the given data are

$$f = g = h_\pm = 0 \tag{4.3.23}$$

but the source $F(x, t)$ is nonzero. In this case, we obtain from (4.3.2) and (4.3.6)

$$U(x, t) = \frac{c}{2\pi} \int \frac{d\omega \, e^{-i\omega t}}{\omega \sin(2\omega L/c)}$$

$$\cdot \int_{-L}^{L} \tilde{F}(x', \omega) \sin\left[\frac{\omega(x_< + L)}{c}\right] \sin\left[\frac{\omega(x_> - L)}{c}\right] dx'. \tag{4.3.24}$$

A first inclination here might be to close the contour of integration again and obtain a solution as a residue sum. However, some care must be taken, because $\tilde{F}$ is a function of ω while f was not. Thus, the analytic continuation to the lower half ω plane of $\tilde{F}$ could introduce new singularities that were not present in the preceding case. Indeed, $\tilde{F}$ could have a *branch point* at any or even all of the poles of the integrand, precluding evaluation by residue sum for those singularities. Consequently, we cannot, in general, calculate the solution by closing contours in the ω plane. However, let us consider two special cases to demonstrate some of the possibilities.

First, let us suppose that

$$F(x, t) = \delta(x - x_0)e^{-\alpha t}, \qquad \alpha > 0. \tag{4.3.25}$$

In this case,

$$\tilde{F}(x', \omega) = i\delta(x' - x_0)/(\omega + i\alpha). \tag{4.3.26}$$

Thus, we see that in addition to the residue sum obtained in the preceding case, we now obtain a residue at the pole introduced by the analytic continuation to the lower half ω plane of the Fourier transform of the source term. The result is

$$U(x, t) = \frac{ce^{-\alpha t}}{\alpha \sinh(2\alpha L/c)} \sinh\left[\frac{\alpha(x_> - L)}{c}\right] \sinh\left[\frac{\alpha(x_< + L)}{c}\right]$$

$$+ \sum_{n=1}^{\infty} 2c \frac{\omega_n \cos \omega_n t - \alpha \sin \omega_n t}{n\pi(\omega_n^2 + \alpha^2)} \sin\left(k_n x + \frac{n\pi}{2}\right) \sin\left(k_n x_0 + \frac{n\pi}{2}\right). \tag{4.3.27}$$

In this equation,

$$x_< = \min(x, x_0), \qquad x_> = \max(x, x_0), \tag{4.3.28}$$

and ω_n and k_n are as defined in (4.3.12).

As a second example, let us consider the source

$$F(x, t) = t^{-1/2} \delta(x - x_0) \sin \alpha t. \tag{4.3.29}$$

In this case,

$$\tilde{F}(x', \omega) = \frac{\sqrt{\pi} e^{-i\pi/4}}{2} \left[\frac{1}{\sqrt{\omega + \alpha}} - \frac{1}{\sqrt{\omega - \alpha}}\right] \delta(x' - x_0). \tag{4.3.30}$$

The square roots here must be analytic functions in the upper half ω plane. Furthermore, both square roots are positive for ω real and larger than α. We make them single valued in the ω plane in the standard manner of introducing *branch cuts* extending from the branch points at $\pm\alpha$ to the point at infinity. The only question that remains is in which direction at infinity these branch cuts should be directed. In fact, all that matters is that they be extended to infinity in such a manner as not to contradict the analyticity and concurrent single-valuedness of each square root in the upper half ω plane. With no loss of generality, we take the branch cuts to be *straight lines* extending *downward* from the branch points to infinity.

We have now given enough information to define both square roots in a single-valued manner in the entire (cut) ω plane. Indeed, we can define a unique square root in the entire ω plane now by prescribing the *angular range* or *argument* of the difference $\omega - \alpha$ (or $\omega + \alpha$). In order that the square root(s) be positive on the far-right real axis and that the branch cuts be vertical, it is necessary that

$$-(\pi/2) \le \arg(\omega \pm \alpha) < 3\pi/2. \tag{4.3.31}$$

Here the only freedom of choice left is which side has equality. In fact, because our transform is defined originally in the upper half plane, this choice has no effect on the solution in the domain of definition. We need only remain consistent when we deform the contour of integration into the lower half plane for analysis of the solution.

We will not carry out the analysis of this example here. However, we note that in closing the contour of integration in the lower half ω plane, it is necessary to include keyholelike contours around the branch points. In Fig. 4.4, we depict an example of this with $\pm \alpha$ lying between the origin and the first poles of the integrand. Thus, the solution to this problem is given as a residue sum plus a sum of two loop contour integrals. See Exercise 4.13.

Now let us consider using the alternative formal method, which required integrating with respect to ω first. Here, however, the dependence of $\tilde{F}$ on ω again makes the procedure not so straightforward as for the initial value problem. We must first rewrite $\tilde{F}$ as an integral with respect to t' as follows:

$$U(x, t) = \frac{c}{2\pi} \int \frac{d\omega\, e^{-i\omega t}}{\omega \sin(2\omega L/c)} \int_{-L}^{L} dx' \int_{0}^{\infty} dt'\, e^{i\omega t'} F(x', t')$$

$$\cdot \sin\left[\frac{\omega(x_< + L)}{c}\right] \sin\left[\frac{\omega(x_> - L)}{c}\right]. \tag{4.3.32}$$

As above, we use (4.3.16) and rewrite the sine functions in the numerator as

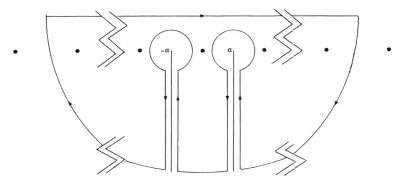

Fig. 4.4. An example of the closed contour for (4.3.24) and (4.3.30).

complex exponentials to obtain

$$U(x, t) = \sum_{n=0}^{\infty} V_n(x, t);$$

$$V_n(x, t) = \frac{c}{4\pi i} \int \frac{d\omega}{\omega} \exp\left(\frac{-i\omega t_n + 2i\omega L}{c}\right) \int_0^{\infty} dt' \exp(i\omega t')$$

$$\cdot \int_{-L}^{L} \left[\exp\left(\frac{i\omega[|x - x'| - 2L]}{c}\right) + \exp\left(\frac{-i\omega[|x - x'| - 2L]}{c}\right) \right.$$

$$\left. - \exp\left(\frac{i\omega[x + x']}{c}\right) + \exp\left(\frac{-i\omega[x + x']}{c}\right) \right] F(x', t') \, dx'. \tag{4.3.33}$$

Here t_n is again defined by (4.3.18).

The major differences between this result and (4.3.17) are (i) the integration with respect to t' and (ii) the division by ω under the integral sign. Again, formally, we interchange orders of integration and consider the ω integral first. This requires that we calculate the inverse Fourier transform of the function $1/i\omega$. The reader should note, however, that this is not the principal value function introduced in Section 2.3, since this inverse transform is defined as an integral *above* the singularities in the ω plane. Indeed, we need only use the result that

$$\frac{1}{2\pi i} \int \frac{d\omega}{\omega} e^{-i\omega t} = \begin{cases} -1, & t > 0 \\ 0, & t < 0 \end{cases} = -H(t), \tag{4.3.34}$$

with H the Heaviside function (2.1.10). Thus, we find that

$$V_n = -\frac{c}{2} \int_0^{\infty} dt' \int_{-L}^{L} dx' \, F(x', t')$$

$$\cdot \left[H\left(t_n - t' - \frac{|x - x'|}{c}\right) + H\left(t_n - t' + \frac{|x - x'|}{c} - \frac{4L}{c}\right) \right.$$

$$\left. - H\left(t_n - t' - \frac{[x + x' + 2L]}{c}\right) - H\left(t_n - t' + \frac{[x + x' - 2L]}{c}\right) \right]. \tag{4.3.35}$$

The effect of the Heaviside functions is to define the domain of integration in (x', t'). We shall demonstrate this fact in only one case. The first Heaviside function, with $n = 0$, is positive for

$$c(t - t') > \pm(x - x'), \tag{4.3.36}$$

which is equivalent to the two conditions

$$ct - x > ct' - x', \qquad ct + x > ct' + x'. \tag{4.3.37}$$

In region I of Figure 4.3, these constraints define the domain of influence of the point (x, t). Thus, for this (x, t), the true domain of integration for this term is just the domain of integration depicted in Fig. 4.1. Furthermore, all other Heaviside functions in (4.3.35) have negative argument in region I and are therefore equal to zero. Consequently, the sum reproduces the solution formula of the preceding sections for this special data in region I. In exactly this manner, the Heaviside functions define the domain of integration for all regions of the (x, t) domain.

To complete this discussion, we consider the case in which the only nonzero data are the function h_+ at the right end. In this case, the solution formula (4.3.6) takes on the somewhat simpler form

$$U(x, t) = \frac{1}{2\pi} \int \frac{\sin[\omega(x + L)/c]}{\sin(2\omega L/c)} \tilde{h}_+ e^{-i\omega t}\, d\omega. \tag{4.3.38}$$

As in the case of a source distribution, we cannot predict the nature of the singularities of the analytic continuation of $\tilde{h}_+$ in the lower half ω plane. Thus, just as in the source problems discussed earlier, we anticipate a residue sum plus *other* contributions to the solution formula. Alternatively,

$$U(x, t) = \sum_{n=0}^{\infty} W_n(x, t);$$

$$W_n(x, t) = \frac{1}{2\pi} \int d\omega \exp\left(\frac{-i\omega t_n + 2i\omega L}{c}\right)$$

$$\cdot \left[\exp\left(\frac{-i\omega[x + L]}{c}\right) - \exp\left(\frac{i\omega[x + L]}{c}\right)\right] \tag{4.3.39}$$

$$\cdot \int_0^{\infty} dt'\, h(t') \exp(i\omega t').$$

Here t_n is again defined by (4.3.18).

We evaluate again by interchanging the order of integration and recognizing the integrals in ω as delta functions. For $n = 0$, the result is

$$W_0 = \int_0^{\infty} dt'\, h(t')\left[\delta\left(t - t' - \frac{x - L}{c}\right) - \delta\left(t - t' - \frac{x + 3L}{c}\right)\right]. \tag{4.3.40}$$

This integral is readily evaluated to yield

$$W_0 = h\left(t + \frac{x - L}{c}\right) H\left(t + \frac{x - L}{c}\right) - h\left(t - \frac{x + 3L}{c}\right) H\left(t + \frac{x + 3L}{c}\right). \tag{4.3.41}$$

The first term here characterizes the propagation of the prescribed data $h(t)$ along the characteristic in the direction of decreasing x with increasing time.

The Heaviside function is needed because its argument must be positive in order that the delta function in the integrand in (4.3.40) have its support in the interval of integration. The second term characterizes the reflection from the opposite endpoint, with change in sign, as predicted in the preceding section. Subsequent terms $n > 0$ produce multiple reflections of the prescribed boundary value.

Exercises

4.8 Derive the solution (4.3.5) to the problem (4.3.2).

4.9 Let

$$f_1(\omega) = \int_{-L}^{L} f(x)e^{i\omega x/c}\, dx,$$

with f continuous on $[-L, L]$.

(a) Show that $f_1 e^{i\omega L/c}$ is bounded in the upper half ω plane.
(b) Show that $f_1 e^{-i\omega L/c}$ is bounded in the lower half ω plane.
(c) Is f_1 differentiable with respect to ω? Why?
(d) Summarize the properties of f_1 and explain the relevance to the discussion following (4.3.8) and (4.3.9).

4.10 Verify (4.3.11).

4.11 Verify (4.3.14).

4.12 Verify (4.3.20) and (4.3.21).

4.13 Complete the analysis of the solution U for F given by (4.3.29) by using the method of closing the contour of integration in the lower half ω plane. Assume that α is not one of the values ω_n given by (4.3.12).

4.14 Describe the domains of integration imposed by the Heaviside functions in (4.3.35) for $n = 0$.

4.15 Verify (4.3.34).

4.16 (a) Solve the problem for U when the only nonzero datum is given by the function

$$h_+(t) = \delta(t - t_0), \qquad 0 < t_0 < 2L/c.$$

(b) Replace t_0 in (a) with $4\pi L/c < t_1 < 6\pi L/c$.

4.4 THE ADJOINT EQUATION AND THE RIEMANN FUNCTION

For linear ordinary differential equations, an important tool for generating integral representations of solutions is the *Green's function*. One approach to the development of these functions is via distributions. More precisely, the source in the ordinary differential equation developed for the Green's function is a Dirac delta function. In the next two sections, we shall discuss the analogous technique for the wave equation in one space dimension. However, here there are two alternative approaches, one in which the *source* is taken to be a delta function the other in which part of the *initial data* is taken to be a distribution. The solution to the former type of problem is again called a Green's function, while the solution to the latter is called a *Riemann function*. It is the latter that will be developed in this section; the former will be developed in the next section.

We introduce the general linear hyperbolic operator

$$\mathcal{L}U = U_{xx} - c^{-2}U_{tt} + \alpha U_x + \beta U_t + \gamma U. \tag{4.4.1}$$

In this operator equation, via a simple transformation, it would be possible to make the coefficient of U_{tt} equal to -1. However, the second-order wave operator is the hyperbolic operator of interest throughout, and this transformation would tend to obscure the dependence of the results that follow on the characteristic speed c. We shall assume that c is a function of x only but that α, β, and γ may be functions of x and t.

For the ordinary differential equation, another differential operator $\mathcal{L}^*$, called the *adjoint operator*, is generated and has the property that $V\mathcal{L}U - U\mathcal{L}^*V$ is an exact differential of lower order than $\mathcal{L}$. Then, integration of this difference produces an expression involving only boundary data. Using this as a guideline, we seek an adjoint operator so that the same difference expression produces a *divergence*. Then, integration over an appropriate domain in space–time will produce an expression completely in terms of "boundary data" in space–time. The quotes here are used to remind the reader that part of our boundary might be at a prescribed value of t and that this use of the term *boundary data* is generic.

We calculate that

$$V\mathcal{L}U = \frac{\partial}{\partial x}[VU_x + \alpha VU] + \frac{\partial}{\partial t}\left[-\frac{1}{c^2}VU_t + \beta VU\right] + \gamma VU$$

$$- V_x U_x - (\alpha V)_x U + \frac{1}{c^2}V_t U_t - (\beta V)_t U$$

$$= \frac{\partial}{\partial x}[VU_x - UV_x + \alpha VU]$$

$$+ \frac{\partial}{\partial t}\left[-\frac{1}{c^2}(VU_t - UV_t) + \beta VU)\right]$$

$$+ U\left[V_{xx} - \frac{1}{c^2}V_{tt} - (\alpha V)_x - (\beta V)_t + \gamma V\right]. \tag{4.4.2}$$

The last line in this calculation defines the adjoint operator; that is,

$$\mathscr{L}^*V = V_{xx} - \frac{1}{c^2}V_{tt} - (\alpha V)_x - (\beta V)_t + \gamma V \qquad (4.4.3)$$

and

$$V\mathscr{L}U - U\mathscr{L}^*V = \frac{\partial}{\partial x}[VU_x - UV_x + \alpha VU]$$

$$+ \frac{\partial}{\partial t}\left[-\frac{1}{c^2}(VU_t - UV_t) + \beta VU\right]. \qquad (4.4.4)$$

We shall now consider the integral of this latter expression over the trapezoidal domain D_1 in Fig. 4.5 bounded by the points $(x_3, 0)$, $(x_4, 0)$, (ξ_2, τ), (ξ_1, τ). As in Fig. 4.1, we draw the back characteristics from the points of interest as if c were constant. However, for the present, we shall allow c to be a function of x. It should be noted that the inner domain of this figure is exactly the domain D of Fig. 4.1. Also, the normal derivatives are determined as in (4.1.10) and (4.1.11). We calculate that

$$I = \int_{D_1} [V\mathscr{L}U - U\mathscr{L}^*V]\,dx\,dt$$

$$= \int_{(x_4,0)}^{(\xi_2,\tau)}\left[-\frac{V}{c}\frac{dU}{d\sigma} + \frac{U}{c}\frac{dV}{d\sigma} + \frac{\alpha + \beta c}{\sqrt{1 + c^2}}VU\right]d\sigma$$

$$- \int_{(x_3,0)}^{(\xi_1,\tau)}\left[\frac{V}{c}\frac{dU}{d\sigma} - \frac{U}{c}\frac{dV}{d\sigma} + \frac{-\alpha + \beta c}{\sqrt{1 + c^2}}VU\right]d\sigma$$

$$+ \int_{x_3}^{x_4}\left[\frac{1}{c^2}[VU_t - UV_t] - \beta VU\right]\Bigg|_{t=0}dx$$

$$- \int_{\xi_1}^{\xi_2}\left[\frac{1}{c^2}[VU_t - UV_t] - \beta VU\right]\Bigg|_{t=\tau}dx. \qquad (4.4.5)$$

Let us suppose that the following problem is prescribed for U:

$$\mathscr{L}U = F(x, t), \qquad t > 0, \quad -\infty < x < \infty;$$

$$U(x, 0) = f(x), \qquad U_t(x, 0) = g(x). \qquad (4.4.6)$$

We are now prepared to prescribe a problem for V so as to make this function a Riemann function and transform (4.4.5) into an integral representation for U in terms of this Riemann function. We want the boundary integral over the line $t = \tau$ to produce the value $U(\xi, \tau)$, and we shall prescribe data for V appropriately. Furthermore, we want the integral over the line at $t = 0$ to

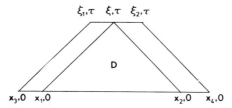

Fig. 4.5. The trapezoidal domain of integration for (4.4.5) with the inner triangular domain denoted by *D*.

produce the expected dependence of U on its initial values. Consequently, the data at $t = \tau$ will be *final data* rather than *initial data*, and the problem for V will be one that is "backward" in time. Thus, we require that

$$\mathcal{L}^*V = 0, \qquad t < \tau, \quad -\infty < x < \infty;$$

$$V(x, \tau; \xi, \tau) = 0, \qquad V_t(x, \tau; \xi, \tau) = -c^2\delta(x - \xi). \tag{4.4.7}$$

Here we have given V four arguments to emphasize its dependence on (ξ, τ). Indeed, the discussion of the domain of influence in Section 4.1 leads us to conclude that, in fact, V is zero on the characteristics connecting $(x_3, 0)$ with (ξ_1, τ) and $(x_4, 0)$ with (ξ_2, τ). Furthermore, V is also zero on the intervals (x_3, x_1) and (x_2, x_4). However, for V_t, more care is required, since the data for V_t is a *distribution*. Whether we think of the distribution itself or its implications as regards a discontinuity in V across $t = \tau$, it is evident that the singular behavior of V at (ξ, τ) *must propagate backward in time along the characteristics*. Thus, the integral of V_t through the points x_1 and x_2 must take account of this distributional behavior. We may "shrink" the interval of integration (x_3, x_4) onto the shorter interval $(x_1 -, x_2 +)$ so long as we remember that the appending $\pm$ here is meant to remind us to retain all distributional contributions right at the endpoints.

Using all of this information in (4.4.5), we find that

$$U(\xi, \tau) = \int_{x_1 -}^{x_2 +} \left[\frac{1}{c^2}[VU_t - UV_t] - \beta UV \right]\bigg|_{t=0} dx - \int_D VF \, dx \, dt. \tag{4.4.8}$$

This is a solution formula for U in terms of the Riemann function V. This formalism does not provide us with a means of finding V. However, it demonstrates that one need only solve this one problem (4.4.7) in order to have a means for solving *all problems* of the type (4.4.6).

We shall now consider some special cases in which one can obtain a closed-form analytical solution for V. First, let us suppose that $\mathcal{L}$ is the wave operator itself,

$$\mathcal{L}U = U_{xx} - c^{-2}U_{tt}, \tag{4.4.9}$$

with c constant. In this case, $\alpha = \beta = \gamma = 0$ in (4.4.1), and then

$$\mathscr{L}^*V = V_{xx} - c^{-2}V_{tt}, \tag{4.4.10}$$

with c constant. That is, $\mathscr{L}^* = \mathscr{L}$ for this example, and $\mathscr{L}$ is *self-adjoint*. The problem (4.4.7) for V now becomes

$$V_{xx} - c^{-2}V_{tt} = 0, \qquad t < \tau, \quad -\infty < x < \infty;$$
$$V(x, \tau; \xi, \tau) = 0, \qquad V_t(x, \tau; \xi, \tau) = -c^2\delta(x - \xi). \tag{4.4.11}$$

If instead we solve the problem

$$W_{xx} - c^{-2}W_{tt} = 0, \qquad 0 < t, \quad -\infty < x < \infty;$$
$$W(x, 0) = 0, \qquad W_t(x, 0) = c^2\delta(x) \tag{4.4.12}$$

and then set

$$V(x, t; \xi, \tau) = W(\xi - x, \tau - t), \tag{4.4.13}$$

we obtain a solution to (4.4.11). For this problem, we simply write down the d'Alembert solution, that is, the first line of (4.1.16):

$$W(x, t) = \frac{c}{2} \int_{x-ct}^{x+ct} \delta(x') \, dx'$$

$$= \begin{cases} \dfrac{c}{2}, & -ct < x < ct \\ 0, & \text{otherwise} \end{cases}$$

$$= \frac{c}{2}[H(x + ct) - H(x - ct)]$$

$$= \frac{c}{2}H(ct - |x|)$$

$$= \frac{c}{2}H(c^2t^2 - x^2). \tag{4.4.14}$$

Each of these representations is useful at times. Stated succinctly, the Riemann function is equal to $c/2$ in the domain of influence of the delta function and zero outside that domain. Here we continue with the third representation and set

$$V(x, t; \xi, \tau) = W(x - \xi, \tau - t)$$

$$= \frac{c}{2}\{H(x - \xi + c[\tau - t]) - H(x - \xi - c[\tau - t])\}. \tag{4.4.15}$$

This solution is to be substituted into (4.4.8). The significance of the outside limits on the x integral can now be seen, since for this Riemann function,

$$V_t(x, 0; \xi, \tau) = -\frac{c^2}{2} [\delta(x - \xi + c\tau) + \delta(x - \xi - c\tau)]. \quad (4.4.16)$$

Consequently, the solution obtained from this substitution is exactly the result (4.1.16) except with a minor change in notation.

As a second example, let us suppose that

$$\alpha = \beta = 0, \qquad \gamma = -(b^2/c^2) \quad (4.4.17)$$

Again the operator $\mathscr{L}$ is self-adjoint with

$$\mathscr{L}^*V = \mathscr{L}V = V_{xx} - c^{-2}V_{tt} - (b^2/c^2)V. \quad (4.4.18)$$

The partial differential equation for U or V is called the *Klein–Gordon equation*. The spatial and time shifts introduced in (4.4.13) can be used here as well. Thus, we consider the following problem for W:

$$W_{xx} - c^{-2}W_{tt} - (b^2/c^2)W = 0,$$
$$W(x, 0) = 0, \qquad W_t(x, 0; 0, 0) = c^2\delta(x). \quad (4.4.19)$$

To solve this problem, we introduce the one-sided Fourier transform u_+ in (2.2.12), denoted here by $w(x, \omega)$. The problem for w is

$$\frac{d^2w}{dx^2} + \frac{\omega^2 - b^2}{c^2}w = -\delta(x). \quad (4.4.20)$$

We seek a solution that remains analytic in some upper half ω plane, that is, above all apparent singularities of w in the ω plane. Furthermore, in accordance with Exercise 2.1, w must be a solution of the homogeneous equation for all nonzero x, continuous at $x = 0$, but with a first derivative that is *discontinuous*, with the magnitude of that discontinuity equal to 1.

The solutions of the homogeneous equation (4.4.20) are

$$w_\pm(x, \omega) = e^{\pm ikx}. \quad (4.4.21)$$

Here

$$k = \sqrt{\omega^2 - b^2}/c. \quad (4.4.22)$$

Since k must be defined in the complex ω plane, it is necessary to be more precise here. Our task is much like that of the preceding section in which it was necessary to define the square roots of $\omega \pm \alpha$. Indeed, we could use the same definitions for each of the square roots here. However, we shall proceed slightly differently for reasons which will become clear later.

We require that

$$0 \leq \arg(\omega \pm b) < 2\pi. \tag{4.4.23}$$

This defines each square root in the ω plane, with branch cuts along the real axis. For the product of these square roots appearing in k, the overlapping cuts to the right of b actually nullify in the sense that the function itself and its derivative have the same limits from above and below the dual cut. Therefore, this part of the cut can be eliminated, leaving only a *finite* branch cut from $-b$ to b.[†] For these defined ranges of the argument of each factor, we find that when

$$0 < \arg(\omega \pm b) < \pi,$$
$$\tag{4.4.24}$$
$$0 < \arg\sqrt{\omega^2 - b^2} = \tfrac{1}{2}[\arg(\omega - b) + \arg(\omega + b)] < \pi.$$

That is, the imaginary part of k is positive in the upper half ω plane. Thus, in the two solutions $w_\pm$, w_+ has a negative real part and w_- has a positive real part for x positive and ω in its upper half plane. Consequently, the former solution decays exponentially while the latter grows if either $x \to \infty$, Im $\omega > 0$, or $x > 0$, Im $\omega \to \infty$. Therefore, for x positive, w_+ is the analytic solution bounded in the upper half ω plane, while for x negative, w_- is. Consequently, the solution to (4.4.20) must be proportional to w_+ for x positive and to w_- for x negative.

The solution to (4.4.20) having the correct discontinuity at $x = 0$ is

$$w(x, \omega) = -\frac{e^{ik|x|}}{2ik}, \tag{4.4.25}$$

and the solution to (4.4.20) is

$$W = -\frac{1}{4\pi i} \int \frac{d\omega}{k} e^{ik|x| - i\omega t}. \tag{4.4.26}$$

Here the path of integration is above the branch points with Re ω ranging from $-\infty$ to ∞.

For t negative, the contour of integration can be closed in the upper half ω plane to verify that $W = 0$. For t positive, we could still close the contour in the upper half plane so long as $|x| > ct$. That is, the solution remains zero outside the domain of influence of the origin. Therefore, in what follows we consider only $|x| < ct$.

We seek an alternative representation of W in terms of "familiar" or "special" functions. First, we seek some insight by appealing to an entirely different problem, namely, the Helmholtz equation (3.1.3) in two spatial

---

[†] This function is discussed in more detail by Henrici [1974].

variables x and y, with source a delta function with support at the origin. With t replaced by iy and b^2 replaced by ω^2, that equation bears some similarity to the one at hand. We remarked in Section 3.1 that the two-dimensional problem models shallow water waves. The particular equation we have introduced here would model the response to a point source at the origin. Anyone who has ever dropped a pebble in the water has seen the waves spread out *radially*. That is, the solution to the problem is cylindrically symmetric or is a function of $x^2 + y^2$. Similarly, here we might now anticipate that the solution for W will be a function of $x^2 - c^2 t^2$. Thus, we might be led to seek a transformation that expresses W in terms of this variable (or its negative).

We set

$$t = \rho \cosh \phi, \qquad x = c\rho \sinh \phi; \qquad \rho = \sqrt{t^2 - x^2/c^2} \; ; \quad (4.4.27)$$

in which case, W can be rewritten as

$$W = -\frac{c}{4\pi i} \int \frac{d\omega}{k} \exp(i\rho[ck \sinh \phi - \omega \cosh \phi]). \qquad (4.4.28)$$

This transformation introduces through ρ the variable of interest. The particular choice relating t to cosh and x to sinh ensures that ct is greater than $|x|$ for real ϕ. We anticipate that W is independent of ϕ. To check this, we shall differentiate with respect to ϕ. However, before we do so, it is necessary to deform the contour of integration in such manner as to assure convergence of the differentiated integral. We deform into the lower half ω plane outside of the branch points, so that the real part of the exponent approaches $-\infty$ as $|\omega| \to \infty$ on the contour. We now differentiate with respect to ϕ and find that

$$\frac{dW}{d\phi} = -\frac{1}{4\pi i} \int \left(ic\rho \left[\cosh \phi - \frac{\omega}{ck} \sinh \phi \right] \right)$$

$$\cdot \exp(i\rho[ck \sinh \phi - \omega \cosh \phi]) \, d\omega$$

$$= \frac{1}{4\pi i} \int \frac{d}{d\omega} \exp(i\rho[ck \sinh \phi - \omega \cosh \phi]) \, d\omega = 0. \quad (4.4.29)$$

The last result follows because the integrand is an exact differential of an expression that is zero at the endpoints of integration. Thus, W is independent of ϕ and is a function of ρ alone, as predicted.

The integrand in (4.4.28) can now be greatly simplified by setting $\phi = 0$. The result is

$$W = -\frac{c}{4\pi i} \int \frac{e^{-i\rho\omega}}{\sqrt{\omega^2 - b^2}} \, d\omega. \qquad (4.4.30)$$

We again view the contour of integration as the limit of integrals over finite paths closed on a semicircle in the lower half ω plane. However, in this example, although we can justify this by taking the limit as the radius of the semicircle approaches infinity, we will deform the contour of integration in just the opposite manner! That is, we think of "shrinking" the contour around the finite branch cut connecting $-b$ to b; see Fig. 4.6. We remark that if we had used semi-infinite branch cuts extending down to $-i\infty$ in the complex plane, then this closed contour would have extended onto a *lower Riemann sheet* of the multivalued square root function. Although this would not have affected the evaluation at all, it would have affected the geometrical depiction of the contour integral. It was in anticipation of this step that we chose a finite branch cut for this example.

On the circular part of the contour in the figure, the square root is of order $1/\sqrt{d}$, where d is the distance from the branch point to the contour. The length of path is $2\pi d = O(d)$. The remainder of the integrand is $O(1)$ in d. Multiplying these estimates together reveals that the integrals on the circles are each $O(\sqrt{d})$ and, hence, approach zero with d. This is the analog in the complex plane of the fact that inverse square root is an integrable singularity on the real line. As for the part of the contour along the cut, on the upper half of the branch cut, the square root in the integrand has argument $i\pi/2$; on the lower half, its argument is the negative of that. Thus, the integral on the upper and lower sides of the branch cut sum to just twice the value on the upper half of the branch cut. Therefore,

$$
W(x, t) = \frac{c}{2\pi} \int_{-b}^{b} \frac{e^{-i\omega\rho}}{\sqrt{b^2 - \omega^2}} \, d\omega
$$

$$
= \frac{c}{2\pi} \int_{-1}^{1} \frac{e^{-i\omega\rho b}}{\sqrt{1 - \omega^2}} \, d\omega \tag{4.4.31}
$$

$$
= \frac{c}{2} J_0 \left[b \sqrt{\frac{t^2 - x^2}{c^2}} \right] H(c^2 t^2 - x^2).
$$

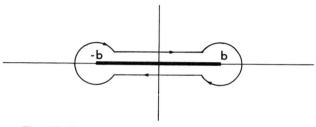

Fig. 4.6. The contour of integration around the branch cut.

Here in the middle equation, we have rescaled ω by $\bar{v}$. In the last equation, we have identified the integral with the Bessel function of the first kind J_0. (See, for example, Gradshteyn and Ryzhik [1965, Section 8.41, Eq. 10].) We have further reintroduced the Heaviside function here, which characterizes the fact that W is given by this function only inside the domain of influence of the point $x = 0$, $t = 0$ and is zero outside. As in the preceding example, this is particularly important in evaluating the integral over the initial line $t = 0$ in (4.4.8), since the Heaviside function has a delta function as its derivative, which again produces two contributions at the endpoints x_1- and x_2+. In order to evaluate $V_t(x, t; \xi, \tau)$, we proceed as follows:

(i)   Use (4.4.13) to express V in terms of W;

(ii)   use the fourth representation of the Heaviside function in (4.4.14); it is easiest to differentiate with respect to t; and

(iii)   use the fact that $J_0' = -J_1$. It then follows that

$$\frac{\partial V}{\partial t} = -\frac{c}{2}\delta\left(\tau - t - \frac{|\xi - x|}{c}\right)$$

$$\cdot\frac{cb(\tau - t)J_1(b\sqrt{(\tau - t)^2 - (\xi - x)^2/c^2})}{2\sqrt{(\tau - t)^2 - (\xi - x)^2/c^2}}H\left(\tau - t - \frac{|\xi - x|}{c}\right). \quad (4.4.32)$$

In this result, we have evaluated J_0 at the zero of the delta function.

We use this result in (4.4.8) along with the data in (4.4.6) to obtain the following representation of the solution:

$$U(\xi, \tau) = \frac{1}{2}[f(\xi - c\tau) + f(\xi + c\tau)]$$

$$-\frac{1}{2c}\int_{\xi-c\tau}^{\xi+c\tau}\left[\frac{b\tau f(x)}{\sqrt{\tau^2 - (\xi - x)^2/c^2}}J_1(b\sqrt{\tau^2 - (\xi - x)^2/c^2})\right.$$

$$\left.- g(x)J_0(b\sqrt{\tau^2 - (\xi - x)^2/c^2})\right]dx$$

$$-\frac{c}{2}\int\int_D F(x, t)J_0(b\sqrt{(\tau - t)^2 - (\xi - x)^2/c^2})\,dx\,dt. \quad (4.4.33)$$

As a check, setting $b = 0$ again leads to the former result.

Exercises

4.17   If γ is replaced by its negative in (4.4.17), then the effect is to replace b by ib throughout the subsequent analysis and, hence, in (4.4.31) to replace J_0 by I_0, the *modified Bessel function of the first kind* of order zero. Confirm this result by using the same method as was used to obtain (4.4.31) to determine the Riemann function in this case. Take care to define the square root of $\omega^2 + b^2$ properly and to calculate the inverse transform as an integral *above* all singularities in the complex ω plane.

4.18 (a) Suppose that in (4.4.1), all coefficients are constant. Introduce the transformation

$$V = We^{(\alpha x - \beta c^2 t)/2},$$

and show that the problem for V [(4.4.7)] leads to the following problem for W:

$$\mathscr{L}^*W = 0, \qquad t < \tau, \quad -\infty < x < \infty;$$

$$W(x, \tau; \xi, \tau) = 0, \qquad W_t(x, \tau; \xi, \tau) = \tfrac{1}{4}c^2 e^{-(\alpha\xi + \beta c^2 \tau)/2} \delta(x - \xi),$$

with new values of α, β, and γ,

$$\tilde{\alpha} = \tilde{\beta} = 0, \qquad \tilde{\gamma} = \gamma + \tfrac{1}{4}(\beta^2 c^2 - \alpha^2).$$

(b) Show that for consistency of dimensions in (4.4.1), α must have the dimension of inverse length, β the dimension of time over length square, and γ the dimension of inverse length square.

(c) Show that $\tilde{\gamma}$ has the same dimension as γ.

4.19 Use the preceding exercise to determine the solution $V(x, t, \xi, \tau)$ to (4.4.7) for constant coefficients in terms of the solution of Exercise 4.17 or in terms of the solution (4.4.31).

4.5 THE GREEN'S FUNCTION

We continue the discussion of the preceding section by developing *the Green's function*. We consider again the identity (4.4.4), with the auxiliary function V denoted instead by G. The domain D_1 used in (4.4.5) is to be slightly modified as well with the upper boundary above the time τ, say, at $\tau+$.

Given the problem (4.4.6) for U, we introduce the following problem for G:

$$\mathscr{L}^*G = -\delta(x - \xi)\delta(t - \tau), \qquad t < \tau+, \quad -\infty < x < \infty;$$

$$G(x, \tau; \xi, \tau) = G_t(x, \tau; \xi, \tau) = 0. \tag{4.5.1}$$

The arguments of the preceding section regarding the domain in which V is nonzero follow exactly along the same lines for G. Thus, by using (4.5.1) in (4.4.5) with V replaced by G, we obtain the representation

$$U(\xi, \tau) = \int_{x_1-}^{x_2+} \left[\frac{1}{c^2}[GU_t - UG_t] - \beta UG \right]\Bigg|_{t=0} dx - \int_D GF \, dx \, dt. \tag{4.5.2}$$

This is the solution formula for U in terms of the Green's function G. We remark that the formula is identical with (4.4.8). Thus, it would be reasonable to expect that, in fact, G and V are one and the same function. To check this, let us consider the following problem for U:

$$\mathscr{L}U = -\delta(x - x')\delta(t - t'), \qquad U(x, 0) = U_t(x, 0) = 0. \qquad (4.5.3)$$

For this problem, the solutions to (4.4.8) and (4.5.2) are, respectively,

$$U(\xi, \tau) = V(x', t'; \xi, \tau), \qquad U(\xi, \tau) = G(x', t'; \xi, \tau). \qquad (4.5.4)$$

By uniqueness of the solution, we conclude that G and V are the same.

The question arises then of why there are two approaches leading to the same result. It is, to this user, a matter of point of view. In dealing with a problem in which the main structure arises through initial data, we prefer to think in terms of the Riemann function. When the structure arises through source and/or boundary data, we prefer to think in terms of the Green's function.

Note in this derivation that we could have formally integrated over all space and time and eliminated the boundary terms "at infinity" by arguments based on *causality*, that is, that the response to data in a finite part of space could not have propagated to infinity in a finite time.

We shall use causality in analyzing the *initial boundary value problem*

$$\mathscr{L}U = F, \qquad 0 < t, \quad -L < x < +L;$$
$$U(x, 0) = f(x), \qquad U_t(x, 0) = g(x), \qquad U(\pm L, t) = h_\pm(t). \qquad (4.5.5)$$

This is a generalization of the problem considered in Section 4.2.

We begin by considering the integral of the identity (4.4.4) over the domain $(-L, L)$ and all positive time, with V replaced by G:

$$\int_{-L}^{L} dx \int_{0}^{\infty} dt \, [G\mathscr{L}U - U\mathscr{L}^*G]$$

$$= \int_{0}^{\infty} [GU_x - UG_x + \alpha GU] \Big|_{-L}^{L} dt$$

$$+ \int_{-L}^{L} \left[\frac{1}{c^2} [GU_t - UG_t] - \beta GU \right] \Big|_{t=0} dx. \qquad (4.5.6)$$

The problem for G now is a modification of the problem (4.5.1) on $(-\infty, \infty)$ or in *free space*. The range of x in (4.5.1) is now restricted to the interval $(-L, L)$ and at these endpoints; in order to eliminate the dependence of the solution on the data we *do not know*, we require that

$$G(\pm L, t; \xi, \tau) = 0. \qquad (4.5.7)$$

Now substituting (4.5.1) and (4.5.7) into (4.5.6) yields the solution formula

$$
U(\xi, \tau) = -\int_0^\infty UG_x \Big|_{-L}^{L} dt + \int_{-L}^{L} \left[\frac{1}{c^2} [GU_t - UG_t] - \beta GU \right] \Big|_{t=0} dx
$$

$$
- \int_{-L}^{L} dx \int_0^\infty dt \, FG. \tag{4.5.8}
$$

Here we rely on the domain of influence of the source for the Green's function to restrict the domains of integration so as to agree with the domains of Section 4.2. Indeed, for (ξ, τ) in region I of Fig. 4.2, the support of the Green's function is just the domain of influence of Fig. 4.1, and the integral over the boundaries in this result are, in fact, zero. Thus, this equation provides an expression for the solution to the general initial boundary value problem with U itself prescribed on the boundary. If, instead of U, U_x were prescribed on one of the boundaries, then it can be seen from (4.5.6) that at the corresponding boundary, it would be necessary to set $G_x - \alpha G = 0$.

Exercises

4.20 In the initial boundary value problem (4.5.5), suppose that $U_x - KU$ is prescribed at one of the endpoints. Show that the Green's function must satisfy

$$
G_x - (\alpha + K)G = 0.
$$

4.21 Find the Green's function for the problem (4.3.2).

4.22 Find the Green's function for the Klein–Gordon operator (4.4.17) on the interval $(-L, L)$ with $G = 0$ at the endpoints. Proceed in the following manner.

(a) Use the Fourier transform technique of Section 4.3 to obtain an inhomogeneous ordinary differential equation. Follow the method of that section to find the solution.

(b) Expand all trigonometric functions as was done to obtain (4.3.17). Take care that the multiplier of $2L/c$ is now k instead of ω.

(c) Identify the integrals of the series with (4.4.25) and then use (4.4.30) to conclude that

$$
V = \sum_{n=0}^{\infty} \sum_{j=1}^{j=4} (-1)^j J_0 \left(b \sqrt{t^2 - \frac{x_{jn}^2}{c^2}} \right);
$$

$$
x_{1n} = x + \xi + (4n + 2)L, \qquad x_{2n} = |x - \xi| + 4nL,
$$

$$
x_{3n} = x + \xi - (4n - 2)L, \qquad x_{4n} = |x - \xi| - (4n + 4)L.
$$

4.23   Show that for the coefficients α, β, and γ in (4.4.1) equal to constants, the Green's function solution to (4.4.6) is in convolution form.

4.6   ASYMPTOTIC SOLUTION OF THE KLEIN–GORDON EQUATION

We consider the following problem for U:

$$U_{xx} - c^{-2}U_{tt} - (b^2/c^2)U = 0,$$

$$U(x, 0) = f(x), \qquad U_t(x, 0) = g(x),$$

(4.6.1)

with b and c constant. At this point, a number of methods of solution are available. First, there is the Green's function/Riemann function approach used in the preceding two sections. Second, there is the technique of using Fourier transform in x or t and solving the resulting ordinary differential equation in the complementary variable. Third, it is observed in Exercise 4.23 that the solution can be written as a product of Fourier transforms. We shall deduce a solution of this last type by transform methods. Our purpose is to produce a representation that lends itself to asymptotic analysis and interpretation.

Define $u(k, \omega)$ by

$$u(k, \omega) = \int_{-\infty}^{\infty} dx \int_{0}^{\infty} dt\, U(x, t)e^{-i\{kx - \omega t\}}.$$

(4.6.2)

Here ω is restricted to some upper half plane whose boundary is to be determined. By applying this integral transform to (4.6.1), we obtain

$$\left[-k^2 + \frac{\omega^2}{c^2} - \frac{b^2}{c^2}\right] u = -\frac{\tilde{g}(k)}{c^2} + i\omega\frac{\tilde{f}(k)}{c^2}.$$

(4.6.3)

In this equation, we have used the tilde to denote the Fourier transform of the data.

We solve for u here and express the inverse transform $U(x, t)$ in terms of that result

$$U(x, t) = \frac{1}{(2\pi)^2} \int_{-\infty}^{\infty} dk \int d\omega\, \frac{i\omega\tilde{f}(k) - \tilde{g}(k)}{\omega^2 - c^2k^2 - b^2} e^{i\{kx - \omega t\}}.$$

(4.6.4)

The ω integral is again an integral parallel to the real axis above all singularities of the integrand, that is, in the upper half plane of analyticity of the integrand. The reader who has completed Exercise 4.23 will recognize this solution as the product of the Fourier transform of the Green's function with the Fourier transform of the data. For t negative, the representation yields zero in the usual manner. For t positive, we close the ω contour in the lower half plane and calculate the integral as a sum of residues at the zeros

$$\omega = \pm\sqrt{c^2 k^2 + b^2}. \tag{4.6.5}$$

The result is

$$U(x, t) = \frac{1}{4\pi}\sum_{\pm}\int_{-\infty}^{\infty} dk\left[\tilde{f}(k) \pm \frac{i\tilde{g}(k)}{\sqrt{c^2 k^2 + b^2}}\right]$$
$$\cdot \exp(i\{kx \mp \sqrt{c^2 k^2 + b^2}\, t\}). \tag{4.6.6}$$

We interpret this result as follows. The solution is made up of a super-position of "plane waves" in which the wave number is defined by k and the attendant frequency is given by a function $\omega(k)$ defined by one of the *dispersion relations* (4.6.5). The amplitude of the wave at each wave number k is a function of the *wave number spectral densities* of the initial data $\tilde{f}(k)$, $\tilde{g}(k)$ given by

$$\frac{1}{4\pi}\left[\tilde{f}(k) \pm \frac{i\tilde{g}(k)}{\sqrt{c^2 k^2 + b^2}}\right] dk.$$

The points of constant phase of each of these waves travel at the *phase speed* given by

$$v_{\text{phase}} = \omega(k)/k = \pm\sqrt{c^2 k^2 + b^2}/k, \tag{4.6.7}$$

which can be seen to vary with k from a minimum magnitude of the charac-teristic speed c to a maximum value of infinity. Thus, we see that the wave does not travel as a function of $x \pm ct$, as it did for the wave equation, but that the initial data propagate at differing speeds in accordance with their decomposition in the spatial Fourier domain.

We remark that consistent with x and t having the dimensions of length and time, respectively, the dimensions of k and ω are inverse length and time. Let us contemplate for a moment introducing dimensionless variables

$$\eta = ck/b, \qquad \lambda = bt, \qquad \theta = x/ct. \tag{4.6.8}$$

In this case, the phases in the integral (4.6.6) become

$$kx \mp \sqrt{c^2 k^2 + b^2}\, t = \lambda[\eta\theta \mp \sqrt{\eta^2 + 1}\,].$$

Thus, for $\lambda = bt \gg 1$, the integral in (4.6.6) is of the type (2.6.1) to which the asymptotic theory of Chapter 2 may be applied. This criterion for applying asymptotic methods may be viewed as *large time* in the units of the inverse of the frequency b. For example, in electromagnetic wave propagation in plasmas modeled by the Klein–Gordon equation, $b = O(10^{10}/\text{sec})$ and bt is large after times measured in fractions of a microsecond.

Being aware of the large parameter that justifies asymptotic analysis, we prefer to proceed formally to apply our asymptotic method to the integral (4.6.6) in dimensional variables with "large parameter" one. Thus, we consider the two phase functions appearing in (4.6.6),

$$\Phi_{\pm}(k) = kx \mp \sqrt{c^2k^2 + b^2}\, t, \tag{4.6.9}$$

and differentiate with respect to k in anticipation of applying the method of stationary phase as discussed in Section 2.7.

$$\frac{d\Phi_{\pm}}{dk} = x \mp \frac{c^2k}{\sqrt{c^2k^2 + b^2}} t, \qquad \frac{d^2\Phi_{\pm}}{dk^2} = \mp \frac{b^2c^2t}{\{c^2k^2 + b^2\}^{3/2}}. \tag{4.6.10}$$

The stationary points occur when the first derivative here is zero. Let us suppose first that x is positive. Then Φ_{+} will have a stationary point at some positive value of k, and Φ_{-} will have a stationary point at $-k$. For x negative, the stationary point of Φ_{-} is at some positive value of k, and Φ_{+} will have a stationary point at the negative of that value. Thus, each phase has a stationary point for either sign of x.

We could now solve for k and substitute into the stationary phase formula (2.7.18). However, instead of this, we prefer to write down a parametric solution in terms of a *positive* parameter k consistent with the stationary phase requirement. Thus, we write

$$U(x,t) \sim \frac{\{c^2k^2 + b^2\}^{3/4}}{2bc\sqrt{2\pi t}} \sum_{\pm} \left[\tilde{f}(k) \pm \frac{i\tilde{g}(k)}{\sqrt{c^2k^2 + b^2}} \right]$$

$$\cdot \exp\left(\pm i \left\{ k|x| - \sqrt{c^2k^2 + b^2}\, t - \frac{\pi}{4} \right\} \right), \tag{4.6.11}$$

$$|x| = \frac{c^2k}{\sqrt{c^2k^2 + b^2}} t.$$

It is left to the reader to verify that in this representation we have properly accounted for the two stationary points, both when x is positive and when x is negative.

As mentioned earlier, one interpretation of this result is that for a given (x, t), we solve the second line for k and insert in the first line to find U. Alternatively, we consider the implications of this representation for fixed k. Then the second line suggests that the solution is observed in some moving reference frame in which the observation point is propagating at a speed $d\omega/dk$ determined by the dispersion relation (4.6.5). The signal propagating at this speed is given by the first line in this last equation. We remind the reader that the stationarity of the phase suggests a constructive interference of the neighboring wave numbers as compared to wave numbers farther away. Thus, the first two lines in (4.6.11) represent the propagation of a *group* of wave numbers in the neighborhood of k that propagate at the approximate speed defined by the third line in this equation. Thus, this speed, given by the k derivative of ω, is called the *group speed*.

We remark that the group speed is less than the characteristic speed c. Thus, the points of constant phase, which travel at the phase speed, which is greater than c, are moving forward through the slower moving packet of waves.

There are further interpretations of this result in the context of *energy conservation*. These will be postponed until such time as we introduce the concept of energy. Suffice it to say, for the present, that the scaling by the second derivative of the phase has an interpretation in terms of *conservation of energy*.

Exercises

4.24 Verify (4.6.11) by separately considering the cases of x positive and x negative in the stationarity condition and substituting into (4.6.6).

4.25 In (4.6.1) consider the special case in which

$$f(x) = 0, \qquad g(x) = c^2 \delta(x).$$

In the formula (4.6.11), eliminate the parameter and show that the solution is the asymptotic expansion of (4.4.31). Use (2.7.20) to write down the asymptotic expansion of this latter result.

4.7 MORE ON ASYMPTOTIC SOLUTIONS

Our objective here is to introduce some elementary methods for developing asymptotic solutions in the limit of high frequency for problems with coefficients that vary with x. Let us suppose again that we are considering a problem for the wave operator

$$\mathscr{L} U = U_{xx} - c^{-2} U_{tt}. \tag{4.7.1}$$

In order to address the question of frequency, let us consider this operator after applying the Fourier transform with respect to t:

$$\mathscr{L}_\omega u(x, \omega) = u''(x, \omega) + (\omega^2/c^2) u(x, \omega). \tag{4.7.2}$$

Here prime denotes x differentiation. We introduce a length scale L and a frequency scale ω_0 by setting $x = L\xi$, $\omega = \omega_0 \eta$, with ξ and η dimensionless variables. We choose L in such a manner that over the length scale L we may introduce an average value of $c(x)$, say, $\bar{c}$, with $c(x) = \bar{c}\theta(\xi)$ and $d\theta/d\xi$ approximately equal to unity.

The operator in (4.7.2) can now be rewritten as

$$\mathcal{L}_\omega u = \frac{1}{L^2}\left[\frac{d^2 u}{d\xi^2} + \lambda^2 \frac{\eta^2}{\theta^2} u\right],$$

with

$$\lambda = \omega_0 L/\bar{c}.$$

By *high frequency*, then, we shall mean frequencies for which the dimensionless parameter λ is large. We remark that $\omega_0/\bar{c}$ is a *wave number* associated with the frequency ω_0 and that

$$\Lambda = 2\pi\bar{c}/\omega_0$$

is the *wave length* of waves with frequency ω_0. Thus,

$$\lambda = 2\pi L/\Lambda$$

is the parameter that must be large in order that asymptotics be justified.

With this discussion in mind, we shall proceed nonetheless by analyzing the *dimensional* operator of (4.7.2). We have already seen that the solutions of the *homogeneous* equation play a crucial role in solving a more general problem for u. Thus, we set

$$\mathcal{L}_\omega u = u'' + (\omega^2/c^2)u = 0. \tag{4.7.3}$$

For c constant, the solutions to this equation are complex exponentials. This motivates the following assumption for the form of the asymptotic solution for variable c:

$$u \sim e^{i\omega\Phi(x)} \sum_{n=0}^{\infty} \frac{A_n(x)}{(i\omega)^n}. \tag{4.7.4}$$

We proceed formally, substituting this series solution into Eq. (4.7.3) to obtain

$$\mathcal{L}_\omega u = e^{i\omega\Phi} \sum_{n=0}^{\infty} \left\{\frac{A^n}{(i\omega)^{n-2}}\left[\Phi'^2 - \frac{1}{c^2}\right]\right.$$
$$\left. + \frac{1}{(i\omega)^{n-1}}[2A_n'\Phi' + A_n\Phi''] + \frac{A_n''}{(i\omega)^n}\right\} = 0. \tag{4.7.5}$$

The coefficient of each power of ω is to be set equal to zero separately. The leading order here is $O(\omega^2)$. Thus,

$$\Phi'^2 = c^{-2}, \tag{4.7.6}$$

from which it follows that

$$\Phi' = \pm \frac{1}{c}, \qquad \Phi = \pm \int^x \frac{dy}{c(y)}. \tag{4.7.7}$$

There are two roots here and, consequently, two formal series solutions to (4.7.3). When necessary, as in (4.7.12) and (4.7.14), we shall distinguish between these two solutions by using the superscripts or subscripts $\pm$. For the present, we continue without the superscripts or subscripts.

When (4.7.6) is satisfied, the first series in (4.7.5) has a multiplier of zero. To order ω, then, we obtain a contribution only from the second series in (4.7.5) with $n = 0$, and therefore we require that

$$2A_0'\Phi' + A_0\Phi'' = 0. \tag{4.7.8}$$

Multiplication by A_0 and integration of the exact differential that results lead to the conclusion

$$A_0^2\Phi' = \text{const.} \tag{4.7.9}$$

In order that A_0 remain real, we must choose the constant positive or negative in accordance with the sign of Φ'. Since an arbitrary constant multiplier is of no concern here, we conclude that

$$A_0 = \sqrt{c(x)} \tag{4.7.10}$$

for either choice of sign in (4.7.7). Thus, A_0 is determined.

For the subsequent coefficients, both the second and third series of (4.7.5) contribute to each order in ω. Then for each n,

$$2A_{n+1}'\Phi' + A_{n+1}\Phi'' = -A_n'', \qquad n \geq 0. \tag{4.7.11}$$

Multiplication by ± 1 and division by $2\sqrt{\pm\Phi'} = 2/\sqrt{c(x)}$ results in an exact differential on the left side and leads to the solution

$$A_{n+1}(x)\sqrt{\pm\Phi'} = B_{n+1}^{\pm} \mp \int^x \frac{A_n''(y)}{2\sqrt{\pm\Phi'(y)}}\,dy. \tag{4.7.12}$$

This leads to the solution

$$A_{n+1}(x) = \sqrt{c(x)}\left[B_{n+1}^{\pm} \mp \frac{1}{2}\int^x A_n''(y)\sqrt{c(y)}\,dy\right]. \tag{4.7.13}$$

In both of these equations, the constant $B_n^{\pm}$ admittedly is redundant because the lower limit of integration has been left open. However, this form emphasizes the presence of an arbitrary additive constant. All of the coefficients in the formal power series (4.7.4) have now been determined, and we obtain to formal asymptotic solutions, say, $u_{\pm}$, corresponding to the two choices of sign in (4.7.7). We remark that the formal Wronskian of these two solutions is nonzero. In particular,

$$u'_+ u_- - u'_- u_+ \sim 2i\omega/c(x) \neq 0. \tag{4.7.14}$$

Thus, the solutions are *asymptotically linearly independent.*

The following theorem can be proven:

Let $c(x)$ be continuous over the x interval of interest. Then the two formal solutions are actually asymptotic to two exact solutions of (4.7.2) as $|\omega| \to \infty$ in a sector in the ω plane in which

$$\text{Re } i\omega/c(x) \neq 0.$$

Thus, there are two linearly independent solutions that have $u_\pm$ as their asymptotic expansions in the upper half ω plane and two linearly independent solutions that have $u_\pm$ as their asymptotic expansions in the lower half ω plane. The theorem does not guarantee (because it is, in general, not true) that the asymptotic expansions approximate the *same* two exact solutions in both half planes. This is a manifestation of a property known as the *Stokes phenomenon* for asymptotic expansions; namely, the analytic continuation of the asymptotic expansion of an exact solution need not be the expansion of the analytic continuation. The boundary across which analytic continuation fails is called the *Stokes line.*

Luckily, we are interested in solutions in an upper half ω plane. Once we have their asymptotic expansion in that domain, the Fourier inversion is to be carried out by using that asymptotic solution. Deformations of contour to be carried out thereafter are done for the given integrand without regard for its origin.

As a simple example of this theory, let us find an asymptotic Green's function; that is, let us solve the inhomogeneous equation

$$u''(x, \omega) + (\omega^2/c^2)u(x, \omega) = -\delta(x - \xi). \tag{4.7.15}$$

In order that the solution to this problem be bounded in the upper half ω plane, we require that the solution be proportional to u_+ for $x \to +\infty$ and proportional to u_- for $x \to -\infty$. Furthermore, the importance of the point ξ motivates us to use this point as the lower limit in all integrals. Also, the requirement that u be continuous at $x = \xi$ leads to the conclusion that the arbitrary constants in the solutions left and right of ξ must be equal; that is,

$$u \sim A\sqrt{c(x)}\, e^{i\omega\Phi} \left\{ 1 + \sum_{n=1}^{\infty} \frac{1}{(i\omega)^n} \left[B_n - \frac{1}{2} \int_{x_<}^{x_>} A''_{n-1}(y)\sqrt{c(y)}\, dy \right] \right\},$$

$$\Phi = \int_{x_<}^{x_>} \frac{dy}{c(y)}, \qquad x_< = \min(x, \xi), \quad x_> = \max(x, \xi). \tag{4.7.16}$$

We denote by $[u']$ the *jump* in u' across $x = \xi$. By differentiating (4.7.16), we find that

$$[u'] = \frac{2i\omega A}{\sqrt{c(\xi)}}\left[1 + \sum_{n=1}^{\infty}\frac{B_n}{(i\omega)^n}\right] - Ac(\xi)\sum_{n=1}^{\infty}\frac{A''_{n-1}(\xi)}{(i\omega)^n}. \qquad (4.7.17)$$

This jump must be equal to -1 (Exercise 2.1c). Thus, the leading order term in this equation is set equal to -1, and the jump to all lower orders in ω is set equal to zero. This leads to the conclusion that

$$A = -\sqrt{c(\xi)}/2i\omega, \qquad B_1 = 0,$$
$$B_{n+1} = \tfrac{1}{2}[c(\xi)]^{3/2}A''_{n-1}(\xi), \qquad n = 1, 2, \ldots. \qquad (4.7.18)$$

To leading order, then,

$$u(x, \omega) \sim -\frac{\sqrt{c(x)c(\xi)}}{2i\omega}\,e^{i\omega\Phi}. \qquad (4.7.19)$$

When $c = $ const, only the first term of the asymptotic series is nonzero, and it yields the *exact* solution for the Fourier transform of the Green's function. Furthermore, if it were valid to use this result over the entire Fourier domain to invert this transform, the result would be

$$\overline{U}(x, t) = -\frac{\sqrt{c(x)c(\xi)}}{2}\,H(t - \Phi). \qquad (4.7.20)$$

We remark that the difference between the exact solution and the leading order asymptotic solution (4.7.19) is $O(\omega^{-2})$. Each reciprocal power of ω can be viewed as integration with respect to t. In this sense, the difference between the exact Green's function and (4.7.20) is "smoother" by one integration than the Heaviside function. Retaining subsequent terms in the asymptotic expansion would leave a progressively smoother error. In any case, (4.7.20) is a representation that retains the "right" discontinuity of the Green's function for this problem.

Let us turn now to the Klein–Gordon operator and consider in the Fourier domain

$$\mathcal{L}_\omega u = u'' + k^2(x)u = 0, \qquad k(x, \omega) = \frac{\sqrt{\omega^2 - b^2(x)}}{c(x)}. \qquad (4.7.21)$$

Here for each fixed x, the branch of the square root is as in the constant coefficient case. Furthermore, we shall assume that $b(x)$ is monotonically increasing from a lower limit of b_1 at $-\infty$ to an upper limit of b_2 at $+\infty$, with $b'(x) > 0$ for all finite x. Although we are interested in a solution in an upper half ω plane, we consider the solutions on the real ω axis, which we view as having been reached as a limit of solutions from the upper half ω plane. For $\omega > b_1$, the solution technique is as above. To leading order, we obtain

$$u_{\pm} \sim k^{-1/2} e^{i\phi}, \qquad \phi = \int^{x} k(y, \omega) \, dy. \qquad (4.7.22)$$

Now let us suppose that ω is in the range $b_1 < \omega < b_2$. Then there is a point, say, x_{ω}, at which

$$\omega^2 = b^2(x_{\omega}). \qquad (4.7.23)$$

This point is called a *turning point*. For the moment, we are, in some sense, "tied" to the real ω axis because, otherwise, there is no turning point as x varies and the structure of the solution changes radically. We proceed, then, under the assumption that the Fourier inversion that we are ultimately to carry out is valid as an integral on the real ω axis, with perhaps branches of square roots to be defined and integrals "through" poles to be interpreted as principal value integrals *minus* half residues.

There is also some question here as to the validity of using asymptotics near the turning point. Indeed, we must take the point of view that the solutions we find will be used only for $|k|L$ "large" and not in the transition region around the tunring point. Here L denotes distance from the turning point.

We continue the analysis with all of these disclaimers in mind. Proceeding as in the analysis of (4.7.3), we find now that there are two linearly independent solutions $v_{\pm}$ for $x > x_{\omega}$ having real exponentials in their asymptotic representation

$$v_{\pm} \sim |k(x, \omega)|^{-1/2} e^{\mp \psi}, \qquad \psi = \int_{x_{\omega}}^{x} \frac{\sqrt{b^2(y) - \omega^2}}{c(y)} \, dy \qquad (4.7.24)$$

and the two solutions $u_{\pm}$ (4.7.22) for $x < x_{\omega}$ having oscillatory solutions (4.7.22) with fixed limit x_{ω}.

It has already been noted that one cannot simply take the analytic continuation of the asymptotic expansion to determine the asymptotic expansion of the analytic continuation for any of these solutions. The problem of determining *connection formulas* relating the oscillatory solutions to the exponential solutions has been extensively studied. (See, for example, Erdélyi [1956] and Olver [1974].) The solution to this problem is associated with the names Wentzel, Kramers, Brillouin, and Jeffreys and is called

WKBJ connection formulas. In particular, let us consider the two solutions $v_\pm$ in (4.7.24). The connection formulas for these two functions are

$$
v_+(x, \omega) \sim \begin{cases} |k|^{-1/2} e^{-\psi}, & x > x_\omega, \\ 2k^{-1/2} \sin[\phi + \pi/4], & x < x_\omega, \end{cases}
$$
$$
v_-(x, \omega) \sim \begin{cases} |k|^{-1/2} e^{\psi}, & x > x_\omega, \\ 2k^{-1/2} \cos[\phi + \pi/4], & x < x_\omega. \end{cases}
\qquad (4.7.25)
$$

Here ψ is as defined by (4.7.24), and ϕ, now with fixed limit, is defined by

$$
\phi = \int_x^{x_\omega} k(y, \omega)\, dy. \qquad (4.7.26)
$$

These are asymptotic expansions of two linearly independent solutions on the real ω axis according to whether k^2 is positive or negative. They are valid for $|k|$ large under the assumption that x_ω is a *simple* zero. The connection formulas change with the order of vanishing of k^2 (see Olver [1974]).

There are two major approaches to deriving these connection formulas. Both can be found in Olver [1974]. On one of the approaches, the analysis of a differential equation with a *simple* turning point is reduced to the study of the "canonical" problem with this character, namely,

$$
u'' - xu = 0.
$$

Here *large argument* means large magnitude of the independent variable x. This equation is *Airy's differential equation*. The asymptotic solutions of (4.7.21) are expressible as multiples of the solutions of Airy's equation. These solutions are called *Airy functions*. Two specific solutions of Airy's equation, denoted by $Ai(x)$ and $Bi(x)$, have the property that they respectively decay or grow exponentially as $x \to \infty$. Asymptotic expansions of these functions for x negative and large yield the connection formulas stated in (4.7.25).

The asymptotic expansions in (4.7.25) are, in fact, specializations to the real ω axis of expansions valid in sectors in the complex pla..e. Indeed, one need only define $|k|$ in a natural way to extend those expansions off the axis. We remark that neither of the solutions whose asymptotic expansions are given in (4.7.25) remains bounded in the upper half ω plane. However, neither solution is the Fourier transform of a solution to an initial value problem. On the other hand, the Green's function, derived in Exercise 4.26 below, *is* the Fourier transform of the solution to an initial value problem, and it *does* remain bounded in the upper half ω plane.

For $\omega < b_2$, the two linearly independent solutions are of the type $v_\pm$ for all x, with the fixed limit in the definition of ψ no longer specified.

Exercises

4.26 The object of this exercise is to find the asymptotic Green's function for the Klein–Gordon operator (4.7.21). Therefore, consider the following equation:

$$\mathcal{L}_\omega u = -\delta(x - \xi).$$

(a) For $\omega > b_2$, show that the Green's function is given by

$$u(x, \xi) \sim \tfrac{1}{2} i u_+(x_>, \omega) u_-(x_<, \omega), \qquad x_< = \min(x, \xi), \qquad x_> = \max(x, \xi)$$

and $u_\pm$ given by (4.7.21).

(b) For $b_1 < \omega < b_2$, show that

$$u(x, \xi) \sim \tfrac{1}{2} v_+(x_>, \omega) v_-(x_<, \omega),$$

with $v_\pm$ given by (4.7.25).

(c) For $\omega^2 < b_1^2$, show that

$$u(x, \xi) \sim \frac{1}{2\sqrt{|k(x, \omega) k(\xi, \omega)|}} \exp\left[-\int_{x_<}^{x_>} |k(y, \omega)| \, dy \right].$$

(d) Suppose in (b) that ξ and x are less than x_ω. Write the Green's function in terms of exponentials. Identify the solution as a sum of a wave incident on the barrier, where $x = x_\omega$ and a wave reflected from the same barrier. Here, define the direction of the wave as bring to the *right* when $\partial\phi/\partial x$ is positive and to the *left* when this derivative is negative.

4.27 Formally replace k^2 in the connection formulas by $f(x, \omega)$ to obtain these formulas more generally for solutions to the equation

$$u'' + f(x, \omega)u = 0,$$

under the assumption that f has a simple zero at x_0 and is an increasing function of x.

4.28 Consider the problem for u defined by

$$u'' + \frac{\omega^2}{c^2(x)} u = 0, \qquad x < x_0; \qquad u'' + \frac{\omega^2}{c_1^2(x)} u = 0, \qquad x > x_0.$$

Require that the solution u and its derivative u' be continuous at the point x_0. Define

$$u_\pm \sim \frac{e^{\pm i\omega\phi(x)}}{\sqrt{c(x)}}, \qquad \phi(x) = \int_{x_0}^{x} \frac{dy}{c(y)}, \qquad x < x_0,$$

$$u_{1\pm} \sim \frac{e^{\pm i\omega\phi_1(x)}}{\sqrt{c_1(x)}}, \qquad \phi_1(x) = \int_{x_0}^{x} \frac{dy}{c_1(y)}, \qquad x > x_0.$$

(a) Let u be the solution that has the asymptotic expansion

$$u = \begin{cases} u_+ + Ru_-, & x < x_0, \\ Tu_{1\pm}, & x > x_0. \end{cases}$$

This solution models the problem of a wave incident from the left, reflected from the discontinuity (interface) and partially transmitted. Show that

$$R = \frac{c_1(x_0) - c(x_0)}{c_1(x_0) + c(x_0)}, \qquad T = \frac{2\sqrt{c_1(x_0)c(x_0)}}{c_1(x_0) + c(x_0)}.$$

(b) Often, in modeling the wave propagating to the right, the discontinuity is ignored, and one uses the formula for the continuous solution

$$u = \begin{cases} u_+, & x < x_0, \\ u_{1\pm}, & x > x_0. \end{cases}$$

The percentage error in this approximation of the *transmitted* wave is given by the transmission coefficient T in (a). Set

$$c(x_0) = c, \qquad c_1(x_0) = c + \Delta c,$$

and show that

$$T = 1 - \tfrac{1}{8}[\Delta c/c]^2 + O((\Delta c/c)^3).$$

That is, the error in the transmitted wave is quadratic in the percentage change in the propagation speed.

References

Coddington, E. A., and Levinson, N. [1955]. "Theory of Ordinary Differential Equations." McGraw-Hill, New York.

Erdélyi, A. [1956]. "Asymptotic Expansions." Dover, New York.

Garabedian, P. R. [1964]. "Partial Differential Equations." Wiley, New York.

Gradshteyn, I. S., and Ryzhik, I. M. [1965]. "Table of Integrals, Series and Products." Academic Press, New York. [Corrected and enlarged edition, 1980.]

Henrici, P. [1974]. "Applied Computational Complex Analysis," Vol. 1. Wiley, New York.

Ince, E. L. [1956]. "Ordinary Differential Equations." Dover, New York.

Jeffreys, H. [1965]. "Asymptotic Approximations." Clarendon Press, Cambridge.

John, F. [1982]. "Partial Differential Equations," 4th ed. Springer-Verlag, New York.

Olver, F. W. J. [1974]. "Asymptotics and Special Functions." Academic Press, New York.

Titchmarsh, E. C. [1962]. "Eigenfunction Expansions Associated with Second Order Differential Equations," Part 1. Oxford University Press, Oxford.

Weinberger, H. F. [1965]. "A First Course in Partial Differential Equations." Blaisdell, New York.

Whitham, G. B. [1974]. "Linear and Nonlinear Waves." Wiley, New York.

5 THE WAVE EQUATION
IN TWO AND THREE DIMENSIONS

We shall now discuss the wave equation in two and three dimensions. As noted in the Preface, we shall be more concerned in this chapter with certain qualitative features of the wave equation than with solution techniques. Thus, we shall discuss ill-posedness and the propagation of discontinuities, uniqueness, and energy conservation; Green's functions; and the representations of solutions of more general problems in terms of them and then, finally, scattering problems.

5.1 CHARACTERISTICS AND ILL-POSED CAUCHY PROBLEMS

We begin again with the m-dimensional generalization of (4.1.1), namely,

$$\nabla^2 U - c^{-2} U_{tt} = F(\mathbf{x}, t, U, \nabla U, U_t). \tag{5.1.1}$$

Here the gradient and $(\mathbf{x})$ are either two- or three-dimensional. [The discussion here will really be $(m > 1)$-dimensional.]

As in Section 4.1, we begin by considering the possibility of introducing a new independent variable ξ in such a manner that the second derivative with respect to ξ does not appear in the equation. For this problem,

$$\xi = \phi(\mathbf{x}, t), \tag{5.1.2}$$

and the analog of equation (4.1.4) is

$$\nabla^2 U - c^{-2} U_{tt} = V_{\xi\xi}[(\nabla\phi)^2 - c^{-2}\phi_t^2] + \cdots. \tag{5.1.3}$$

Thus, there is no second derivative with respect to ξ in the equation if

$$(\nabla\phi)^2 - c^{-2}\phi_t^2 = 0. \tag{5.1.4}$$

If, as in Exercise 1.7, we assume that ϕ is of the form

$$\phi(\mathbf{x}, t) = \psi(\mathbf{x}) - c_0 t, \tag{5.1.5}$$

then the equation for ϕ reduces to the *eikonal equation* for ψ:

$$(\nabla\psi)^2 = n^2. \tag{5.1.6}$$

Consider the case $m = 2$. Let us suppose that U_ξ is discontinuous along some curve in (x, y) at $t = 0$. Then that curve becomes *Cauchy data* for the eikonal equation (5.1.6), and the solution describes a surface in space–time or, equivalently, a curve moving in space as time progresses along which the discontinuity in U_ξ propagates. Of course, if the concept of solution is extended as in the preceding chapter to allow for one-sided derivatives and differentiable solutions on either side of a surface in space–time, then the discontinuity could be in U itself. See Section 4.1. The discontinuity might exist only at a point, in which case the surface of discontinuity is the conoidal solution for ϕ emanating from that point.

It need not be the case that the discontinuity is originally defined at a fixed time. It could be that the curve of discontinuity is a curve in space–time or a moving locus of points in space. Nonetheless, the subsequent propagation of that discontinuity must be along a surface in space–time satisfying (5.1.4) or (5.1.5) and (5.1.6).

When there are three space dimensions, the initial manifold of discontinuity may be a point, a curve, or a surface. In any case, the discontinuity propagates in space according to the eikonal equation, as in the discussion following (5.1.6).

We shall now turn to the question of Cauchy problems that admit exponential solutions. For the wave equation itself

$$\nabla^2 U - c^{-2} U_{tt} = 0, \tag{5.1.7}$$

we seek solutions that contain real exponentials in a variable,

$$\Lambda = \sum_{j=1}^{m} \lambda_j x_j - \lambda_0 t, \tag{5.1.8}$$

with $\lambda_0, \ldots, \lambda_m$ real. If such solutions can be found, then data on $\Lambda = 0$ can be set to construct an example of an exponentially growing solution, as was done in Section 3.2. We try a solution of the form

$$U(\mathbf{x}, t) = e^{\Lambda + iM}, \qquad M(\mathbf{x}, t) = \sum_{j=1}^{m} \mu_j x_j - \mu_0 t, \tag{5.1.9}$$

with the μ_j's real. Substitution of this solution into (5.1.7) yields a complex equation for $\lambda_0, \ldots, \lambda_m, \mu_0, \ldots, \mu_m$ equivalent to the two real equations

$$\sum_{j=1}^{m} [\lambda_j^2 - \mu_j^2] - \frac{1}{c^2}[\lambda_0^2 - \mu_0^2] = 0, \qquad \sum_{j=1}^{m} \lambda_j \mu_j - \frac{1}{c^2}\lambda_0\mu_0 = 0. \quad (5.1.10)$$

If $\lambda_0 = 0$, one can always find nontrivial values of λ_j's and produce exponential solutions. Indeed, this was the type of example constructed in Section 3.2. Therefore, consider the case $\lambda_0 \neq 0$. Then solve in the second line of this equation for μ_0 and substitute into the first line

$$\sum_{j=1}^{m} [\lambda_j^2 - \mu_j^2] - \frac{\lambda_0^2}{c^2} + \frac{c^2}{\lambda_0^2}\left[\sum_{j=1}^{m} \lambda_j \mu_j\right]^2. \quad (5.1.11)$$

We now apply the *Cauchy–Schwarz inequality*

$$\left[\sum_{j=1}^{m} \lambda_j \mu_j\right]^2 \leq \left[\sum_{j=1}^{m} \lambda_j^2\right]\left[\sum_{j=1}^{m} \mu_j^2\right]$$

to find that

$$0 \leq \sum_{j=1}^{m} [\lambda_j^2 - \mu_j^2] - \frac{\lambda_0^2}{c^2} + \frac{c^2}{\lambda_0^2}\left[\sum_{j=1}^{m} \lambda_j^2\right]\left[\sum_{j=1}^{m} \mu_j^2\right]$$

$$= \left[\sum_{j=1}^{m} \lambda_j^2 - \frac{\lambda_0^2}{c^2}\right]\left[1 + \frac{c^2}{\lambda_0^2}\sum_{j=1}^{m} \mu_j^2\right]. \quad (5.1.12)$$

In order for the right side to be nonnegative, the first factor containing a difference of squares must be nonnegative; that is,

$$\frac{\lambda_0^2}{c^2} \leq \sum_{j=1}^{m} \lambda_j^2. \quad (5.1.13)$$

The case $\lambda_0 = 0$ is a special case of this result. When this inequality is satisfied, we can create Cauchy problems with exponentially growing solutions. For example, the Cauchy problem for U in which

$$U(\mathbf{x}, t) = 1, \qquad \frac{\partial U}{\partial n} = \left[\sum_{j=1}^{m+1} \lambda_j^2\right]^{1/2} \quad (5.1.14)$$

on the hyperplane $\Lambda = 0$ has solution

$$U(\mathbf{x}, t) = e^\Lambda \cos \Lambda. \quad (5.1.15)$$

This solution grows exponentially with increasing Λ. It is also possible to construct solutions that grow exponentially with $\pm\Lambda$.

The condition (5.1.13) can be rewritten as

$$\frac{1}{c^2}\left[\frac{\partial \Lambda}{\partial t}\right]^2 \leq \sum_{j=1}^{m}\left[\frac{\partial \Lambda}{\partial x_j}\right]^2. \tag{5.1.16}$$

In this form, the constraint can be viewed as being applicable *pointwise* to an arbitrary function $\Lambda(\mathbf{x}, t)$ for which the parameters $\lambda_0, \ldots, \lambda_m$ are the components of the normal vector to the surface at a point. When equality holds here, the tangent to this surface is also tangent to the characteristic cone centered at that point; compare (5.1.16) with (5.1.4), and see Fig. 5.1. For inequality, the normal to the initial surface lies outside the cone and the surface itself cuts the characteristic cone. In this case, the initial surface is called *timelike*. The coordinate planes of the spatial variables are timelike surfaces. It is for these surfaces then that the Cauchy problem is ill posed.

On the other hand, let us consider the case in which (5.1.16) is violated:

$$\sum_{j=1}^{m}\left[\frac{\partial \Lambda}{\partial x_j}\right]^2 < \frac{1}{c^2}\left[\frac{\partial \Lambda}{\partial t}\right]^2. \tag{5.1.17}$$

The space–time surface $t = 0$ is an example of a surface satisfying this criterion. In this case, one *cannot* construct exponentially growing solutions to the wave equation, and it can be shown that the Cauchy problem is well conditioned. Surfaces on which (5.1.17) holds are called *spacelike*. See Fig. 5.2.

Let us suppose that we seek a solution for the wave equation with a moving point source, such as an airplane in flight. This problem is equivalent to one in which the equation is homogeneous but nonzero Cauchy data are

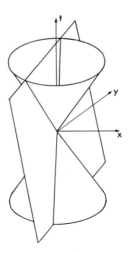

Fig. 5.1. A timelike initial surface for $m = 2$.

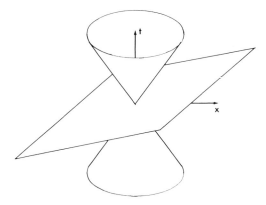

Fig. 5.2. A spacelike initial surface for $m = 2$.

prescribed on some space–time curve. If the speed of the source is less than the characteristic speed, the initial curve is spacelike and the problem is well posed. If the speed of the source is greater than the characteristic speed, the initial curve is timelike and the problem is ill posed.

As a second example, let us consider the seismic inverse problem mentioned in Chapter 3. A signal is propagated into the earth, and the return signal at the earth's surface is observed. The objective is to solve the wave equation downward in space and backward in time to determine from where the upward propagating wave came. This is a Cauchy problem with data given on a plane, say, $z = 0$, which is timelike. Hence, the problem is ill posed. For some purposes, the data recorded at different points are processed as though they arrived at delayed times varying linearly across the array of receivers. Thus, the space–time surface of the observations is exactly of the form $\Lambda = 0$. So long as (5.1.13) is satisfied, the problem remains ill posed. It should be noted that this criterion can be expressed in terms of a "speed" of propagation across the array. The criterion is then written as

$$\frac{d|x|}{dt} = \frac{\partial \Lambda/\partial t}{[\sum_{j=1}^{m} [\partial \Lambda/\partial x_j]^2]^{1/2}} \le c. \tag{5.1.18}$$

Often in practice, the receiver array gathering data for an inverse problem is moving, for example, an antenna array in an airplane or a towed array in water. However, in both of these cases, the speed of propagation of the observation curve is so much lower than the characteristic speed of the medium (light speed in the first example, sound speed in water in the second) that the motion of the array can be neglected. However, were sound waves to be recorded by an airplane, this would no longer be the case, and the effects of the motion of the observation surface would be of interest.

Exercises

5.1   Verify that (5.1.15) is a solution to the wave equation with the Cauchy data (5.1.14).

5.2   Solve the wave equation with the Cauchy data

$$U = 1, \qquad \partial U/\partial n = 0, \qquad \text{for} \quad \Lambda = 0.$$

Here Λ is defined by (5.1.8).

5.2   THE ENERGY INTEGRAL, DOMAIN OF DEPENDENCE, AND UNIQUENESS

We shall consider here the Klein–Gordon operator

$$\mathscr{L}U = \nabla^2 U - c^{-2}U_{tt} - (b^2/c^2)U, \tag{5.2.1}$$

and introduce the product

$$U_t\mathscr{L}U = U_t\left[\nabla^2 U - \frac{1}{c^2}U_{tt} - \frac{b^2}{c^2}U\right]$$

$$= \nabla\cdot[U_t\nabla U] - \frac{1}{2}\frac{\partial}{\partial t}\left[\frac{1}{c^2}[U_t^2 + b^2 U^2] + (\nabla U)^2\right]. \tag{5.2.2}$$

We shall analyze the integral of this quantity (5.2.2) over a domain D depicted in Fig. 5.3 (in two space dimensions and time). The upper boundary of D is a disk in space (sphere in three dimensions) at time t_2. The lower surface is a disk at time t_1. The boundary surface between these two planes denoted by B is assumed to be spacelike or, at worst, on the boundary of the class of spacelike surfaces. That is, if the normal to this surface at each point is denoted by

$$(\boldsymbol{\lambda}, \lambda_0) = (\lambda_1, \ldots, \lambda_m, \lambda_0), \tag{5.2.3}$$

then from Section 5.1 [in particular, (5.1.17)],

$$\lambda_0^2/c^2 \geq \boldsymbol{\lambda}\cdot\boldsymbol{\lambda} = \lambda^2. \tag{5.2.4}$$

We set

$$I = \int_D U_t\mathscr{L}U \, dV \, dt. \tag{5.2.5}$$

Here dV is the *differential content* in m dimensions, that is, differential volume when $m = 3$ or differential surface area when $m = 2$. By applying

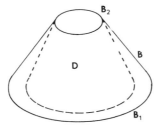

Fig. 5.3. The domain of integration D with boundary B, B_1, B_2 (solid line) and inner domain of dependence B_1 (broken line).

the divergence theorem, we find that

$$I = -\frac{1}{2}\int\left[\frac{1}{c^2}[U_t^2 + b^2U^2] + (\nabla U)^2\right]\Bigg|_{B_1}^{B_2} dV$$

$$+ \int_B\left\{U_t\boldsymbol{\lambda}\cdot\nabla U - \frac{\lambda_0}{2}\left[\frac{1}{c^2}[U_t^2 + b^2U^2] + (\nabla U)^2\right]\right\} dS. \quad (5.2.6)$$

Here dS is the differential content on the side boundary:

$$dS = \frac{\sqrt{\lambda^2 + \lambda_0^2}}{\lambda_0}\begin{cases} dx_1\,dx_2, & m = 2\\ dx_1\,dx_2\,dx_3, & m = 3\end{cases}. \quad (5.2.7)$$

Our objective now is to obtain an upper bound on the second line in (5.2.6). To do so, we first use the Cauchy–Schwarz inequality to estimate $|\boldsymbol{\lambda}\cdot\nabla U| \leq \lambda|\nabla U|$. Then use the fact that $|ab| \leq \frac{1}{2}[a^2 + b^2]$, which follows from $[a \pm b]^2 = a^2 \pm 2ab + b^2 \geq 0$, to deduce

$$|U_t\boldsymbol{\lambda}\cdot\nabla U| \leq |U_t|\lambda|\nabla U| = c\lambda\left|\frac{U_t}{c}\right||\nabla U| \leq \frac{c\lambda}{2}\left[\frac{U_t^2}{c^2} + (\nabla U)^2\right].$$

This result is used in the second line of (5.2.6) to deduce that

$$I \leq -\frac{1}{2}\int\left[\frac{1}{c^2}[U_t^2 + b^2U^2] + (\nabla U)^2\right]\Bigg|_{B_1}^{B_2} dV$$

$$+ \int_B\left[\frac{c}{2}\left[\lambda - \frac{\lambda_0}{c}\right]\left[\frac{U_t^2}{c^2} + (\nabla U)^2\right] - \frac{\lambda_0 b^2U^2}{2c^2}\right] dS. \quad (5.2.8)$$

As a consequence of (5.2.4), the first term of the second line is nonpositive. Then, since the second term is also nonpositive, the second line in (5.2.8) cannot be positive and I is bounded by the first line

$$I \leq -\frac{1}{2}\int\left[\frac{1}{c^2}[U_t^2 + b^2U^2] + (\nabla U)^2\right]\Bigg|_{B_1}^{B_2} dV. \quad (5.2.9)$$

Let us suppose now that U is a solution of an initial value problem, with

$$\mathscr{L}U = F(\mathbf{x}, t). \tag{5.2.10}$$

Then

$$I = \int_D U_t F(\mathbf{x}, t)\, dV\, dt. \tag{5.2.11}$$

By setting the right sides of (5.2.9) and (5.2.11) equal to one another and isolating the integral over B_1, we conclude that

$$\frac{1}{2} \int \left[\frac{1}{c^2} [U_t^2 + b^2 U^2] + (\nabla U)^2 \right]\Bigg|_{B_1} dV$$

$$\geq \frac{1}{2} \int \left[\frac{1}{c^2} [U_t^2 + b^2 U^2] + (\nabla U)^2 \right]\Bigg|_{B_2} dV + \int_D U_t F(\mathbf{x}, t)\, dV\, dt. \tag{5.2.12}$$

In the limit of equality in (5.2.4), the domain D is the *domain of dependence* of the upper region B_2. Therefore, if $F = 0$ on the interior and the Cauchy data at $t = t_1$ are also zero,[†] then the integral over B_2 is zero. Since the integrand is nonnegative, this could only be the case if U itself were identically zero on B_2. Thus, we have proven for this equation in m dimensions that the solution in a bounded spatial region at a prescribed time depends only on the data in the domain of dependence of that region. In the limit, when B_2 shrinks to a point, it follows that the solution at a point depends only on the data in the domain of influence of that point, *interpreted as a limit from the exterior.* This last comment again will cause us to include the "full strength" of distributions on the characteristic conoid.

This same estimate provides a proof of uniqueness of the solution. To see why this is so, let us suppose that there are two solutions U_1 and U_2, with U being their difference. Then U is a solution of the problem with zero data. We have just verified that the only solution to that problem is $U = 0$. Therefore, $U_1 = U_2$ and the solution is unique.

If B_1 were a more general *spacelike* initial domain, then it would be necessary to estimate the integral over B_2 exactly as was done for the integral over B. The same conclusion would follow.

The integrand in (5.2.12)

$$c^{-2}[U_t^2 + b^2 U^2] + (\nabla U)^2$$

is the mathematical energy density for this problem. It differs from the physical energy density only in a scale factor. For example, if U were the displacement of an area or a volume element, then the appropriate multiplier

---

[†] The Cauchy data U provide ∇U on the initial surface by differentiation.

of U_{tt} would be the mass density rather that c^{-2}. This rescaling would not change the essential character of the conclusion.

In the absence of external forces ($F = 0$), with B_1 and B_2 extending over all space, and assuming that the integral on the lower surface is finite, we conclude that the energy at time t_2 is bounded by the energy at time t_1. In particular, when the initial data are bounded to a finite region, the integral over B is zero for B_2 large enough. In this case, equality holds and *energy is conserved*. It is in this sense that we identify the Klein–Gordon equation as energy conserving. We have already seen in the case of one space dimension that it is dispersive, hence, the identification of the Klein–Gordon equation as an energy-conserving dispersive hyperbolic equation. Of course, the limit $b = 0$ leads to the conclusion that the wave equation itself is energy conserving.

Exercises

5.3 Verify (5.2.7).

5.4 Suppose that

$$\mathscr{L}U = \nabla^2 U - c^{-2}U_{tt} - \boldsymbol{\alpha} \cdot \nabla U + \beta U_t + \gamma U = 0, \qquad \boldsymbol{\alpha} = (\alpha_1, \ldots, \alpha_m).$$

Substitute

$$U = W \exp[(\beta c^2 t + \boldsymbol{\alpha} \cdot \mathbf{x})/2],$$

and show that energy is conserved for W if

$$\gamma - \tfrac{1}{4}(\alpha^2 - \beta^2 c^2) \leq 0.$$

5.3 THE GREEN'S FUNCTION

We consider the following problem for the Klein–Gordon equation (with the wave equation resulting from the special case $b = 0$):

$$\mathscr{L}U = \nabla^2 U - c^{-2}U_{tt} - (b^2/c^2)U = F(\mathbf{x}, t), \qquad (5.3.1)$$

with Cauchy data U and U_t prescribed at $t = 0$. Here the spatial dimensions are unbounded; that is, this is a *free-space* problem. A solution formula expressing U in terms of a Green's function will be derived here. However, the discussion will be much briefer than analogous discussions of earlier sections, relying in part on insights gained from the discussion of the one-dimensional problem in the preceding chapter and on the discussion of the domain of dependence in the preceding section.

Thus, let us define the Green's function as a solution of the following problem:

$$\mathscr{L}\, G(\mathbf{x}, t; \boldsymbol{\xi}, \tau) = -\delta(\mathbf{x} - \boldsymbol{\xi})\delta(t - \tau), \qquad t < \tau+;$$

$$G(\mathbf{x}, \tau; \boldsymbol{\xi}, \tau) = G_t(\mathbf{x}, \tau; \boldsymbol{\xi}, \tau) = 0. \tag{5.3.2}$$

We set

$$I = \int_D [G\mathscr{L}U - U\mathscr{L}G]\, dV\, dt. \tag{5.3.3}$$

Here the domain D should contain the *domain of influence* "backward in time" of the source for the Green's function. The domain of Fig. 5.3 will suffice for this purpose if we require that on $B_2\ t > \tau$. By using the divergence theorem, this integral is expressible as an integral in terms of the "boundary data" on the boundary $t = 0$ of D:

$$I = \int_{B_1} \frac{1}{c^2} [GU_t - UG_t]\, dV. \tag{5.3.4}$$

Here B_1 is as shown in Fig. 5.3. On the other hand, using (5.2.1) and (5.2.2) yields an alternative expression for the volume integral. Therefore,

$$I = \int_D GF\, dV\, dt + U(\boldsymbol{\xi}, \tau). \tag{5.3.5}$$

Solving for U here yields

$$U(\boldsymbol{\xi}, \tau) = \int_{B_1} \frac{1}{c^2} [GU_t - UG_t]\, dV - \int_D GF\, dV\, dt. \tag{5.3.6}$$

As in earlier discussions, the domain D can be shrunk here to the characteristic conoid of the point $(\boldsymbol{\xi}, \tau)$. The domain B_1 is then shrunk to the domain of influence of the source point of the Green's function, so long as one interprets this domain as a limit from the outside and takes into account the "full strength" of all distributions on the boundary of the domain.

Let us suppose now that U is to satisfy (5.3.1) in the exterior of some domain D_0 with boundary B_0. Then, when the divergence theorem is applied, it must be done in the new domain, all space with D_0 deleted for all time. Thus,

$$\int_D [G\mathscr{L}U - U\mathscr{L}G]\, dV\, dt = \int_{B_1} \frac{1}{c^2} [GU_t - UG_t]\, dV$$

$$= \int_0^\tau \int_{B_0} \left[U\frac{\partial G}{\partial n} - G\frac{\partial U}{\partial n} \right] dS. \tag{5.3.7}$$

If the prescribed data on B_0 is the value of

$$\cos \psi (\partial U / \partial n) - \sin \psi U,$$

then we require that G satisfy

$$\cos \psi (\partial G / \partial n) - \sin \psi G = 0. \tag{5.3.8}$$

In this case, we can again solve for U in (5.3.7) in terms of this enlarged set of data:

$$U(\xi, \tau) = \int_{B_1} \frac{1}{c^2} [GU_t - UG_t] \, dV - \int_0^\tau \int_{B_0} G \left[\frac{\partial U}{\partial n} - \tan \psi U \right] dS$$

$$- \int_D GF \, dV \, dt. \tag{5.3.9}$$

It is hoped that the reader has drawn the following conclusion from the discussion thus far in this and preceding chapters:

Given a problem consisting of a differential equation and "data," there is a fundamental or *canonical* problem to be addressed, namely, the problem for the Green's function, for which the (homogeneous) data should be deduced from the data for the given problem. If the problem for the Green's function does not lend itself to solution by a particular approach, it is not likely that the given problem will be solvable by that method either. On the other hand, solution for the Green's function provides the means to determine the general solution.

We consider now the determination of the free-space Green's functions in two and three dimensions when b and c are constant. First, let us set $b = 0$ in (5.3.2) in order to find the Green's function for the wave equation. We employ the time reversal $\tau - t \rightarrow t$ and the spatial shift $\mathbf{x} - \xi \rightarrow \mathbf{x}$ to replace the problem for G by the following problem:

$$\mathcal{L} G = -\delta(t) \delta(\mathbf{x}), \quad t > 0; \qquad G = G_t = 0, \quad t = 0. \tag{5.3.10}$$

We introduce the multifold Fourier transform of G, denoted by g and defined by (2.4.3), with its inversion given by (2.4.4). The Fourier transformation of (5.3.10) leads to the solution

$$g = -[c^2 / (\omega^2 - c^2 k^2)]. \tag{5.3.11}$$

For $m = 3$, the Fourier inversion is carried out in Section 2.4, Eqs. (2.4.6)–(2.4.13), to yield the solution

$$G(\mathbf{x}, t; \xi, \tau) = \frac{\delta(\tau - t - r/c)}{4\pi r}, \qquad r = |\mathbf{x} - \xi|, \quad m = 3. \tag{5.3.12}$$

Similarly, from Exercise 2.13,

$$G(\mathbf{x}, t; \xi, \tau) = \frac{H(\tau - t - r/c)}{\pi\sqrt{(\tau - t)^2 - r^2/c^2}}, \qquad m = 2. \qquad (5.3.13)$$

For $b \neq 0$ and the problem temporally and spatially shifted as above,

$$g = -\frac{c^2}{\omega^2 - (c^2k^2 + b^2)}. \qquad (5.3.14)$$

As in Section 2.4, we calculate the ω integral by residues to obtain

$$G = \frac{-ic^2}{2(2\pi)^m} \int dk^m \, e^{i\mathbf{k}\cdot\mathbf{x}} \left[\frac{e^{i\omega t} - e^{-i\omega t}}{\omega}\right],$$
$$\omega = \omega(k) = \sqrt{c^2k^2 + b^2}. \qquad (5.3.15)$$

Let us first consider the case $m = 3$. We introduce polar coordinates as in Section 2.4 and carry out the angular integrals as described there to obtain in this case

$$G = -\frac{c^2}{8\pi^2 r} \int_0^\infty \frac{k\,dk}{\omega} [e^{ikr} - e^{-ikr}][e^{i\omega t} - e^{-i\omega t}]. \qquad (5.3.16)$$

For large k, the modulus of the integrand is nearly unity, so that the integral makes sense only as a distribution. Indeed, the case $b = 0$ reduces to the result (5.3.12) after undoing the coordinate shifts. Preferring to deal with a convergent integral, we set

$$G = \frac{i}{8\pi^2 r} \frac{\partial}{\partial r} G_1;$$

$$G_1 = \int_0^\infty \frac{c^2\,dk}{\omega} [e^{ikr} + e^{-ikr}][e^{i\omega t} - e^{-i\omega t}] \qquad (5.3.17)$$

$$= \int_{-\infty}^\infty \frac{c^2\,dk}{\omega} e^{ikr}[e^{i\omega t} - e^{-i\omega t}].$$

This integral is now in a form in which it can be analyzed as was (4.4.25). That is, we can show that I is a function only of ρ introduced in (4.4.26), with x now the magnitude of a vector. We leave the details to the exercises and state the result

$$G_1 = 2\pi i J_0(b\rho/c)H(ct - r), \qquad \rho = \sqrt{c^2t^2 - r^2}, \qquad r^2 = \mathbf{x}\cdot\mathbf{x}. \quad (5.3.18)$$

From this result, it follows by using (5.3.17) that

$$G = \frac{-c}{4\pi r} \frac{\partial}{\partial r} \left[J_0 \left(\frac{b\rho}{c} \right) H(ct - r) \right]$$

$$= \frac{c}{4\pi r} \delta(ct - r) + \frac{b}{4\pi \rho} J_1 \left(\frac{b\rho}{c} \right) H(ct - r). \tag{5.3.19}$$

As a check, note that for $b = 0$, this result reduces to the preceding one.

For $m = 2$, we return to (5.3.15) and introduce polar coordinates in k with the polar angle measures from $\mathbf{x}$. Thus,

$$G = -\frac{ic^2}{8\pi^2} \int_0^\infty k \, dk \int_0^{2\pi} d\phi \, e^{ikr \cos\phi} \left[\frac{e^{i\omega t} - e^{-i\omega t}}{\omega} \right]. \tag{5.3.20}$$

The angular integral here can be recognized as a multiple of the Bessel function of the first kind and order zero, (2.7.19) with $n = 0$. This leads to the result

$$G = \frac{c^2}{2\pi} \int_0^\infty \frac{k \, dk}{\omega} J_0(kr) \sin \omega t. \tag{5.3.21}$$

The analysis of this integral is by no means straightforward (see Watson [1966, Section 13.47]). However, the integral is really a Bessel transform of a sine function, or with a change of variable of integration, it is a Fourier sine transform of the Bessel function. Consequently, it can be found in a number of appropriate tables, such as those from Erdélyi [1954, Vol. 1, p. 113, Eq. (47) or Vol. 2, p. 9, Eq. (25)], Gradshteyn and Ryzhik [1965, p. 736, Section 6.667, Eq. (1)], and Oberhettinger [1972, p. 12, Eq. (2.53)]. The reader is cautioned to use multiple sources when possible because of the possibility of there being typographical errors in transcription. In this regard, it should be noted that Gradshteyn and Ryzhik are courteous enough to provide sources. For the particular example, they cite Erdélyi as the source and therefore do not provide an independent check on that source.

Using any of these references, we find that

$$G = \frac{c}{2\pi} \frac{\cos[b\sqrt{t^2 - r^2/c^2}]}{\sqrt{c^2 t^2 - r^2}} H(ct - r). \tag{5.3.22}$$

Exercises

5.5 For the integral (5.3.17), repeat the analysis (4.4.25) to (4.4.28) to confirm that G is a function of ρ only.

5.6 (a) Suppose that the domain of interest for the Klein–Gordon equation, $m = 3$, is the region $x_3 \geq 0$. Suppose further that G must satisfy (5.3.8) with $\psi = 0$. Show that the solution G is now the difference between the response to a source at $\xi_3 > 0$ and a source at the point $-\xi_3$.

(b) For $\psi = \pi/2$, show that the difference in (a) should be replaced by a sum.

(c) Specialize each of these results to the wave equation.

5.4 SCATTERING PROBLEMS

A class of problems that arises in mathematical physics (and is of interest to the author) can be characterized as follows. In some "local" region in space, there is an obstacle or *inhomogeneity* of the medium characterized by variable coefficients in the governing equation. A wave is transmitted "from a distance" toward this region and "reacts" with it. The consequence of this interaction is to change the shape and character of the wave from what it would have been in the absence of local variations.

Problems of this sort were described in Chapter 1, in which it was argued that $u(\mathbf{x}) - c_0 t$, with u a solution of the eikonal equation, represented the location in space at time t of a *wave front* associated with acoustic, electromagnetic, and elastic waves among others. In the examples presented there, we introduced the idea of an *incident wave* that as time progressed, impacted on the local region, giving rise to other waves that were collectively called *scattered waves* and individually called *reflected, refracted,* and *diffracted waves*.

Let us consider the problem of a wave incident on an *obstacle* or *scatterer*, such as in the example beginning with the plane wave (1.5.15). In that problem, the plane wave is incident from the right on a circular cylinder centered at the origin. The question that arises is how one is to set up an appropriate problem for the full wave equation, modeling the same situation as was modeled more primitively in that example. First, the obstacle itself should be characterized by its location and a boundary condition describing how solutions of the wave equation interact with it. Thus, let us denote the domain occupied by the obstacle by D and its boundary by B. Then, for example, we might be considering a solution of the wave equation

$$\nabla^2 U - c^{-2} U_{tt} = 0, \qquad \mathbf{x} \quad \text{outside} \quad D, \qquad (5.4.1)$$

satisfying a boundary condition, say,

$$U(\mathbf{x}, t) = 0, \qquad \mathbf{x} \quad \text{on} \quad B. \qquad (5.4.2)$$

The total solution is to consist of a plane wave incident from the right on the obstacle and the consequences of the presence of the obstacle. We can check that

$$U_1(\mathbf{x}, t) = \delta(x - ct) \tag{5.4.3}$$

is a solution of the wave equation. However, if the origin of the coordinate system is inside of D, then this solution of the equation does not also satisfy the boundary condition for t near zero. On the other hand, if the obstacle extends in the negative x direction no more than a distance L, then U_1 *does* satisfy the boundary condition for times less than $-L/c$ since this solution is identically zero on B in that time range.

With this added insight, it is fairly straightforward to set down *initial conditions* at an earlier time, say, $t = -2L/c$, in order that the solution U be given by (5.4.3) up to time $-L/c$. Those conditions are

$$U(\mathbf{x}, 0) = \delta(x + 2L), \qquad U_t(\mathbf{x}, 0) = -c\delta'(x + 2L). \tag{5.4.4}$$

Now, (5.4.1), (5.4.2), and (5.4.4) define an initial boundary value problem for U.

An alternative approach to this problem more widely in use is as follows. We set

$$U = U_1 + U_S \qquad \text{outside} \quad D \tag{5.4.5}$$

and proceed to set down an initial boundary value problem for the scattered wave U_S; namely,

$$\nabla^2 U_S - \frac{1}{c^2} \frac{\partial^2}{\partial t^2} U_S = 0 \qquad \text{outside} \quad D, \qquad t > -2L/c,$$

$$U_S = \frac{\partial U_S}{\partial t} = 0, \qquad t = -2L/c, \tag{5.4.6}$$

$$U_S = -U_1 \qquad \text{on} \quad B,$$

the last equation following from (5.4.2) and (5.4.5). We see that U_S satisfies an initial boundary value problem with homogeneous initial data and inhomogeneous boundary data.[†] Furthermore, the functions U_1 and U_S are in some sense unphysical in that (i) U_1 exists for all time as if the obstacle D were not in place and, therefore, (ii) the *mathematical* scattered field U_S must carry the "burden" of *negating* this unphysical incident field in regions where it is "blocked" by the obstacle.

This approach to scattering problems is a *departure* from the approach in Chapter 1. There the incident wave existed only in that region exterior to D covered by the *characteristics* or *rays* of the incident wave front. Thus, the

---

[†] Contrast with U itself.

incident field did not exist on the "back side" of the scatterer *as it does here.* Nonetheless, this approach leads to a more tractable mathematical problem.

Now let us consider inhomogeneity problems. To be more precise, let us suppose that we are considering the wave equation (5.4.1) again, with c variable. Here a variation on the above theme for the problem of scattering by an obstacle is again possible. For example, let us suppose that c has two constant values in two complementary domains D_1 and D_2. Let us suppose further that a wave is incident on the boundary between these domains from within D_1. Then in D_1 we set $U = U_I + U_S$, but in D_2 we set $U = U_T$, a *transmitted* wave, which is really another form of scattered wave. In this case, we impose appropriate *continuity* conditions on the total field at the boundary. Thus, U_I is a discontinuous solution existing only in a part of space, namely, D_1. However, U_S and U_T are also discontinuous, while the total solution is not.

As a second example, let us suppose that c varies "smoothly" in some finite domain from some constant reference value c_0 "at infinity." Then we could proceed as in the first problem to write U as a sum, with

$$\nabla^2 U_I - \frac{1}{c_0^2} \frac{\partial^2}{\partial t^2} U_I = 0,$$

$$\nabla^2 U_S - \frac{1}{c^2} \frac{\partial^2}{\partial t^2} U_S = \left[\frac{1}{c^2} - \frac{1}{c_0^2} \right] \frac{\partial^2}{\partial t^2} U_I.$$

(5.4.7)

For most problems of interest, closed-form analytical solutions are not available. Known analytical solutions are usually arrived at after applying Fourier transform to the time domain. Thus, discussion of such exact solutions is best postponed until after the discussion of the Helmholtz equation in the next chapter.

Much insight into the nature of exact solutions can be obtained form *approximate* or, more precisely, *asymptotic* solutions. Here again, though, the most natural large parameter is a dimensionless scale connoting high frequency.

The class of problems being described is collectively known as *direct* scattering problems to distinguish them from another class, the so-called *inverse* scattering problems. In this class of problems, one observes the scattered field, and the objective is to determine the parameters of the scattering mechanism. That is, the scattered field might have been the result of the scattering of a (known) incident wave from an obstacle, semitransparent or solid, or might be the result of scattering from an inhomogeneity.

One other problem of a slightly different nature is included in the class of inverse problems, namely, the *inverse source problem.* Here a wave produced by an unknown source is observed. The objective is to determine the source.

This problem is fundamentally different from the others as can be seen from a simple count of degrees of freedom or independent variables of the solution versus the data. In both cases, we might contemplate observing the scattered field in two space dimensions (over a surface) for all time, three independent variables. In the former problem, the medium parameters to be determined are functions of three spatial variables, and the number of degrees of freedom of the data agrees with the number of degrees of freedom of the source, albeit at the expense of an interchange of space and time. In the latter problem, the source may be a function of time as well as all spatial variables, thus having more degrees of freedom than the given (observed) data. Hence, in general, we should expect nonuniqueness in the inverse source problem and uniqueness only under severe restrictions on the source—enough to reduce the number of independent variables by one at the very least. On the other hand, there are many situations in which the inverse scattering problem has unique solutions, in general, or at least in a subclass of "physically reasonable" parameter characterizations. Such problems will be discussed in Chapter 9.

Exercises

5.7 (a) Consider the following problem for $U(\mathbf{x}, t)$ in three spatial dimensions:

$$\nabla^2 U - c^{-2}U_{tt} = -\delta(\mathbf{x} - \mathbf{x}_0)\delta(t), \qquad x \neq 0,$$

$$\mathbf{x}_0 = (x_0, 0, 0), \qquad x_0 < 0, \qquad c = \begin{cases} c_0, & x < 0, \\ c_1, & x > 0, \end{cases}$$

$$U(\mathbf{x}, 0) = U_t(\mathbf{x}, 0) = 0,$$

with both U and U_x continuous at $x = 0$. Write U as

$$U = \begin{cases} U_I + U_S, & x < 0, \\ U_T, & x > 0. \end{cases}$$

Use for U_I the result (2.4.13). Write down the initial boundary value problems for U_S and U_T, with the values of U_I and its normal derivative at $x = 0$ explicitly stated in the boundary conditions.

(b) Repeat (a) when the continuity in normal derivative is replaced by continuity of c times the normal derivative.

5.8 The purpose of this exercise is to develop a Fourier transform needed in the next exercise. Let G be the free-space Green's function with source point at the origin given by

$$G(\mathbf{x}, t; \mathbf{0}, 0) = \frac{\delta(t - r/c)}{4\pi r}, \qquad r = |\mathbf{x}|.$$

Define

$$g(k_1, k_2, z, \omega)$$
$$= \int_{-\infty}^{\infty} dx_1 \int_{-\infty}^{\infty} dx_2 \int_{0}^{\infty} dt \, G(\mathbf{x}, t; \mathbf{0}, 0) \exp\{-i[k_1 x + k_2 y - \omega t]\}.$$

Show that

$$g(k_1, k_2, z, \omega) = \frac{ie^{ik_3|z|}}{2k_3}.$$

Here

$$k_3 = [\omega^2/c^2 - k_1^2 - k_2^2]^{1/2}.$$

In the ω plane, the square root is defined to be positive when $\omega \to +\infty$. Since k_3 must be analytic in the upper half ω plane, this is sufficient to define k_3 everywhere. We note that k_3 is negative when $\omega \to -\infty$ and is purely imaginary between the two branch points $\pm \sqrt{k_1^2 + k_2^2}$, with argument $\pi/2$. Consequently, g is bounded and analytic in the upper half ω plane for any choice of z. *Hint:* Start from the problem (5.3.10) for G and apply Fourier transform in x, y, and t to obtain a problem for g. Solve the problem for g under the requirement that the solution must be analytic in the upper half ω plane.

5.9 The purpose of this exercise is to present a case in which the inverse source problem admits a unique solution. Suppose that U is a solution of the problem

$$\nabla^2 U - c^{-2} U_{tt} = -\delta(t)F(\mathbf{x}), \qquad t > 0;$$
$$U(\mathbf{x}, 0) = U_t(\mathbf{x}, 0) = 0.$$

(a) Use (5.3.9) and (5.3.12) to show that

$$U(\mathbf{x}, t) = \frac{1}{4\pi} \int F(\mathbf{x}') \frac{\delta(t - |\mathbf{x} - \mathbf{x}'|/c)}{|\mathbf{x} - \mathbf{x}'|} dV'.$$

Here the integral is over all space.

(b) Suppose that the source is confined to some sphere, say, of radius less than L. Suppose further that U is observed everywhere on the surface $z = L$. Introduce the same threefold Fourier transform as was introduced in the preceding exercise and show that

$$u(k_1, k_2, L, \omega) = \frac{i}{2k_3} \int_{-\infty}^{\infty} f(k_1, k_2, z) e^{ik_3(L - z')} dz'.$$

Here k_3 is as defined in the preceding exercise.

(c)   Conclude that

$$f(k_1, k_2, k_3) = -2ik_3 u(k_1, k_2, L, \omega)e^{-ik_3 L},$$

$$-\infty < k_1, k_2 < \infty, \qquad \begin{cases} -\infty < k_3 < 0, & -\infty < \omega < -\sqrt{k_1^2 + k_2^2}, \\ 0 < k_3 < \infty, & \sqrt{k_1^2 + k_2^2} < \omega < \infty. \end{cases}$$

The function F is determined by Fourier inversion over real values of k_1, k_2, k_3.

(d)   Suppose that two source functions of this same type and confined to some finite region to one side of $z = L$ produce the same observed field. Conclude that the difference between the two sources must have Fourier transform zero. More precisely, the Fourier transform of the source function is uniquely determined in the spectrum of the observations.

5.10   Suppose that $F(\mathbf{x}, t)$ is a function with two derivatives, all of its arguments, vanishing outside of some finite x domain, say, $|\mathbf{x}| \le 0$. Suppose further that U is a solution of the problem

$$\nabla^2 U - c^{-2} U_{tt} = \nabla^2 F - c^{-2} F_{tt};$$

$$U(\mathbf{x}, 0) = F(\mathbf{x}, 0), \qquad U_t(\mathbf{x}, 0) = F_t(\mathbf{x}, 0).$$

Conclude that $U = F$ and explain the implications as regards the inverse source problem.

References

Courant, R., and Hilbert, D. [1964]. "Methods of Mathematical Physics," Vol. 2, Partial Differential Equations. Wiley, New York.

Erdélyi, A. (ed.) [1954]. "Tables of Integral Transforms," Vols. 1 and 2. McGraw-Hill, New York.

Friedlander, F. G. [1973]. An inverse problem for radiation fields, *Proc. London Math. Soc.* **3**, 551–576.

Garabedian, P. R. [1964]. "Partial Differential Equations." Wiley, New York.

Gradshteyn, I. S., and Ryzhik, I. M. [1965]. "Table of Integrals, Series and Products." Academic Press, New York. [Corrected and enlarged edition, 1980.]

John, F. [1982]. "Partial Differential Equations," 4th ed. Springer-Verlag, New York.

Oberhettinger, F. [1972]. "Tables of Bessel Transforms." Springer-Verlag, New York.

Watson, G. N. [1966]. "A Treatise on the Theory of Bessel Functions." Cambridge University Press, Cambridge.

6 THE HELMHOLTZ EQUATION AND OTHER ELLIPTIC EQUATIONS

In this chapter, we shall discuss primarily the *Helmholtz* or *reduced wave equation* obtained by Fourier transform in time of the wave equation. Some aspects of the theory will be developed for other elliptic equations having the Laplacian as highest-order operator. Many of the qualitative features of solutions of problems with this operator carry over to more general elliptic equations. We shall not discuss such generalizations here. The reader is referred to the literature for generalizations and rigorous theory.

We have already seen some features of elliptic problems. First, in Section 3.2, the Hadamard example was presented [Eqs. (3.2.12) and (3.2.13)]. This example demonstrated that the Cauchy problem is an ill-posed problem for Laplace's equation in two dimensions. From the discussion in Section 5.1, if we set $\lambda_0 = 0$ in (5.1.8) and $\mu_0 = 0$ in (5.1.9), the resulting solution demonstrates the ill-posedness of the Cauchy problem for Laplace's equation in m dimensions.

Exercise 3.1 demonstrates that the homogeneous Helmholtz equation with zero boundary data can have nonzero solutions under prescribed circumstances. Thus, this particular elliptic equation, as well as the class of equations for which it is prototypical, can have nonunique solutions. The parameter values (λ_{mn}^2 in Exercise 3.1) that allow these special solutions are *eigenvalues* or *characteristic values*, while the solutions themselves are *eigenfunctions* or *characteristic functions*.

Equation (4.3.2) is a one-dimensional inhomogeneous Helmholtz equation. In Section 4.3, the eigenvalues and eigenfunctions associated with the problem (4.3.2) arose as poles in the ω plane of the solution formula for u; see the discussion pertaining to Eqs. (4.3.12) and (4.3.13) as well as the digression on eigenfunction expansions in Section 4.3.

We shall present some general remarks on elliptic equations in the next section and then proceed to the class of problems of interest.

6.1   GREEN'S IDENTITIES AND UNIQUENESS RESULTS

Let us define the operator

$$\mathscr{L}u(\mathbf{x}) = \nabla^2 u(\mathbf{x}) - qu(\mathbf{x}). \tag{6.1.1}$$

We shall concern ourselves with problems in two or three independent variables, although much of what is said is readily generalized to more independent variables. For $q = 0$, $\mathscr{L}$ is the Laplace operator. As already noted, problems for u involving the operator $\mathscr{L}$ have quite different features for q positive or negative, with $q = 0$ being more like the limit from positive values of q.

When q is negative, we shall think of the operators having arisen from the wave operator in the time domain after Fourier transform. Thus, in that case,

$$-q = k^2 = \omega^2/c^2. \tag{6.1.2}$$

Often, the solutions we examine will have singularities in ω for real ω. When the problem of interest can be viewed as arising from an initial value problem, these solutions should be viewed as the limit from above of solutions in the upper half ω plane. This will be discussed further later.

In the problems that we shall consider, u will be required to satisfy

$$\mathscr{L}u(\mathbf{x}) = f(\mathbf{x}), \qquad \mathbf{x} \quad \text{in} \quad D, \tag{6.1.3}$$

subject to a boundary condition on the boundary of D, denoted by ∂D. There are three types of boundary conditions of interest. There is the *Dirichlet boundary condition*, or the *boundary condition of the first kind*, in which u itself is prescribed,

$$u(\mathbf{x}) = g(\mathbf{x}), \qquad \mathbf{x} \quad \text{on} \quad \partial D; \tag{6.1.4}$$

the *Neumann boundary condition*, or the *boundary condition of the second kind*, in which the normal derivative of u is prescribed,

$$\partial u(\mathbf{x})/\partial n = g(\mathbf{x}), \qquad \mathbf{x} \quad \text{on} \quad \partial D; \tag{6.1.5}$$

and the *mixed boundary condition*, or the *boundary condition of the third kind*,

$$[\partial u(x)/\partial n] + \alpha u(\mathbf{x}) = g(\mathbf{x}), \qquad \mathbf{x} \quad \text{on} \quad \partial D. \tag{6.1.6}$$

The problems associated with each type of boundary condition are identified in the same manner, that is, Dirichlet, Neumann, or mixed problem, respectively. It is important to note here that the normal derivative we use points out of the domain D on which the differential equation is to hold.

Two important results from multidimensional calculus used in the study of these problems with elliptic operators are the *Green's identities*. The following is *Green's first identity* for two functions $u(\mathbf{x})$ and $v(\mathbf{x})$ having the necessary derivatives explicit in the identity

$$\int_D [v\,\nabla^2 u + \nabla v \cdot \nabla u]\,dV = \int_{\partial D} v\,\frac{\partial u}{\partial n}\,dS. \tag{6.1.7}$$

Here dV denotes the *differential content*, which means differential volume in three dimensions or differential surface area in two dimensions. Similarly, dS denotes differential surface area of the boundary of D in three dimensions or counterclockwise differential arc length on the boundary of D in two dimensions. The normal derivative in this equation is to be interpreted as a directional derivative in the normal direction *outward* from D. If we interchange u and v and subtract, we obtain *Green's second identity* or *Green's theorem*:

$$\int_D [u\,\nabla^2 v - v\,\nabla^2 u]\,dV = \int_{\partial D} \left[u\,\frac{\partial v}{\partial n} - v\,\frac{\partial u}{\partial n}\right]dS. \tag{6.1.8}$$

We restate these results for the operator $\mathscr{L}$:

$$\int_D [v\mathscr{L}u + \nabla v \cdot \nabla u + qvu]\,dV = \int_{\partial D} v\,\frac{\partial u}{\partial n}\,dS \tag{6.1.9}$$

and

$$\int_D [u\mathscr{L}v - v\mathscr{L}u]\,dV = \int_{\partial D} \left[u\,\frac{\partial v}{\partial n} - v\,\frac{\partial u}{\partial n}\right]dS. \tag{6.1.10}$$

We shall exploit (6.1.9) to obtain some results about uniqueness of solutions of the problems defined by (6.1.3) and one of the conditions (6.1.4), (6.1.5), or (6.1.6). To study uniqueness, we consider problems with zero data, that is, with $f = g = 0$. If such problems have only the zero solution, then whenever a problem with nonzero data has a solution, it must indeed be unique. For q nonnegative, we shall consider only real solutions. Let us set $v = u$ in (6.1.9) and use (6.1.3) with $f = 0$. In this manner, we obtain

$$\int_D [(\nabla u)^2 + qu^2]\,dV = \int_{\partial D} u\,\frac{\partial u}{\partial n}\,dS. \tag{6.1.11}$$

For either the Dirichlet or Neumann problem, the right side is equal to zero here. If q is positive, then the only way for the left side to be zero is for u to be zero in D. Thus, for either of these problems, the solution to the inhomogeneous problem is unique. When $q = 0$ (Laplace's equation), we can only conclude from the vanishing of the left side in (6.1.11) that the gradient of u vanishes in D. Thus, u must be a constant. For the Dirichlet problem, that constant is zero on the boundary and, hence, zero in the interior as well. Therefore, we conclude that when $q = 0$ and $\mathscr{L}$ is just the Laplacian, the Dirichlet problem has a unique solution whenever it has a solution. For the Neumann problem, the constant cannot be determined, since we know only that the normal derivative is zero on the boundary, which is true for all constant solutions. Thus, the Neumann problem for Laplace's equation has solutions that are at best unique *only up to a constant*.

We remark that in problems in which Laplace's equation arises, the solution represents a *potential function*—a velocity potential, electric potential, etc.—and only the *difference* in function values at two points or the *gradient* of the potential is a function of physical relevance. Thus, uniqueness up to a constant suffices for these applications.

Let us now consider homogeneous data for the mixed problem. In this case, we rewrite (6.1.11) as

$$\int_D [(\nabla u)^2 + qu^2]\, dV = -\int_{\partial D} \alpha u^2\, dS. \tag{6.1.12}$$

Now for q nonnegative and α positive, the right side is nonpositive while the left side is nonnegative. Thus, they must both be zero; u is a constant function that is zero on the boundary; u must be identically zero, and again the solution is unique.

We consider now the case $q = -\omega^2/c^2$. Clearly, in this case, for real values of ω, knowing that the left side of (6.1.11) is zero will not allow us to draw any conclusions about u or its gradient in D. On the other hand, let us consider complex q. Now we must allow the solutions to be complex, too. Thus, we again consider the homogeneous problem but in (6.1.9), set v equal to the complex conjugate of u ($v = u^*$) and take the imaginary part of the result

$$\operatorname{Im} \omega^2 \int_D \frac{|u|^2}{c^2}\, dV = \operatorname{Im} \int_{\partial D} u^* \frac{\partial u}{\partial n}\, dS. \tag{6.1.13}$$

Here we see that for zero boundary data for either the Dirichlet or Neumann problem, the right side is zero. Hence, the left side is zero as well, which makes $u = 0$ in the interior for $\operatorname{Im} \omega$ nonzero. Consequently, for either of these problems, the solution is unique for ω in the upper half or lower half plane. As noted earlier, we have already demonstrated through examples that this equation admits eigenvalues and eigenfunctions, that is, nontrivial solutions with zero data. We can now conclude that when such eigenvalues exist, they must lie on the Re ω axis. In terms of q then, we conclude that the eigenvalues occur only for certain negative values of q.

Let us now consider the mixed problem (6.1.6) for this case. For zero data, we rewrite (6.1.11) as

$$\text{Im } \omega^2 \int_D \frac{|u|^2}{c^2} \, dV = -\text{Im} \int_{\partial D} \alpha |u|^2 \, dS. \qquad (6.1.14)$$

If the two sides here were of opposite signs, then we could conclude that, in fact, both are zero. However, the sign of the left side is different in the first and second quadrants of the complex ω plane. Thus, there is no chance to conclude uniqueness here by the proposed method unless α also depends on ω, which we shall now assume.

We consider first the case of ω in the upper half plane and introduce the *impedance boundary condition*

$$[\partial u(x)/\partial n] - i\omega Z u(\mathbf{x}) = g(\mathbf{x}), \qquad \mathbf{x} \text{ on } \partial D, \quad Z > 0. \qquad (6.1.15)$$

We remark that such a boundary condition in frequency domain implies a balance between spatial and temporal derivatives in the domain. For this boundary condition, we again consider the case $g = 0$ in order to study uniqueness. In this case, we write down both the real and imaginary parts of (6.1.12)

$$\int_D \left[|\nabla u|^2 - \frac{\text{Re } \omega^2}{c^2} |u|^2 \right] dV = -\text{Im } \omega \int_{\partial D} Z |u|^2 \, dS,$$

$$-\text{Im } \omega^2 \int_D \frac{|u|^2}{c^2} \, dV = \text{Re } \omega \int_{\partial D} Z |u|^2 \, dS. \qquad (6.1.16)$$

In the second line here, we see that for $0 < \arg \omega < \pi/2$, the left side is nonpositive and the right side nonnegative. Hence, both must be zero in this range of values of ω. For $\pi/2 < \arg \omega < \pi$, the left side is nonnegative while the right side is nonpositive. Hence, again, both must be zero. Thus, for either of these ranges of ω, u is zero in D. Finally, let us suppose that ω is on the positive imaginary axis. In this case, let us consider the first line in (6.1.16). The left side is nonnegative while the right side is nonpositive. As for the operator (6.1.1) with q positive, we conclude that u must be zero in D. Consequently, the Helmholtz equation with impedance boundary condition (6.1.15) will have a unique solution whenever it has a solution for ω in the upper half plane. When the minus sign in (6.1.15) is replaced by a plus sign, we obtain in similar fashion uniqueness in the lower half plane *for that boundary condition*. We remark that this situation corresponds to considering the Helmholtz equation as having arisen from taking the Fourier time

Table 6.1

Uniqueness Results

q	Type	q	Type
≥ 0	Dirichlet	≥ 0	Mixed, $\alpha > 0$
> 0	Neumann		$\left\{\begin{array}{l}\text{Dirichlet}\end{array}\right.$
$= 0$	Neumann	$-\omega^2/c^2$, Im $\omega \neq 0$	Neumann
	up to const		$\left.\text{Mixed, } \alpha = \mp i\omega Z, Z > 0, \pm \text{Im } \omega > 0\right.$

transform with opposite sign in the exponent. Hence, opposite sign in the boundary condition has the same meaning as regards differentiation with respect to t. Thus, this latter case is not really a new uniqueness result, but only one equivalent to the result with Z in (6.1.15) positive and ω in the upper half plane.

These conclusions are summarized in Table 6.1.

Exercises

6.1 Suppose that $q > 0$ and that we are given a Dirichlet boundary condition on part of the boundary and a Neumann boundary condition on the remainder of the boundary. Prove that if a solution exists, it is unique.

6.2 Consider the following eigenvalue problem in one independent variable:

$$u'' + \lambda^2 u = 0, \qquad |x| < 1, \qquad \pm(du/dx) - i\lambda Zu = 0, \qquad x = \pm 1.$$

Here Z is positive and not equal to 1.

(a) Show that up to an arbitrary multiplicative constant, the differential equation has as general solution

$$u = e^{i\lambda(x - \alpha)} + e^{-i\lambda(x - \alpha)},$$

with α arbitrary.

(b) Apply the boundary conditions and conclude that the problem for u will have nontrivial solutions when

$$\lambda = \frac{1}{2i}\log\left|\frac{1 + Z}{1 - Z}\right| + \frac{n\pi}{2} \qquad \text{and} \qquad \alpha = \frac{n\pi}{2\lambda}, \qquad n \text{ an integer.}$$

(c) Locate the eigenvalues in the complex λ plane and discuss the limits $Z \to 0$, $Z \to \infty$.

(d) Consider the case $Z = 1$. Give physical arguments to explain why there are no eigenvalues for this choice of Z.

6.3   Now consider the following eigenvalue problem on the square $|x| < 1$, $|y| < 1$:

$$\nabla^2 u + \lambda^2 u = 0, \qquad \begin{cases} \pm(\partial u/\partial x) - i\lambda Z u = 0, & x = \pm 1, \\ \pm(\partial u/\partial y) - i\lambda Z u = 0, & y = \pm 1. \end{cases}$$

(a)   Show that

$$u = [e^{i\lambda \cos \beta(x-\alpha)} + e^{-i\lambda \cos \beta(x-\alpha)}][e^{i\lambda \sin \beta(y-\gamma)} + e^{-i\lambda \sin \beta(y-\gamma)}].$$

Here, α, β, and γ are to be determined along with λ.

(b)   Show that $\cos \beta = \sin \beta$ and that this problem has nontrivial solutions when

$$\lambda = \frac{1}{i\sqrt{2}} \log \left| \frac{1 + Z\sqrt{2}}{1 - Z\sqrt{2}} \right| - \frac{n\pi}{\sqrt{2}}.$$

6.2   SOME SPECIAL FEATURES OF LAPLACE'S EQUATION

We shall describe here some noteworthy features of Laplace's equation. Thus, we consider (6.1.1) with $q = 0$:

$$\mathscr{L}u = \nabla^2 u. \tag{6.2.1}$$

Let us first integrate $\mathscr{L}u$ over some finite domain D and apply the divergence theorem

$$\int_D \nabla^2 u \, dV = \int_{\partial D} \frac{\partial u}{\partial n} \, dS. \tag{6.2.2}$$

Given a Neumann problem in D [(6.1.3) and (6.1.5)], we see from (6.2.2) that the data cannot be chosen arbitrarily. It is necessary that

$$\int_D f(\mathbf{x}) \, dV = \int_{\partial D} g(\mathbf{x}) \, dS. \tag{6.2.3}$$

This condition can be understood in terms of a physical phenomenon that it models. Let us suppose that D is a volume or surface of an incompressible fluid in *steady-state* or time independent motion. Then u represents a *potential* function whose gradient is the velocity vector at each point. The function f represents a source/sink distribution of fluid in the interior of D, while g represents the density of the influx or efflux of fluid through the boundary. Equation (6.2.3) has the interpretation that for an incompressible fluid, the sources of the fluid in the interior must be balanced by the inflow or

outflow of fluid through the boundary. This is the mathematical manifestation of the principle of conservation of mass for an incompressible fluid. The
special case $f = 0$ has the interpretation that the net flow into or out of D
must be zero.

Let us now consider the problem of determining the "singular part" of the
Green's function, which is also the *free-space Green's function*; this is the
Green's function in unbounded space. Thus, we consider the equation

$$\nabla^2 G = -\delta(\mathbf{x} - \boldsymbol{\xi}). \tag{6.2.4}$$

We remark that translation leaves the equation unaltered. Therefore, we
might as well take $\boldsymbol{\xi} = \mathbf{0}$. Also, introduction of a new set of coordinates that
is a rotation of $\mathbf{x}$ about the source point leaves the equation unaltered. Thus,
the equation can depend only on the radial variable r in n dimensions.
Therefore, we consider instead of (6.2.4) the equation

$$\frac{1}{r^{n-1}} \frac{d}{dr}\left[r^{n-1} \frac{dG}{dr} \right] = -\delta(\mathbf{x}), \tag{6.2.5}$$

which applies to g only the radial part of the Laplace operator in n dimensions. Away from the origin, G satisfies a homogeneous equation, which on
integration once with respect to r becomes

$$dG/dr = c_n/r^{n-1} \tag{6.2.6}$$

with c_n to be determined. Another integration yields

$$G(r) = c_2 \log r, \quad n = 2, \qquad G(r) = -\frac{c_n}{(n-2)r^{n-2}}, \quad n > 2. \tag{6.2.7}$$

In order to determine c_n, we integrate (6.2.4) on a small sphere about the
source point (at the origin) and apply the divergence theorem to obtain

$$-1 = \int_{r=\varepsilon} \frac{dG}{dr} \, dS = \int_{r=\varepsilon} \frac{c_n}{r^{n-1}} \, dS = \int_{r=\varepsilon} c_n \, d\Omega_n. \tag{6.2.8}$$

Here $d\Omega_n$ denotes the differential solid angle in n dimensions. For example,

$$d\Omega_2 = d\theta, \qquad d\Omega_3 = \sin\theta \, d\theta \, d\phi.$$

The integration in (6.2.8) can now be carried out to yield

$$-1 = c_n \Omega_n = \begin{cases} 2\pi c_2, & n = 2, \\ 4\pi c_3, & n = 3, \\ \dfrac{2\pi^{n/2} c_n}{\Gamma(n/2)}, & n \geq 3. \end{cases} \tag{6.2.9}$$

Here Ω_n denotes the *content*—arc length in two dimensions, surface area in three—of the surface of the unit sphere in n dimensions. Derivation of this result is outlined in Exercise 6.4.

We conclude from (6.2.9) that

$$c_n = -\Gamma(n/2)/2\pi^{n/2}, \qquad n \geq 2. \tag{6.2.10}$$

We denote the Green's function that satisfies (6.2.4) by $G(\mathbf{x}; \xi)$. Then

$$G(\mathbf{x}; \xi) = \begin{cases} -\dfrac{1}{2\pi}\log r, & n = 2, \\[2mm] \dfrac{1}{4\pi r}, & n = 3, \\[2mm] \dfrac{\Gamma(n/2)}{2(n-2)\pi^{n/2}r^{n-2}}, & n \geq 3. \end{cases} \tag{6.2.11}$$

Let us now define the function

$$G_1(\mathbf{x}; \xi) = \begin{cases} G(\mathbf{x}; \xi) + \dfrac{1}{2\pi}\log a, & n = 2, \\[2mm] G(\mathbf{x}; \xi) - \dfrac{1}{\Omega_n a^{n-2}}, & n \geq 3, \end{cases} \tag{6.2.12}$$

so that $G_1 = 0$ on the surface of the sphere of radius a. Now we apply *Green's theorem* (6.1.8) to u and G_1, where the domain D is the sphere of radius a centered at ξ and u a solution of the homogeneous Laplace equation. The result is

$$u(\xi) = \frac{1}{\Omega_n a^{n-1}} \int_{r=a} u\, dS = \frac{1}{\Omega_n} \int_{r=a} u\, d\Omega_n. \tag{6.2.13}$$

Here $r = |\mathbf{x} - \xi|$. We see that $u(\xi)$ is the *mean* of its values on a sphere centered at ξ so long as the sphere is in a region D in which u is a well-behaved solution of Laplace's equation. Furthermore, for any point in D, we now conclude that such a point could not be an isolated maximum or minimum. If it were, it could not be the mean of its values on a surrounding sphere. Indeed, more generally, u could only attain a maximum at an interior point if it were constant. This result provides another approach to the question of uniqueness and provides a means for assuring continuous dependence on the boundary data. Suppose, for example, that we consider two problems for which the boundary values differ by a small amount. The difference of these solutions can be no larger in the interior than on the boundary. The same is true for the difference in opposite order. Hence, the absolute difference is bounded by the absolute difference of the boundary values. This constitutes continuous dependence on the boundary data for the Dirichlet problem for Laplace's equation.

Exercises

6.4   The purpose of this exercise is to derive the result for Ω_n stated in (6.2.9). Let V_n be the content—volume for $n = 3$, area for $n = 2$—of the interior of the n-dimensional unit sphere. Then

$$V_n = \int_{r \le 1} r^{n-1} \, dr \, d\Omega_n .$$

(a)   Introduce the variable

$$\rho = \sqrt{x_1^2 + \cdots + x_{n-1}^2}$$

and show that

$$V_n = 2 \int_{\rho \le a} \rho^{n-2} \, d\rho \, d\Omega_{n-1} \int_0^{\sqrt{1-\rho^2}} dx_n .$$

(b)   Conclude from the first representation that

$$V_n = \Omega_n / n$$

and from the second representation that

$$V_n = 2\Omega_{n-1} \int_0^a \rho^{n-2} \sqrt{1 - \rho^2} \, d\rho = \Omega_{n-1} B\left[\frac{n-1}{2}, \frac{3}{2} \right]$$

$$= \Omega_{n-1} \frac{\Gamma((n-1)/2)\sqrt{\pi}}{n\Gamma(n/2)} .$$

Here B denotes the beta function.

(c)   Equate the two values of V_n, obtain a recursion relation for Ω_n, and deduce the result

$$\Omega_n = 2\pi^{n/2} / \Gamma(n/2) .$$

6.5   In (6.2.12), let the radius of the sphere a be variable. Multiply by a^{n-1} and integrate from zero to b to obtain the second mean value theorem, namely,

$$u(\xi) = \frac{n}{b_n \Omega_n} \int_{r \le b} u(\mathbf{x}) \, dV, \qquad r = |\mathbf{x} - \xi| .$$

That is, $u(\xi)$ is the mean of its values in the surrounding spherical volume centered at ξ for any sphere in D.

6.6   Let u be a solution of

$$\nabla^2 u - qu = 0$$

in some domain D with q positive in D.

(a)   Show that at any point in D at which u is positive, at least one term of the Laplacian of u must be positive. Then u cannot have a positive maximum in D; explain. Similarly, u cannot have a negative minimum in D; explain.

(b)   Use (a) to prove uniqueness for the Dirichlet or mixed problem with α positive or for the Neumann problem up to a constant.

6.3   GREEN'S FUNCTIONS

We shall now develop the Green's function representation of the solution to the class of problems for which the differential equation is given by (6.1.3). We begin from (6.1.10), the generalization of Green's theorem applied to the operator $\mathscr{L}$. For the present, we do not require that q or α be constant, although we shall not go much farther than theoretical development for nonconstant values of these parameters. We choose for v a function satisfying the equation

$$\mathscr{L}v = -\delta(\mathbf{x} - \boldsymbol{\xi}). \tag{6.3.1}$$

Now, with u satisfying (6.1.3) and $\boldsymbol{\xi}$ in D, (6.1.10) becomes

$$-u(\boldsymbol{\xi}) - \int_D v(\mathbf{x}; \boldsymbol{\xi}) f(\mathbf{x})\, dV = \int_{\partial D} \left[u(\mathbf{x}) \frac{\partial v(\mathbf{x}; \boldsymbol{\xi})}{\partial n} - v(\mathbf{x}; \boldsymbol{\xi}) \frac{\partial u(\mathbf{x})}{\partial n} \right] dS$$

$$= \int_{\partial D} \left\{ u(\mathbf{x}) \left[\frac{\partial v(\mathbf{x}; \boldsymbol{\xi})}{\partial n} - \alpha v(\mathbf{x}; \boldsymbol{\xi}) \right] \right.$$

$$\left. - v(\mathbf{x}; \boldsymbol{\xi}) \left[\frac{\partial u(\mathbf{x})}{\partial n} - \alpha u(\mathbf{x}) \right] \right\} dS. \tag{6.3.2}$$

The choice of boundary data to be imposed on v depends on the *type* of problem satisfied by u. For each problem type, we require that v be a solution of the *same* type of problem with *homogeneous* boundary data. Thus,

$$v(\mathbf{x}; \boldsymbol{\xi}) = 0, \qquad \mathbf{x} \quad \text{on} \quad \partial D \qquad \text{(Dirichlet)},$$

$$\frac{\partial v(\mathbf{x}; \boldsymbol{\xi})}{\partial n} = 0, \qquad \mathbf{x} \quad \text{on} \quad \partial D \qquad \text{(Neumann)}, \tag{6.3.3}$$

$$\frac{\partial v(\mathbf{x}; \boldsymbol{\xi})}{\partial n} - \alpha v(\mathbf{x}; \boldsymbol{\xi}) = 0, \qquad \mathbf{x} \quad \text{on} \quad \partial D \qquad \text{(mixed)}.$$

Now in (6.3.2) we may solve for u; we delete the arguments under the integrals for brevity:

$$u(\xi) = -\int_D fv \, dV - \int_{\partial D} u \frac{\partial v}{\partial n} \, dS \qquad \text{(Dirichlet)},$$

$$u(\xi) = -\int_D fv \, dV + \int_{\partial D} v \frac{\partial u}{\partial n} \, dS \qquad \text{(Neumann)}, \qquad (6.3.4)$$

$$u(\xi) = -\int_D fv \, dV - \int_{\partial D} \left[\frac{\partial u}{\partial n} - \alpha u \right] v \, dS \qquad \text{(mixed)}.$$

We have now reduced the problem of finding a solution to a general problem to one of finding a Green's function. This is of both theoretical and practical interest. From the theoretical point of view, existence is reduced to the question of existence of a Green's function. From the practical point of view, any proposed solution technique for a given problem ought first to be tried on the Green's function. If the solution technique is approximate—and most are whether they are numerical or analytical—then it had better produce the Green's function with accuracy and stability as desired before proceeding to employ the technique more generally.

We consider now the question of determining the Green's functions for constant q. We consider first the determination of a solution of the equation having the "right" singularity at the source point. The Green's function that does not satisfy any "particular" boundary condition will again be denoted by G, as in the preceding section.

We remark that for q constant, (6.3.1) is rotationally symmetric. That is, as in the preceding section, the solution must be a function of $r = |\mathbf{x} - \xi|$ alone, and therefore we set $G = G(r)$. Indeed, since the most singular part of $\mathscr{L}$ near the source point must come from the Laplacian, the analysis of the behavior of G near this point must be exactly as it was in the preceding section. Thus, we shall consider the equation for G only away from this critical point and seek solutions that are singular at this point. Therefore, we consider the equation

$$\frac{1}{r^{n-1}} \frac{d}{dr} \left[r^{n-1} \frac{dG}{dr} \right] - qG = 0 \quad \text{or} \quad \frac{d^2G}{dr^2} + \frac{n-1}{r} \frac{dG}{dr} - qG = 0. \quad (6.3.5)$$

The case $q = 0$ was solved in Section 6.2; see (6.2.11). For q nonzero, the reader familiar with the method of Frobenius[†] can readily verify that the equation has a regular singular point at the origin and an irregular singular

[†] See, for example, Hildebrand [1962] or Coddington and Levinson [1955].

point at infinity. Thus, the solutions to this equation are related to *Bessel functions.* For q positive, the transformations

$$G(r) = r^{(2-n)/2} G_1(s), \qquad s = \sqrt{q}\, r, \tag{6.3.6}$$

lead to the differential equation for G_1

$$\frac{d^2 G_1}{ds^2} + \frac{1}{s} \frac{dG_1}{ds} - \left[1 + \frac{(n-2)^2}{4s^2} \right] G_1 = 0. \tag{6.3.7}$$

This is the equation for the modified Bessel functions of order $(n-2)/2$ and argument s. (See, for example, Abramowitz and Stegun [1965].) Two solutions of this equation are denoted by I and K. The first of these is a function that is regular at the origin and exponentially growing at infinity. Hence, this solution is not of interest to us here. The second solution is singular at the origin and decays exponentially at infinity. This is the solution we seek; that is,

$$G_1 = C_n K_p(s), \qquad G = C_n r^{-p} K_p(\sqrt{q}\, r), \qquad p = (n-2)/2, \tag{6.3.8}$$

with C_n to be determined.

It is important for us to know the asymptotic behavior of $K_p(\sqrt{q}\, r)$ in the neighborhood of the origin. We express this result in terms of the function G rather than in terms of K_p itself:

$$G(r) = \begin{cases} -C_0 \log r, & n = 2, \\ \dfrac{C_n}{2r^{n-2}} \Gamma\left[\dfrac{n-2}{2}\right] \left[\dfrac{2}{\sqrt{q}}\right]^{(n-2)/2}, & n > 2, \end{cases} \qquad r \to 0. \tag{6.3.9}$$

Also,

$$G(r) \sim \frac{C_n}{r^{(n-1)/2}} \frac{\sqrt{\pi}}{q^{1/4}} \exp(-\sqrt{q}\, r), \qquad r \to \infty. \tag{6.3.10}$$

In order to determine C_n as we determined c_n in the preceding section [in the discussion following (6.2.7)], we also need the result

$$\frac{dG}{dr} \sim -\frac{C_n}{r^{n-1}} \Gamma\left[\frac{n}{2}\right] \left[\frac{2}{\sqrt{q}}\right]^{(n-2)/2}, \qquad n \geq 2. \tag{6.3.11}$$

We now are prepared to repeat the calculation (6.2.8) for this case:

$$-1 = \int_{r=\varepsilon} \frac{dG}{dr} dS = -\int_{r=\varepsilon} \frac{C_n}{r^{n-1}} \Gamma\left[\frac{n}{2}\right] \left[\frac{2}{\sqrt{q}}\right]^p dS$$

$$= -C_n \sqrt{q} \left[\frac{2\pi}{\sqrt{q}}\right]^{n/2}. \tag{6.3.12}$$

From this result, we conclude that

$$C_n = \left[\frac{\sqrt{q}}{2\pi}\right]^{n/2} \frac{1}{\sqrt{q}}$$

and then

$$G(r) = \begin{cases} \dfrac{\sqrt{q}}{2\pi} K_0(\sqrt{q}\, r), & n = 2, \\[2ex] \dfrac{q^{1/4}}{(2\pi)^{3/2}\sqrt{r}} K_{1/2}(\sqrt{q}\, r), & n = 3, \\[2ex] \dfrac{1}{2\pi}\left[\dfrac{\sqrt{q}}{2\pi r}\right]^{p} K_p(\sqrt{q}\, r), & n \geq 2, \quad p = (n-2)/2. \end{cases} \tag{6.3.13}$$

We turn now to the Helmholtz equation; that is, we take q to be given by (6.1.2) and G to satisfy

$$\nabla^2 G + (\omega^2/c^2)G = -\delta(\mathbf{x} - \boldsymbol{\xi}). \tag{6.3.14}$$

As above, we consider the case of constant c and conclude that G is a function of radial distance only. Thus, writing $G = G(r)$, we find that

$$\frac{1}{r^{n-1}}\frac{d}{dr}\left[r^{n+1}\frac{dG}{dr}\right] + \frac{\omega^2}{c^2}G = 0$$

or

$$\frac{d^2G}{dr^2} + \frac{n-1}{r}\frac{dG}{dr} + \frac{\omega^2}{c^2}G = 0. \tag{6.3.15}$$

The transformation

$$G(r) = r^{(2-n)/2}G_1(z), \qquad z = (\omega/c)r, \tag{6.3.16}$$

leads to the differential equation

$$\frac{d^2G_1}{dz^2} + \frac{1}{z}\frac{dG_1}{dz} + \left[1 - \frac{(n-2)^2}{4z^2}\right]G_1 = 0. \tag{6.3.17}$$

This is the differential equation for the ordinary Bessel functions of order $(n-2)/2$ and argument z. The solution that is regular at the origin is denoted by J, the solutions that are singular at the origin are denoted by Y or N and $H^{(1)}$ and $H^{(2)}$. It is these latter two that are of interest to us here. In particular, to leading order, asymptotically,

$$H_p^{(1)}(z) \sim \sqrt{2/(\pi z)}\, e^{i(z - p\pi/2 - \pi/4)}, \qquad -\pi < \arg z < 2\pi, \ |z| \to \infty,$$

$$H_p^{(2)}(z) \sim \sqrt{2/(\pi z)}\, e^{-i(z - p\pi/2 - \pi/4)}, \qquad -2\pi < \arg z < \pi, \ |z| \to \infty, \tag{6.3.18}$$

and

$$H_0^{(1)}(z) \sim -H_0^{(2)}(z) \sim \frac{2i}{\pi} \log z, \qquad\qquad |z| \to 0, \quad -\pi < \arg z \le \pi,$$

$$H_p^{(1)}(z) \sim -H_p^{(2)}(z) \sim -\frac{i}{\pi} \Gamma(p) \left[\frac{z}{2}\right]^{-p}, \quad p \ne 0, \quad |z| \to 0, \quad -\pi < \arg z \le \pi.$$

$$(6.3.19)$$

For z given by (6.3.16), the solution that decays in the upper half ω plane is $H_p^{(1)}$, while the other solution grows in the upper half ω plane and is therefore unacceptable. Therefore,

$$G(r) = C_n r^{-p} H_p^{(1)}(\omega r/c), \qquad p = (n-2)/2, \qquad (6.3.20)$$

with C_n to be determined as it was for the result (6.3.9).

We leave it to the reader to carry out the computation for C_n for this case. The result is

$$C_n = \frac{i}{4} \left[\frac{\omega}{2c\pi}\right]^{(n-2)/2}. \qquad (6.3.21)$$

We conclude then that

$$G(r) = \frac{i}{4} \left[\frac{\omega}{2\pi rc}\right]^p H_p^{(1)}(\omega r/c), \qquad n \ge 2, \quad p = (n-2)/2. \quad (6.3.22)$$

The cases $n = 2$ or 3 are of greatest interest to us. In particular,

$$G(r) = \frac{i}{4} H_0^{(1)}(\omega r/c), \qquad n = 2. \qquad (6.3.23)$$

For $n = 3$, the order of the Hankel function reduces to a half integer. For such values, these functions are expressible in terms of trigonometric functions and powers. Using a standard reference on Bessel functions, we can show that

$$G(r) = \frac{e^{i\omega r/c}}{4\pi r}, \qquad n = 3. \qquad (6.3.24)$$

We have now found the singular part or free-space Green's function for constant q for (6.1.1). Qualitatively, for each n and any choice of q, all of the Green's functions have the same singularity at the source point. Indeed, since this feature arises from the r derivatives of the Laplacian, the same must be true for the case of variable q.

The Green's function v used in the representation of a solution (6.3.4) must also satisfy a boundary condition (6.3.3). Because this problem is linear, let us set

$$v(\mathbf{x}; \boldsymbol{\xi}) = G(\mathbf{x}; \boldsymbol{\xi}) + w(\mathbf{x}; \boldsymbol{\xi}). \qquad (6.3.25)$$

Here in a slight abuse of notation, we have returned to writing G as a function of $\mathbf{x}$ and ξ rather than r because this discussion applies equally to the case of constant and variable q. Since G satisfies the inhomogeneous equation (6.3.1), w must satisfy the homogeneous equation

$$\mathscr{L}w = 0. \qquad (6.3.26)$$

Furthermore, from the boundary condition, one of (6.3.3), for v, w must satisfy the boundary condition

$$w(\mathbf{x};\xi) = -G(\mathbf{x};\xi), \qquad \mathbf{x} \text{ on } \partial D \qquad \text{(Dirichlet)},$$

$$\frac{\partial w(\mathbf{x};\xi)}{\partial n} = -\frac{\partial G(\mathbf{x};\xi)}{\partial n}, \qquad \mathbf{x} \text{ on } \partial D \qquad \text{(Neumann)},$$

$$\frac{\partial w(\mathbf{x};\xi)}{\partial n} - \alpha w(\mathbf{x};\xi) = -\frac{\partial G(\mathbf{x};\xi)}{\partial n} + \alpha G(\mathbf{x};\xi), \qquad \mathbf{x} \text{ on } \partial D \qquad \text{(mixed)}.$$

$$(6.3.27)$$

The question of the existence of a solution to the general boundary value problem (6.1.3) with (6.1.4) or (6.1.5) or (6.1.6), with constant q, is now reduced to the existence of the solution of the homogeneous equation (6.3.26) for w, with one of the boundary conditions in (6.3.27). The proof of existence either for w or for u itself would take us very far afield of the objectives of this text. Therefore, we will not address this matter but will refer the reader to the literature (in particular, Garabedian [1964], Courant and Hilbert [1962], and John [1982]) for extensive discussion of the subject.

Existence proofs that lend themselves to consideration of data in the broadest class of functions are based on results for integral equations. The solution, say, for $w(\mathbf{x};\xi)$, is represented as an integral over the boundary of the Green's function $G(\mathbf{x};\mathbf{x}')$ multiplied by an unknown function. When the observation point $\mathbf{x}$ is moved to the boundary, a *Fredholm integral equation of the second kind* is obtained for the unknown integrand. Existence proofs based on iteration or eigenfunction expansions are then possible for data of a type general enough to encompass models of realistic physical problems.

Exercises

6.7 (a) Use a standard reference on modified Bessel functions and show that for $n = 3$; the result (6.3.13) is

$$G(r) = [\exp(-\sqrt{q}\,r)]/4\pi r.$$

(b) Derive this result by directly solving the differential equation (6.3.5) with $n = 3$.

6.8 The purpose of this exercise is to show how the Green's function representation (6.3.2) can be deduced without the use of distributions. Characterize the Green's function $v(\mathbf{x}; \xi)$ as a solution of the equation

$$\mathscr{L}v(\mathbf{x}; \xi) = 0, \qquad \mathbf{x} \neq \xi,$$

having the singular behavior as $r = |\mathbf{x} - \xi| \to 0$:

$$v(r) \sim \frac{-1}{2\pi} \log r, \qquad\qquad \text{2 dimensions},$$

$$v(r) \sim \frac{\Gamma(m/2)}{2(m-2)\pi^{m/2}r^{m-2}}, \qquad m > 2 \text{ dimensions},$$

$$\frac{\partial v}{\partial r} \sim \frac{-\Gamma(m/2)}{2\pi^{m/2}r^{m-1}}, \qquad m \geq 2.$$

(a) Replace the finite domain D by D', created from D by excluding a small sphere S_ε around $\mathbf{x} = \xi$, assumed to be an interior point of D. Assume that u satisfies (6.1.3) in D. Apply (6.1.10) to u and v on the domain D' to conclude that

$$-\int_{D'} v(\mathbf{x}; \xi) f(\mathbf{x}) \, dV = \int_{\partial D} \left[u(\mathbf{x}) \frac{\partial v(\mathbf{x}; \xi)}{\partial n} - v(\mathbf{x}; \xi) \frac{\partial u(\mathbf{x})}{\partial n} \right] dS$$

$$- \int_{\partial S_\varepsilon} \left[u(\mathbf{x}) \frac{\partial v(\mathbf{x}; \xi)}{\partial n} - v(\mathbf{x}; \xi) \frac{\partial u(\mathbf{x})}{\partial n} \right] dS.$$

(b) Assume that u and its normal derivative are bounded and smooth in D. Use the estimates in part (a) and the result of Exercise 6.4c to conclude the result (6.3.2) by taking the limit as the radius of the sphere S_ε approaches zero.

6.4 PROBLEMS IN UNBOUNDED DOMAINS AND THE SOMMERFELD RADIATION CONDITION

In Section 5.4, we introduced the concept of scattering problems for which the domain of interest was unbounded. In this section, we shall discuss such problems in unbounded domains for functions that are solutions of the inhomogeneous Helmholtz equation

$$\mathscr{L}u(\mathbf{x}; \omega) = \nabla^2 u(\mathbf{x}; \omega) + (\omega^2/c^2)u(\mathbf{x}; \omega) = f(\mathbf{x}; \omega). \qquad (6.4.1)$$

Here the source term f will be assumed to be nonzero at most in some finite domain; that is, f has *compact support* in space. The question arises of how one assures uniqueness in an unbounded domain. In all of the examples discussed up to this point, we have done so by designing a problem in the time domain for which the solution started at some finite time; that is, the solution was *causal*. Then in the frequency domain, it was required that the solution be analytic in some upper half ω plane.

Indeed, this condition will suffice to make the solution of the Helmholtz equation in an unbounded domain with data—source and boundary values confined to a finite domain—unique. However, this condition is, in some sense, *unphysical*. It is possible to develop a criterion for uniqueness more closely tied to the property of "radiation" or propagation outward from the finite domain of the data for ω "large enough" but real. This is the *Sommerfeld radiation condition*. It is this condition (actually, two conditions) that will be developed and discussed in this section.

Let us begin by considering the free-space Green's function in three dimensions (6.2.23) and note that the solution in the time domain will be a superposition over frequencies of the function

$$\bar{G}(r, t) = G(r)e^{-i\omega t} = \frac{e^{i\omega(r/c - t)}}{4\pi r}. \tag{6.4.2}$$

We note that the surfaces of constant phase for this function are spheres whose radii increase with time. Indeed, Fourier inversion yields the distribution (2.4.13), which is nonzero only on these same expanding spheres. Thus, this function in the time domain certainly satisfies our notion of a function that is *outgoing*; that is, a function that represents a wave propagating outward toward infinity. The alternative to this solution was rejected in the discussion following (6.3.18) on the basis that it was not analytic in an upper half ω plane. We remark that, in the time domain, this second solution would represent a wave that propagates inward for time in the range $-\infty < t < 0$, collapsing on the origin at $t = 0$, and is zero thereafter. Thus, our earlier criterion did pick out the outward-propagating Green's function and reject the inward-propagating one.

Far away from the domain containing the data, the solution to a problem of the type we defined in the first paragraph of this section should behave qualitatively as this Green's function does. The reason is that any finite domain when measured on the scale of the range to an observation point receding toward infinity shrinks (in relative size) to zero. Thus, it would be reasonable to expect the solution to behave in the same way as the Green's function in (6.4.2) in the sense that it should represent a wave propagating radially outward from any origin in the finite part of the plane. Of course, the amplitude of this radially propagating wave would be expected to have angular dependence arising from the specific properties of the data.

Examination of (6.3.23) or (6.4.2) suggests as one criterion that the solution must decay to zero at the rate of $1/r$ as $r \to \infty$. The second criterion must characterize the direction of propagation, which arises from the r derivative of the phase. In general, it is not practical to think in terms of differentiating only the phase, since that need not be an identifiable expression for the solution of an arbitrary problem. However, we can see that for G, differentiation with respect to r leads to one term of order $1/r$ arising from differentiation of the phase and another of order $1/r^2$ from differentiation of the amplitude. Furthermore, the former term is a multiple of G itself, which for inward-propagating waves would be of *opposite sign*. Thus, we characterize the outward propagation of G by the two criteria

$$u(\mathbf{x}; \omega) = O(1/r), \qquad u_r(\mathbf{x}; \omega) - (i\omega/c)u(\mathbf{x}; \omega) = o(1/r), \qquad r \to \infty. \quad (6.4.3)$$

Here r is the radial variable measured from any point in the finite part of the plane. (See Exercise 6.9.) These equations comprise the Sommerfeld radiation condition in three space dimensions.[†] We state the main conclusion of this section in the following theorem.

Theorem Suppose that u is a solution of (6.4.1) in an unbounded domain D with one or more finite "holes" on the boundary of which either Dirichlet or Neumann boundary conditions are prescribed. Furthermore, we suppose that u satisfies the Sommerfeld condition [(6.4.3)]. Then u is unique.

Corollary For the mixed boundary condition (6.1.15), the solution is also unique. Note here that the normal derivative pointing out of D points *into* each of the finite "scatterers" with boundary ∂D.

We shall prove this theorem and corollary at the end of this section. However, we shall first examine some consequences of the radiation condition, along the way establishing some preliminary results needed for the proof.

We introduce a sphere S_1 centered at the origin of coordinates and containing all of the data for u. Thus, outside of this sphere, u satisfies a homogeneous equation with constant coefficient c. We suppose that u is to be evaluated at some point $\mathbf{x}$ outside this sphere. We introduce a second sphere S_2 large enough to contain $\mathbf{x}$ as well. See Fig. 6.1.

We now apply Green's theorem (6.1.10) to u and G, given by (6.3.23) over the domain between S_1 and S_2. The result is[‡]

$$u(\xi) = \int_{\partial S_1} \left[G(\mathbf{x}; \xi) \frac{\partial u(\mathbf{x})}{\partial n} - u(\mathbf{x}) \frac{\partial G(\mathbf{x}; \xi)}{\partial n} \right] dS$$

$$+ \int_{\partial S_2} \left[G(\mathbf{x}; \xi) \frac{\partial u(\mathbf{x})}{\partial n} - u(\mathbf{x}) \frac{\partial G(\mathbf{x}; \xi)}{\partial n} \right] dS. \quad (6.4.4)$$

[†] Courant and Hilbert [1962] have noted that under an integral form of the radiation condition, we can actually conclude that the term $o(1/r)$ in the second condition here can be shown to be $O(1/r^2)$ in general, as is the case specifically for the Green's function.

[‡] The explicit dependence of u on ω is not important here and is therefore omitted.

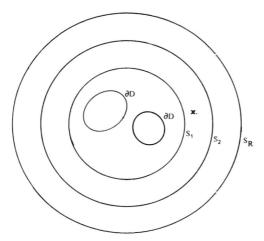

Fig. 6.1. Spheres containing data (S_1) and observation points (S_2); sphere of increasing radius (S_R).

The Sommerfeld radiation condition (6.4.3) guarantees that the second integral here must be zero. This can be seen in the following manner. First, introduce a still larger sphere S_R whose radius R will ultimately be allowed to approach infinity. Because the singularity of the Green's function lies inside of S_2, applying Green's theorem to the region between S_2 and S_R shows the equality of the surface integrals:

$$\int_{S_2} \left[G(\mathbf{x}; \xi) \frac{\partial u(\mathbf{x})}{\partial n} - u(\mathbf{x}) \frac{\partial G(\mathbf{x}; \xi)}{\partial n} \right] dS$$

$$= \int_{\partial S_R} \left[G(\mathbf{x}; \xi) \frac{\partial u(\mathbf{x})}{\partial n} - u(\mathbf{x}) \frac{\partial G(\mathbf{x}; \xi)}{\partial n} \right] dS. \qquad (6.4.5)$$

Here the direction of the normal on the left side is taken to be the same as it was in (6.4.4), that is, outward to S_2, although this is opposite to the direction it would have in the application of Green's theorem. This results in like signs on both sides of this equation when the normal in the second line is taken as the ordinary outward normal. We remark that in these surface integrals, the normal direction is merely the radial direction. Thus,

$$G \frac{\partial u}{\partial n} - u \frac{\partial G}{\partial n} = G \left[\frac{\partial u}{\partial r} - \frac{i\omega}{c} u \right] - u \left[\frac{\partial G}{\partial r} - \frac{i\omega}{c} G \right]$$

$$= o\left(\frac{1}{r^2} \right), \qquad r \to \infty. \qquad (6.4.6)$$

The last part of the equation here follows from applying the Sommerfeld condition (6.4.3). Each term on the right side here contains one factor that is $O(1/r)$ and one factor that is $o(1/r)$; hence, the entire expression is as stated. We note further that the surface area on the sphere S_R is $O(R^2)$. Thus, the integral over ∂S_R is $o(1)$ in R; that is, the integral decays to zero as $R \to \infty$. Consequently, the integral on the left in (6.4.5), being independent of R, must be identically equal to zero.

This leads us to the following integral representation for $u(\mathbf{x})$ outside the region of nonzero data:

$$u(\xi) = \int_{\partial S_1} \left[G(\mathbf{x}; \xi) \frac{\partial u(\mathbf{x})}{\partial n} - u(\mathbf{x}) \frac{\partial G(\mathbf{x}; \xi)}{\partial n} \right] dS. \tag{6.4.7}$$

We remind the reader that in this region outside the support of the data, the Green's function is given by

$$G(\mathbf{x}; \xi) = \frac{e^{i\omega|\mathbf{x} - \xi|/c}}{4\pi|\mathbf{x} - \xi|}. \tag{6.4.8}$$

Let us introduce (ρ, θ, ϕ) as polar coordinates for the point ξ and use the binomial expansion to approximate $|\mathbf{x} - \xi|$ for values of ρ large compared to $|\mathbf{x}| = x$. The expansion is

$$|\xi - \mathbf{x}| = [\rho^2 - 2\xi \cdot \mathbf{x} + x^2]^{1/2} = \rho \left[1 - \frac{2\xi \cdot \mathbf{x}}{r^2} + \frac{x^2}{\rho^2} \right]^{1/2}$$

$$= \rho - \hat{\xi} \cdot \mathbf{x} + \sum_{p=1}^{\infty} \frac{a_p(\theta, \phi; \mathbf{x})}{\rho^p}. \tag{6.4.9}$$

Here in the last line, we use $\hat{\xi}$ to denote a unit vector in the direction of ξ. Also, more explicit information about the coefficients in the sum does not concern us. By inverting this series, we also obtain the result that

$$\frac{1}{|\xi - \mathbf{x}|} = \frac{1}{\rho} \left[1 + \sum_{p=1}^{\infty} \frac{b_p(\theta, \phi; \mathbf{x})}{\rho^p} \right]. \tag{6.4.10}$$

By substituting these series into the representation (6.4.7) and integrating term by term, we obtain the following representation for the outward radiating field:

$$u(\xi) = \frac{e^{i\omega\rho/c}}{4\pi\rho} \sum_{p=1}^{\infty} \frac{f_p(\theta, \phi)}{\rho^p}. \tag{6.4.11}$$

In this equation, the leading coefficient is called the *far-field scattering amplitude* given by

$$f_0 = \int_{\partial S_1} \left[\frac{\partial u}{\partial n} - i\omega\hat{\xi} \cdot \hat{n}u \right] e^{i\omega\hat{\xi} \cdot \mathbf{x}/c} \, dS. \tag{6.4.12}$$

Here we have used the notation $\hat{n}$ to denote a unit outward normal vector to the surface S_1. By substituting the functional form (6.4.11) into the Helmholtz equation in polar coordinates and equating the coefficient of each power of ρ equal to zero, we obtain the following *recursion relation* for the coefficients in the expansion:

$$\frac{2i\omega}{c}(p + 1)f_{p+1} = p(p + 1)f_p + \frac{1}{\sin\theta}\frac{\partial}{\partial\theta}\left[\sin\theta\frac{\partial f_p}{\partial\theta}\right] + \frac{1}{\sin^2\theta}\frac{\partial^2 f_p}{\partial\phi^2},$$

$$p \geq 0. \qquad (6.4.13)$$

An important consequence of this result is that the radiated field will be identically zero if only f_0 is zero. That is, there is no nontrivial outward-radiated field with algebraic decay faster than $1/\rho$! We remark, however, that the k vector might be complex and thus lead to exponential decay; still, the amplitude must be as stated.

Let us now consider an observation point ξ, which is inside the sphere S_1. We shall apply Green's theorem (6.1.10) to the region D', which is the domain D restricted to the interior of S_1. We deduce that the boundary integral over ∂S_1 is zero now, just as was the integral over ∂S_2 earlier. Thus, we conclude that u satisfies the first line of (6.3.2) but in terms of the outward-radiating Green's function G in place of v:

$$-u(\xi) - \int_D G(\mathbf{x};\xi)f(\mathbf{x})\,dV = \int_{\partial D}\left[u(\mathbf{x})\frac{\partial G(\mathbf{x};\xi)}{\partial n} - G(\mathbf{x};\xi)\frac{\partial u(\mathbf{x})}{\partial n}\right]dS. \quad (6.4.14)$$

We do not know both u and its normal derivative on the boundary surfaces. Hence, this equation does not provide a solution formula for u. However, it does provide a point of departure for solution techniques, either by numerical approximation or asymptotic approximation. We shall not discuss the former, but we shall discuss the latter in subsequent sections and chapters.

We turn now to the consideration of the energy in a wave field. We suppose that (6.4.1) arose as the Fourier transform of a problem in the time domain for a function $U(\mathbf{x}, t)$ after Fourier transform as given by

$$u(\mathbf{x};\omega) = \int_0^\infty U(\mathbf{x}, t)e^{i\omega t}\,dt. \qquad (6.4.15)$$

The function U satisfies the equation

$$\nabla^2 U - \frac{1}{c^2}\frac{\partial^2 U}{\partial t^2} = F(\mathbf{x}, t), \qquad (6.4.16)$$

with prescribed initial and boundary data. We remark that the source for (6.4.1) is a linear combination of the Fourier transform of the source F and the initial data; see, for example, (4.3.2).

Again, we use the domain D', D truncated to the interior of S_1, and define the energy at time t, $E(t)$, by the integral

$$E(t) = \frac{1}{2} \int_{D'} \left[\frac{U_t^2}{c^2} + |\nabla U|^2 \right] dV. \qquad (6.4.17)$$

We take the derivative of this expression with respect to t and use (6.4.15) and the divergence theorem as follows:

$$\frac{\partial E}{\partial t} = \int_{D'} \left[\frac{U_t U_{tt}}{c^2} + \nabla U \cdot \nabla U_t \right] dV = \int_{D'} [U_t \nabla^2 U + \nabla U \cdot \nabla U_t - U_t F] \, dV$$

$$= \int_{D'} [\nabla \cdot [U_t \nabla U] - U_t F] \, dV = \int_{\partial D} U_t \frac{\partial U}{\partial n} \, dS$$

$$+ \int_{\partial S_1} U_t \frac{\partial U}{\partial n} \, dS - \int_D U_t F \, dV. \qquad (6.4.18)$$

In the last part of the equation here, we see that the time rate of change of energy is made up of three terms. We interpret the first term as the flux (influx, positive; efflux, negative) of energy through the boundary ∂D; the second term is the flux of energy outward from the finite support of the data; the third term is the flux of energy due to the source. In this last term, we are justified in replacing D' by D because the source is nonzero only in D' and the integrals over these two domains are equal.

Let us focus our attention now on the second term, the radiated energy. Therefore, we set

$$\frac{\partial E_{\text{rad}}}{\partial t} = \int_{\partial S_1} U_t \frac{\partial U}{\partial n} \, dS. \qquad (6.4.19)$$

We note that the normal direction here is radially outward; hence, the normal derivative in this expression is just the radial derivative. By taking the Fourier transform of the Sommerfeld condition (6.4.3), we obtain the following radiation condition in the time domain:

$$U(\mathbf{x}, t) = O(1/r), \qquad U_r(\mathbf{x}, t) = [-U_t(\mathbf{x}, t)/c] + o(1/r), \qquad r \to \infty. \quad (6.4.20)$$

Substitution of this result into (6.4.19) yields

$$\frac{\partial E_{\text{rad}}}{\partial t} = - \int_{\partial S_1} \frac{U_t^2}{c} [1 + o(1)] \, dS, \qquad r \to \infty. \qquad (6.4.21)$$

Thus, for r large enough, we conclude that *the rate of change of energy through the boundary ∂S_1 is negative.* That is, the domain D' is "losing" energy toward infinity when the wave field satisfies the Sommerfeld radiation condition.

Thus, solutions satisfying the Sommerfeld condition behave in a manner that is reasonable on physical grounds.

We close this section with proofs of the theorem on p. 182 and its corollary.

Proof of Theorem   Let us suppose that the problem (6.4.1) and (6.4.3), with all the data nonzero only in a finite domain, has two solutions u_1 and u_2. We define u then to be the difference of these two solutions. Thus, all of the data for u are equal to zero. We again consider the domain D'—D truncated to the interior of a sphere S_1—and apply Green's theorem to the pair u and u^*:

$$0 = \int_{D'} [u^* \mathscr{L}u - u\mathscr{L}u^*]\, dS = \int_{\partial D + \partial S_1} \left[u^* \frac{\partial u}{\partial n} - u \frac{\partial u^*}{\partial n} \right] dS$$

$$= \int_{\partial S_1} \left[u^* \frac{\partial u}{\partial n} - u \frac{\partial u^*}{\partial n} \right] dS. \tag{6.4.22}$$

In going from the first line to the second line, we have used the fact that u has either zero Dirichlet data or zero Neumann data. We apply the Sommerfeld condition (6.4.3) to conclude that

$$0 = \frac{2i\omega}{c} \int_{\partial S_1} u^* u \left[1 + o\!\left(\frac{1}{r^2}\right) \right] dS. \tag{6.4.23}$$

We use the representation (6.4.11) and let $r \to \infty$ to conclude that

$$0 = \int_{\Omega} f_0^* f_0 \, d\Omega. \tag{6.4.24}$$

Here Ω denotes the solid angle on the unit sphere, and we have set $dS = r^2 d\Omega$ in (6.4.23). From this result, we conclude that f_0 must be zero. However, we have already seen that this implies that the wave field is identically zero in the domain outside of S_1 and on the boundary ∂S_1. Thus, both u and its normal derivative are zero on ∂S_1. Now (6.4.14) implies that u is identically equal to zero. This is what we wanted to prove for u; the solution must be unique.

Proof of the Corollary   Now let us suppose that u satisfies a homogeneous mixed boundary condition (6.1.5), with $g = 0$ and $Z > 0$ on $\partial D''$, which is all or part of ∂D. Then from the middle part of (6.4.22) and from (6.4.23), we conclude that

$$0 = \frac{2i\omega}{c} \int_{\partial S_1} u^* u \left[1 + o\!\left(\frac{1}{r^2}\right) \right] dS + 2i\omega Z \int_{\partial D''} u^* u \, dS. \tag{6.4.25}$$

Now as we let the radius of ∂S_1 approach infinity, we see that both terms are imaginary and of the same sign. Thus, in order for the sum to be zero, they must separately be zero; u is zero on ∂D, and the proof is reduced to the preceding case. Thus, u is zero, and again the solution is unique.

Exercises

6.9 (a) Let $r = |\xi - x|$, with x a fixed point and ξ of much larger magnitude ρ than x, the magnitude of x. Show that

$$\partial r/\partial \rho = 1 + O(1/\rho^2) = 1 + O(1/r^2), \qquad \rho \to \infty.$$

(b) What are the implications of this result as regards the origin of coordinates in the Sommerfeld condition (6.4.3)?

6.10 The purpose of this exercise is to demonstrate how the solution of a Dirichlet boundary value problem for the Helmholtz equation is reduced to the solution of a Fredholm integral equation of the second kind. See Fig. 6.2. Let S_1 denote the disc of radius ε in Fig. 6.2, and define w_1 by

$$w_1(\mathbf{x}) = \int_{S_1} g(\mathbf{x}') \frac{\partial G(\mathbf{x}'; \mathbf{x})}{\partial n'} dS.$$

(a) Use the coordinate system of the figure and verify that

$$dS = r \, dr \, d\theta = \mu^2 \sec^2 \phi \tan \phi \, d\theta \, d\phi.$$

(b) Show that

$$w_1(\mathbf{x}) = \frac{1}{4\pi} \int_{S_1} g(\mathbf{x}_0) \cos \phi \, e^{i\omega|\mathbf{x} - \mathbf{x}'|/c}[1 + O(\varepsilon)][1 + O(\mu)] \, d\theta \, d\phi.$$

(c) Conclude from (b) that the one-sided limit

$$w_1(\mathbf{x}_0) \equiv \lim_{\mu \to 0} w_1(\mathbf{x}) = g(\mathbf{x}_0)[1 + O(\varepsilon)].$$

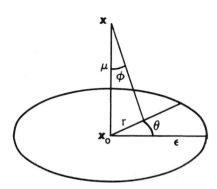

Fig. 6.2. Local coordinate system for Exercise 6.10.

(d) Let D be some exterior domain with a single boundary surface ∂D. Define

$$w(\mathbf{x}) = \int_{\partial D} g(\mathbf{x}') \frac{\partial G(\mathbf{x}'; \mathbf{x})}{\partial n'} dS.$$

Show that w is an outgoing solution of the Helmholtz equation.

(e) For $\mathbf{x}$ "near" ∂D, decompose the boundary into a part S_1 as introduced in (a) and its complement S_2. (This decomposition is only approximate, since S_1 is not flat in this application.) Conclude that

$$w(\mathbf{x}_0) = \frac{g(\mathbf{x}_0)}{2}[1 + O(\varepsilon)] + \int_{S_2} g(\mathbf{x}') \frac{\partial G(\mathbf{x}'; \mathbf{x}_0)}{\partial n} dS.$$

(f) Take the limit as $\varepsilon \to 0$. Explain why the integral remains finite; confirm that the normal derivative is actually $o(1/|\mathbf{x}' - \mathbf{x}_0|^2)$ and hence that the integral remains convergent in this limit. Thus, obtain the integral equation

$$w(\mathbf{x}_0) = \frac{g(\mathbf{x}_0)}{2} + \int_{\partial D} g(\mathbf{x}') \frac{\partial G(\mathbf{x}'; \mathbf{x}_0)}{\partial n} dS.$$

This is the Fredholm integral equation for the unknown function g when Dirichlet data for w are given. A similar equation can be obtained when Neumann data are given. (See Garabedian [1964].)

6.11 (a) The uniqueness proof used in this section for the three-dimensional problem can also be applied to the one-dimensional problem. That is, suppose that u is a solution to the equation

$$u''(x, \omega) + (\omega^2/c^2) u(x, \omega) = f(x, \omega), \qquad x_0 < x < \infty,$$

subject to a Neumann, Dirichlet, or mixed boundary condition at x_0; the last of these of the form

$$-u'(x, \omega) - i\omega Z u(x, \omega) = g(\omega), \qquad x = x_0,$$

and a Sommerfeld condition

$$u(x, \omega) = O(1), \qquad u'(x, \omega) - i\omega u(x, \omega)/c = o(1), \qquad x \to \infty.$$

Suppose further that the data—f, g, and nonconstant c—are confined to some finite domain in D. Then follow the outline of the proof of this section to prove that u is unique. [*Remark*: The choice of sign in the mixed boundary condition is equivalent to an *outward* normal derivative at the boundary point.]

(b) Repeat (a) for a left semi-infinite domain $-\infty < x < x_0$. For this case, the mixed boundary condition and Sommerfeld condition must be replaced by, respectively,

$$u'(x, \omega) - i\omega Z u(x, \omega) = g(\omega), \qquad x = x_0,$$

and

$$u(x, \omega) = O(1), \qquad -u'(x, \omega) - i\omega u(x, \omega)/c = o(1), \qquad x \to \infty.$$

6.12   The purpose of this exercise is to prove uniqueness of the exterior problem in two dimensions under the appropriate Sommerfeld condition. Suppose that u satisfies (6.4.1) in two space dimensions in some exterior domain D with Dirichlet or Neumann boundary conditions and the Sommerfeld radiation condition

$$u = O(r^{-1/2}), \qquad u_r - i\omega u/c = o(r^{-1/2}), \qquad r \to \infty,$$

with r the polar distance from some finite point in two dimensions. For uniqueness, we consider a problem with zero data; the objective is to prove that the solution must then be zero.

(a)   As in the discussion of this section, prove that

$$\lim_{R \to \infty} \int_S u^* u \, ds = 0.$$

Here the path of integration is a circle of radius R, and the integral is with respect to arc length on that circle.

(b)   Outside of a sphere containing the finite boundaries of D, assume that u has a Fourier series solution

$$u = \sum_{p=-\infty}^{\infty} a_p(r) e^{ip\theta}$$

in the polar coordinates r and θ. Show that

$$a_p = b_p H_p^{(1)}(\omega r/c),$$

with b_p constant and $H_p^{(1)}$ the Hankel function of the first kind of order p.

(c)   Use (b) in (a) to deduce that

$$\sum_{p=-\infty}^{\infty} |b_p|^2 = 0$$

and thus that

$$b_p = 0, \qquad \text{all} \quad p.$$

(d)   As in the discussion of this section, deduce now that u must be identically zero and, thus, that the Dirichlet and Neumann problems have unique solutions.

(e)   Extend to the mixed boundary value problem.

6.13   Use spherical harmonics and the method of proof of 6.12 in three dimensions.

6.14   Show that the function

$$u = [\sin(\omega r/c)]/r$$

is a solution of the homogeneous Helmholtz equation in three dimensions.

(a)   Show that this solution does not satisfy the Sommerfeld condition (6.4.3).

(b)   Show that the function

$$u = [\sin(\omega r/c)]/r - [\sin(\omega a/c)]/a$$

is a solution of the Helmholtz equation in the exterior of the sphere of radius a, satisfying a homogeneous Dirichlet condition on the boundary $r = a$.

(c)   What is the relevance of this result as regards the Sommerfeld condition and uniqueness?

(d)   Construct examples like these in two dimensions.

(e)   Define the functions

$$u_p = \left[H_p^{(1)}\left(\frac{\omega r}{c}\right) - \frac{J_p'(\omega a/c)}{H_p^{(1)\prime}(\omega a/c)} J_p\left(\frac{\omega r}{c}\right) \right] e^{ip\theta}.$$

Here the prime denotes differentiation with respect to the argument. Show that for $r \neq 0$ these functions satisfy the homogeneous Helmholtz equation in two dimensions and a homogeneous Neumann boundary condition on the sphere $r = a$.

(f)   Construct a similar class of functions for the Dirichlet boundary condition.

(g)   Repeat the preceding two examples in three dimensions.

6.15   (a)   Verify that

$$(1 + z)^{1/2} = \sum_{k=1}^{\infty} \frac{\Gamma(\frac{1}{2})}{\Gamma(\frac{1}{2} - k)} \frac{z^k}{k!}, \qquad |z| < 1.$$

(b)   Prove that (6.4.9) converges for $x/\rho < 1$ uniformly in θ and ϕ and that

$$\left| \sum_{p=1}^{\infty} \frac{a_p}{\rho^p} \right| \leq \frac{2x}{\rho}.$$

(c)   Prove that, similarly, (6.4.10) converges for $x/\rho < 1$ uniformly in θ and ϕ, with an analogous estimate for the corresponding sum.

(d)   Show that the expansion of (6.4.8) obtained by using (6.4.9) and (6.4.10) converges for $x/\rho \leq \frac{1}{4}$ uniformly in θ, ϕ, and $\mathbf{x}$.

(e)   Conclude that (6.4.11) is convergent for ρ at least four times the diameter of the support of the data for the problem for u.

6.5 SOME EXACT SOLUTIONS

We shall discuss here a few examples of exact solutions[†] of the Helmholtz equation and as part of that discussion, introduce representations of solutions in cylindrical and spherical coordinates. In this context, we shall introduce the special functions appropriate to solutions in these coordinate systems. It is not our intent to be expository in the discussion of these special functions, but only informative. More complete discussions can be found in texts devoted to special functions, such as those of Watson [1966], Wittaker and Watson [1958], and Lebedex [1972]. The last discusses these special functions in the context of applications, as do texts such as those by Jackson [1975] and Stratton [1941], among others. There are also compendia of information on special functions, such as those by Abramowitz and Stegun [1965], Erdélyi *et al.* [1953], and Gradshteyn and Ryzhik [1965].

We shall be concerned in this section with the homogeneous Helmholtz equation

$$\nabla^2 u(\mathbf{x}; \omega) + (\omega^2/c^2)u(\mathbf{x}; \omega) = 0 \qquad (6.5.1)$$

in unbounded space (unless otherwise specified) in two or three space dimensions, with constant c.

PLANE WAVES

We begin by assuming a plane-wave solution to (6.5.1), namely,

$$u = e^{i\mathbf{k}\cdot\mathbf{x}}, \qquad (6.5.2)$$

with the vectors being two- or three-dimensional as dictated by the equation itself. Substitution into (6.5.1) reveals that this is a solution if

$$\mathbf{k}\cdot\mathbf{k} = k^2 = \omega^2/c^2. \qquad (6.5.3)$$

More general solutions can now be obtained by Fourier superposition, that is,

$$u(\mathbf{x}; \omega) = \int_{k^2 = \omega^2/c^2} A(\mathbf{k})e^{i\mathbf{k}\cdot\mathbf{x}} \, dS_k. \qquad (6.5.4)$$

The integral is over (all or part of) a sphere in three dimensions or (all or part of) a circle in two dimensions. The function $A(\mathbf{k})$ is one for which the integral makes sense either as an ordinary function or as a distribution. That this is a formal solution can be checked by direct substitution into (6.5.1). (In fact, that is what I mean by formal solution!)

[†] A useful compendium of exact and asymptotic solutions has been provided by Bowman *et al.* [1969].

Recall that the convention we have adopted for our Fourier transform requires multiplication of a solution in frequency domain by $\exp(-i\omega t)$. Thus, in (6.5.2), the planes of constant phase propagate in the direction of $\mathbf{k}$ sign ω.

Let us consider now a plane wave incident from $x_1 < 0$ on the plane $x_1 = 0$. This requires that sign $k_1 = \text{sgn } \omega$. Let us suppose that c in (6.5.1) is replaced by the value c_1 in $x_1 > 0$ and that the total solution and its normal derivative must be continuous across $x_1 = 0$. Then let us set

$$u = \begin{cases} u_1 + u_R, & x_1 < 0, \\ u_T, & x_1 > 0, \end{cases}$$

$$(6.5.5)$$

$$u_1 = e^{i\mathbf{k}\cdot\mathbf{x}}, \qquad k_1 = \text{sign } \omega \sqrt{(\omega^2/c^2) - k_2^2 - k_3^2}.$$

In the last line, the square root is positive when it is real; otherwise, k_1 is positive imaginary. In this manner, U_I is a wave that either propagates toward the interface or attenuates in the direction of the interface.

We assume that the reflected wave u_R and the transmitted wave u_T are also plane waves:

$$u_R = R e^{i\mathbf{k}^R\cdot\mathbf{x}}, \qquad [k^R]^2 = \frac{\omega^2}{c^2}, \qquad u_T = T e^{i\mathbf{k}^T\cdot\mathbf{x}}, \qquad [k^T]^2 = \frac{\omega^2}{c_1^2}, \quad (6.5.6)$$

with R and T constant. Furthermore, we seek solutions that radiate away from $x_1 = 0$. Thus,

$$\text{sign } k_1^R = -\text{sign } \omega \qquad \text{if} \quad k_1^R \text{ is real}, \qquad \text{Im } k_1^R < 0 \qquad \text{otherwise,}$$

$$\text{sign } k_1^T = \text{sign } \omega \qquad \text{if} \quad k_1^T \text{ is real}, \qquad \text{Im } k_1^T > 0 \qquad \text{otherwise.}$$

$$(6.5.7)$$

The continuity of u and its normal derivative across $x_1 = 0$ leads to

$$\begin{aligned} e^{i\mathbf{k}\cdot\mathbf{x}} + R e^{i\mathbf{k}^R\cdot\mathbf{x}} &= T e^{i\mathbf{k}^T\cdot\mathbf{x}}, \\ k_1 e^{i\mathbf{k}\cdot\mathbf{x}} + k_1^R R e^{i\mathbf{k}^R\cdot\mathbf{x}} &= k_1^T T e^{i\mathbf{k}^T\cdot\mathbf{x}}, \end{aligned} \qquad x_1 = 0. \qquad (6.5.8)$$

The phases here must agree. If they did not, then one could deduce that R and T were nonconstant. (See Exercise 6.16.) This would contradict the basic assumption of the form of solution. Thus, the transverse (with respect to the interface) components of the wave vector must agree:

$$k_2 = k_2^R = k_2^T, \qquad k_3 = k_3^R = k_3^T, \qquad (6.5.9)$$

with the second equation here relevant only to the three-dimensional case.

We now determine k_1^R and k_1^T consistent with these results, the constraints on $[k^R]^2$ and $[k^T]^2$ in (6.5.6), and the conditions (6.5.7). The results are

$$k_1^R = -k_1, \qquad k_1^T = \text{sign } \omega \sqrt{(\omega^2/c_1^2) - k_2^2 - k_3^2}. \qquad (6.5.10)$$

We remark that the second equation here along with (6.5.7) define k_1^T both for real and imaginary values of the square root.

Now (6.5.8) is a pair of equations for R and T:

$$1 + R = T, \qquad 1 - R = (k_1^T/k_1)T, \tag{6.5.11}$$

with solutions

$$R = (k_1 - k_1^T)/(k_1 + k_1^T), \qquad T = 2k_1/(k_1 + k_1^T). \tag{6.5.12}$$

For the special case of normal incidence, $k_2 = k_3 = 0$, $k_1^2 = \omega^2/c^2$, and

$$k_1^T = \omega/c_1, \qquad R = (c_1 - c)/(c_1 + c), \qquad T = 2c_1/(c_1 + c). \tag{6.5.13}$$

CYLINDRICAL WAVES

Let us consider the Helmholtz equation in cylindrical coordinates (ρ, ϕ, z):

$$\frac{1}{\rho}\frac{\partial}{\partial\rho}\left[\rho\frac{\partial u}{\partial\rho}\right] + \frac{1}{\rho^2}\frac{\partial^2 u}{\partial\phi^2} + \frac{\partial^2 u}{\partial z^2} + k^2 u = 0, \qquad k^2 = \frac{\omega^2}{c^2}. \tag{6.5.14}$$

We seek solutions of the form

$$u = R(\rho)\Phi(\phi)Z(z) \tag{6.5.15}$$

and find that

$$\frac{1}{R}\rho\frac{d}{d\rho}\left[\rho\frac{dR}{d\rho}\right] + \frac{1}{\Phi}\frac{d^2\Phi}{d\phi^2} + \frac{\rho^2}{Z}\frac{d^2 Z}{dx^2} + k^2\rho = 0. \tag{6.5.16}$$

The second term is a function of ϕ alone, and the remaining terms are independent of ϕ; thus, each of these must be constant and negative of one another:

$$\frac{1}{\Phi}\frac{d^2\Phi}{d\phi^2} = -n^2; \qquad \frac{1}{R}\rho\frac{d}{d\rho}\left[\rho\frac{dR}{d\rho}\right] + \frac{\rho^2}{Z}\frac{d^2 Z}{dz^2} + k^2\rho^2 = n^2. \tag{6.5.17}$$

We now rewrite the second line in this equation as

$$\frac{1}{R}\frac{d}{d\rho}\left[\rho\frac{dR}{d\rho}\right] + \left[k^2 - \frac{n^2}{\rho^2}\right] = \frac{1}{Z}\frac{d^2 Z}{dz^2} \tag{6.5.18}$$

and use the same argument to conclude that

$$\frac{1}{Z}\frac{d^2 Z}{dz^2} = -m^2; \qquad \frac{1}{\rho R}\frac{d}{d\rho}\rho\frac{dR}{d\rho} + \left[k^2 - \frac{n^2}{\rho^2}\right] = m^2, \tag{6.5.19}$$

with m constant, too.

We rewrite the equations for Φ, Z, and R as

$$\frac{d^2\Phi}{d\phi^2} + n^2\Phi = 0; \tag{6.5.20}$$

$$\frac{d^2Z}{dz^2} + m^2Z = 0; \tag{6.5.21}$$

$$\frac{1}{\rho}\frac{d}{d\rho}\left[\rho\frac{dR}{d\rho}\right] + \left[k^2 - m^2 - \frac{n^2}{\rho^2}\right]R = m^2. \tag{6.5.22}$$

The solutions to (6.5.20) are

$$\Phi_n^{\pm} = e^{\pm in\phi}. \tag{6.5.23}$$

If the solution were required to be single valued in a closed 2π sector, then n would have to be an integer. In wedge-shaped domains of smaller angle, this need not be the case.

The solutions to (6.5.21) are

$$Z_m^{\pm} = e^{\pm imz}. \tag{6.5.24}$$

The *separation constant* m plays the role of k_3 in our earlier discussion of plane waves. In a domain infinite or semi-infinite in z, m might range over a continuum of values. For a domain finite in z, m would be restricted to a discrete set of values by boundary conditions.

Equation (6.5.22) is Bessel's differential equation, with solutions

$$R_{mn} = C_n(\sqrt{k^2 - m^2}\,\rho). \tag{6.5.25}$$

Here $C_n(x)$ denotes any of the Bessel functions of order n:

$$J_n(x), \qquad H_n^{(1)}(x), \qquad H_n^{(2)}(x), \qquad Y_n(x).$$

The solution (6.5.15) now has the form

$$u = C_n(\sqrt{k^2 - m^2}\,\rho)e^{\pm in\phi \pm imz}. \tag{6.5.26}$$

More general solutions can be obtained by Fourier superposition—summation or integration as appropriate—over m and n.

Solutions of the two-dimensional equation in polar coordinates are obtained by assuming that u is z independent in (6.5.14). Equivalently, we set $m = 0$ throughout the discussion following that equation. In particular, (6.5.26) becomes

$$u = C_n(k\rho)e^{\pm in\phi}. \tag{6.5.27}$$

From the asymptotic expansions (6.3.18) for the Hankel functions and the two-dimensional Sommerfeld condition, Exercise 6.12 at the end of Section

6.4, it is straightforward to verify that $H_n^{(1)}$ represents outgoing waves and $H_n^{(2)}$ represents incoming waves.

It is useful to note that the plane wave in two dimensions

$$u = e^{i\mathbf{k}\cdot\mathbf{x}} \tag{6.5.28}$$

has an expansion in the *cylindrical waves* (6.5.27). We write

$$u = e^{ik\rho\cos(\phi-\psi)} = \sum_{n=-\infty}^{\infty} a_n(\rho)e^{in\phi}. \tag{6.5.29}$$

Here ψ is the polar angle of $\mathbf{k}$. The coefficients are then given by

$$a_n(\rho) = \frac{1}{2\pi}\int_{-\pi}^{\pi} e^{ik\rho\cos(\phi-\psi)-in\phi}\,d\phi = J_n(k\rho)e^{-in(\psi+i\pi/2)}. \tag{6.5.30}$$

Thus,

$$e^{i\mathbf{k}\cdot\mathbf{x}} = \sum_{n=-\infty}^{\infty} J_n(k\rho)e^{in(\phi-\psi-\pi/2)} = J_0(k\rho) + 2\sum_{n=0}^{\infty} J_n(k\rho)e^{in(\phi-\psi-\pi/2)}. \tag{6.5.31}$$

This result allows us to write down a series solution to the problem of scattering of a plane wave by a circular cylinder. Let us consider the case of the homogeneous Dirichlet problem in the exterior of the circle of radius a. We set

$$u = u_1 + u_S, \tag{6.5.32}$$

with u_1 given by (6.5.31). Now we must choose u_S as an outgoing solution such that $u_1 + u_S = 0$ on $r = a$. We shall do this term by term in a Fourier series solution. Therefore, we assume that u_S has a series solution terms of the functions (6.5.27), with each C_n being the Hankel function of the first kind. We then need only to pick the coefficients of the series so that u is zero on the boundary. The result is

$$u(\rho, \phi) = J_0(k\rho) - \frac{J_0(ka)}{H_0^{(1)}(ka)} H_0^{(1)}(k\rho)$$

$$+ 2\sum_{n=1}^{\infty}\left[J_n(k\rho) - \frac{J_n(ka)}{H_n^{(1)}(ka)} H_n^{(1)}(k\rho)\right]e^{in(\phi-\psi-\pi/2)}. \tag{6.5.33}$$

The Green's function (6.3.24) also has a Fourier series representation of this type. The derivation is not straightforward. It can be found in the treatise by Watson [1966, Section 11.41]. The result is

$$\frac{i}{4}H_0^{(1)}(k|\mathbf{x}-\mathbf{x}'|) = \frac{i}{4}\left[J_0(k\rho_<)H_0^{(1)}(k\rho_>)\right.$$

$$\left. + 2\sum_{n=0}^{\infty} J_n(k\rho_<)H_n^{(1)}(k\rho_>)\cos n(\phi-\phi')\right]. \tag{6.5.34}$$

In this equation, ρ and ϕ are the polar coordinates of $\mathbf{x}$; ρ' and ϕ' are the polar coordinates of $\mathbf{x}'$ and $\rho_< = \min(\rho, \rho')$, $\rho_> = \max(\rho, \rho')$. Also, this identity remains true when the Hankel functions on both sides of the equation are replaced by Bessel functions of the same type. In the literature, the same and related identities are often written with the argument in its polar form

$$|\mathbf{x} - \mathbf{x}'| = \sqrt{\rho^2 + \rho'^2 - 2\rho\rho'\cos(\phi - \phi')}. \qquad (6.5.35)$$

SPHERICAL WAVES

We consider now the Helmholtz equation in spherical coordinates:

$$\frac{1}{r^2}\frac{\partial}{\partial r}\left[r^2\frac{\partial u}{\partial r}\right] + \frac{1}{r^2\sin^2\theta}\frac{\partial}{\partial\theta}\left[\sin\theta\frac{\partial u}{\partial\theta}\right] + \frac{1}{\sin^2\theta}\frac{\partial^2 u}{\partial\phi^2} + k^2 u = 0. \quad (6.5.36)$$

The angle θ is the polar angle and the angle ϕ the azimuthal angle, as shown in Figure 6.3. As for Eq. (6.5.14) in cylindrical coordinates, we shall first seek separable solutions of the form

$$u = R(r)\Theta(\theta)\Phi(\phi). \qquad (6.5.37)$$

Proceeding as we did following (6.5.15), we obtain the following three equations for R, Θ, and Φ.

$$\frac{d^2\Phi}{d\phi^2} + n^2\Phi = 0, \qquad (6.5.38)$$

$$\frac{1}{r^2}\frac{d}{dr}\left[r^2\frac{dR}{dr}\right] + \left[k^2 - \frac{m(m+1)}{r^2}\right]R = 0, \qquad (6.5.39)$$

$$\frac{1}{\sin\theta}\frac{d}{d\theta}\left[\sin\theta\frac{d\Theta}{d\theta}\right] + \left[m(m+1) - \frac{n^2}{\sin^2\theta}\right]\Theta = 0. \qquad (6.5.40)$$

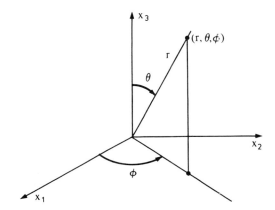

Fig. 6.3. Spherical polar conditions.

The separation constants here are m and n. The first equation of this set is again (6.5.20) with solutions (6.5.23). The second equation (6.5.39) is the equation for the spherical Bessel functions denoted by

$$R_m = c_m(kr) \tag{6.5.41}$$

and related to the ordinary Bessel functions by

$$c_m(z) = \sqrt{\pi/(2z)}\, C_{m+1/2}(z). \tag{6.5.42}$$

Each of these spherical Bessel functions is denoted by the lowercase letters of their counterparts, namely, j_m, y_m, $h_m^{(1)}$, and $h_m^{(2)}$. For integer values of m, these functions are all expressible in terms of ordinary trigonometric functions. In particular,

$$j_0(z) = (\sin z)/z; \qquad y_0(z) = -(\cos z)/z;$$
$$h_0^{(1)}(z) = j_0(z) + iy_0(z) = -ie^{iz}/z. \tag{6.5.43}$$

In (6.5.40), we introduce the new variable $x = \cos\theta$ and rewrite the equation as

$$\frac{d}{dx}\left[(1 - x^2)\frac{d\Theta}{dx}\right] + \left[m(m+1) - \frac{n^2}{1-x^2}\right]\Theta = 0. \tag{6.5.44}$$

As the polar angle θ ranges from 0 to π, the range of values of x is $|x| \le 1$. We consider first the case $n = 0$:

$$\frac{d}{dx}\left[(1 - x^2)\frac{d\Theta}{dx}\right] + m(m+1)\Theta = 0. \tag{6.5.45}$$

This is Legendre's equation. It is known that for noninteger values of m, the solutions are singular at both endpoints. For integer values of m, there is one solution that is regular at both endpoints and, in fact, is a polynomial of degree m, the Legendre polynomial, denoted by P_m:

$$\Theta_m = P_m(x) = P_m(\cos\theta) = \frac{1}{2^m m!}\frac{d^m}{dx^m}(x^2 - 1)^m. \tag{6.5.46}$$

The first few functions here are given by

$$P_0(x) = 1, \qquad P_1(x) = x, \qquad P_2(x) = (3x^2 - 1)/2,$$
$$P_3(x) = (5x^2 - 3x)/2, \qquad P_4(x) = (35x^4 - 30x^2 + 3)/8. \tag{6.5.47}$$

These functions satisfy the *orthogonality relation*

$$\int_{-1}^{1} P_m(x)P_{m'}(x)\, dx = \frac{2}{2m+1}\delta_{mm'}. \tag{6.5.48}$$

Here δ is the Kronecker delta function. These functions are *complete* on the interval $(-1, 1)$. That is, we can represent "arbitrary" functions in terms of these, much like the Fourier sine and cosine series. See Coddington and Levinson [1955] for a discussion of completeness of the eigenfunctions of Sturm–Liouville differential equations.

For n nonzero, the solutions are the *associated Legendre functions* denoted by P_m^n:

$$\Theta(x) = P_m^n(x) = (-1)^n(1 - x^2)^{n/2}\frac{d^n}{dx^n}P_m(x), \qquad n \le m. \qquad (6.5.49)$$

From this formula, it can be seen that these functions are zero for $n > m$. While these functions would seem to have branch points at $x = \pm 1$ for n odd, this is not the case in θ near $\theta = 0$ or π, since $1 - x^2 = \sin^2 \theta$.

What is important to us here is that the functions Φ_n and P_m^n can be combined to define the *spherical harmonics* Y_{mn}:

$$Y_{mn}(\theta, \phi) = \sqrt{\frac{(2m + 1)(m - n)!}{4\pi(m - n)!}}P_m^n(\cos\theta)e^{in\phi}. \qquad (6.5.50)$$

These functions form a *complete orthonormal set* in the class of square integrable functions on the unit sphere $0 \le \theta \le \pi$ and $0 \le \phi < 2\pi$ with respect to integration over solid angle as in the orthonormality condition stated as

$$\int_0^{2\pi} d\phi \int_0^\pi \sin\theta\, d\theta\, Y_{m'n'}^*(\theta, \phi)Y_{mn}(\theta, \phi) = \delta_{mm'}\,\delta_{nn'}. \qquad (6.5.51)$$

We use (6.5.50) and (6.5.42) in (6.5.37) to write the class of solutions

$$u = c_m(kr)Y_{mn}(\theta, \phi). \qquad (6.5.52)$$

Other solutions can be generated by Fourier superposition over m and n.

Let us now consider the expansion of a plane wave in terms of the solutions (6.5.52). We define γ as the angle between $\mathbf{x}$ and $\mathbf{k}$ and begin by writing the plane wave solution to the Helmholtz equation as

$$e^{i\mathbf{k}\cdot\mathbf{x}} = e^{ikr\cos\gamma}. \qquad (6.5.53)$$

This function of $\cos\gamma$ has an expansion in the Legendre polynomials of argument $\cos\gamma$. A derivation of the expansion can be found in Section 11.5 in Watson [1966]. The result is

$$e^{i\mathbf{k}\cdot\mathbf{x}} = \sum_{m=0}^{\infty} i^m(2m + 1)j_m(kr)P_m(\cos\gamma). \qquad (6.5.54)$$

This result can also be expressed in terms of the polar angles of $\mathbf{x}$ (θ and ϕ)

and the polar angles of $\mathbf{k}$ (τ and ψ). The expansion uses the *addition theorem* for spherical harmonics. Derivation of this result has been given by Stratton [1941, Section 7.5] and Erdélyi *et al.* [1953, Section 11.2]. The addition theorem states that

$$P_m(\cos \gamma) = \frac{4\pi}{2m+1} \sum_{n=-m}^{m} Y_{mn}^*(\tau, \psi) Y_{mn}(\theta, \phi);$$

$$\cos \gamma = \frac{\mathbf{k} \cdot \mathbf{x}}{kr} = \sin \phi \sin \psi \cos(\theta - \tau) + \cos \phi \cos \psi.$$

(6.5.55)

By using this result and (6.5.54), we obtain the following representation of the plane wave in terms of spherical harmonics:

$$e^{i\mathbf{k} \cdot \mathbf{x}} = 4\pi \sum_{m=0}^{\infty} i^m j_m(kr) \sum_{n=-m}^{m} Y_{mn}^*(\tau, \psi) Y_{mn}(\theta, \phi).$$

(6.5.56)

The Green's function also has a spherical harmonic expansion

$$\frac{e^{ik|\mathbf{x}-\mathbf{x}'|}}{4\pi|\mathbf{x}-\mathbf{x}'|} = ik \sum_{m=0}^{\infty} j_m(kr_<)h_m^{(1)}(kr_>) \sum_{n=-m}^{m} Y_{mn}^*(\theta', \phi') Y_{mn}(\theta, \phi).$$

(6.5.57)

This result has also been derived by Watson [1966, Section 11.41].

As in the problem of scattering of a plane wave by a cylinder (6.5.33), we can use these results to derive a series solution for the scattering of a plane wave by a sphere. We shall take the direction of an incident plane wave to be the positive z direction so that $\gamma = \phi$, and we assume that the total solution (6.5.32) satisfies a homogeneous Dirichlet condition on the sphere of radius a. We leave to the exercises the derivation of the solution

$$u(r, \theta, \phi) = \sum_{m=0}^{\infty} i^m(2m+1)\left[j_m(kr) - \frac{j_m(ka)}{h_m^{(1)}(ka)} h_m^{(1)}(kr) \right] P_m(\cos \gamma).$$

(6.5.58)

Exercises

6.16 (a) Suppose in (6.5.8) that the phases did not agree. Show that then we could solve uniquely for R and T in terms of k vectors so long as $k_1^R \neq k_1^T$ and, in this case, that R and T are functions of the transverse x variables.

(b) If $k_1^R = k_1^T$, show that a solution would still exist if they also both equaled k_1 but that at least one of the coefficients R and T would then have to be x dependent.

6.17 Consider the problem of scattering of a plane wave by a circular cylinder of radius a. Suppose that the boundary condition on the cylinder is given by

$$\Omega u = \frac{\partial u(r, \phi)}{\partial r} - i\omega Z u(r, \phi)\bigg|_{r=a} = 0.$$

Then show that

$$u(\rho, \phi) = J_0(k\rho) - \frac{\Omega J_0}{\Omega H_0^{(1)}} H_0^{(1)}(k\rho)$$

$$+ 2 \sum_{n=1}^{\infty} \left[J_n(k\rho) - \frac{\Omega J_n}{\Omega H_n^{(1)}} H_n^{(1)}(k\rho) \right] e^{in(\phi - \psi - \pi/2)}.$$

6.18 Verify (6.5.58) by starting from series (6.5.54), adding to it a series of outgoing waves with arbitrary coefficients, and determining those coefficients from the boundary condition at $r = a$.

6.19 Repeat Exercise 6.17 for scattering by a sphere of radius a, now under the boundary condition

$$\Omega u = \frac{\partial u(r, \theta, \phi)}{\partial r} - i\omega Z u(r, \theta, \phi)\bigg|_{r=a} = 0.$$

Show that

$$u(r, \theta, \phi) = \sum_{m=0}^{\infty} i^m (2m + 1) \left[j_m(kr) - \frac{\Omega j_m}{\Omega h_m^{(1)}} h_m^{(1)}(kr) \right] P_m(\cos \gamma).$$

6.20 Let us suppose now that a point source in two dimensions is located at the point $\mathbf{x}_0$, with polar coordinates (ρ_0, ϕ_0), $\rho_0 > a$. Use (6.5.34) to represent the incident field on a circular cylinder of radius a on which the boundary condition in Exercise 6.17 is to be satisfied. Find the scattered wave.

6.21 Repeat Exercise 6.20 for the three-dimensional case by using (6.5.57) for the incident field and the boundary condition in Exercise 6.19.

6.22 (a) Suppose that u is a solution of the inhomogeneous Helmholtz equation

$$\nabla^2 u + k^2 u = f(r, \theta, \phi; k),$$

with f nonzero only for $r \leq a$. Use (6.5.57) to show that for $r > a$,

$$u(r, \theta, \phi; k) = -ik \sum_{m=0}^{\infty} h_m^{(1)}(kr) Y_{mn}(\theta, \phi) \int_{r'=0}^{a} j_m(kr') r' \, dr' \, f_{mn}(r'; k).$$

Here

$$f_{mn}(r'; k) = \int_0^{2\pi} d\phi' \int_0^{\pi} Y_{mn}(\theta', \phi') \sin \theta' \, d\theta' \, f(r', \theta', \phi'; k).$$

(b)   Find a function f for which f_{00} is nonzero but u is zero for $r > a$.

(c)   State a criterion for nonzero f for which u is zero for $r > a$.

Such sources are called *nonradiating sources*.

(d)   In Section 5.4, we introduced the concept of an inverse source problem. What are the implications of this exercise as regards uniqueness of solutions to the inverse source problem?

6.23   Let u be a solution of the problem

$$\nabla^2 u + (\omega^2/c^2)u = -\delta(x_1 - h)\delta(x_2)\delta(x_3), \qquad x_1 < 0, \quad h < 0,$$

$$\nabla^2 u + (\omega^2/c_1^2)u = 0, \qquad\qquad\qquad\qquad x_1 > 0,$$

with u outgoing and satisfying continuity of function and normal derivative across $x_1 = 0$. Write u as a sum of an incident plus reflected field for $x_1 < 0$ and as a transmitted field for $x_1 > 0$. Introduce Fourier transform in the transverse variables and conclude that

$$u_I(\mathbf{x}; \omega) = \frac{i}{(2\pi)^2} \int_{-\infty}^{\infty} dk_2 \int_{-\infty}^{\infty} \frac{dk_3}{2k_1} \exp(i[k_1|x_1 - h| + k_2 x_2 + k_3 x_3]),$$

$$u_R(\mathbf{x}; \omega) = \frac{i}{(2\pi)^2} \int_{-\infty}^{\infty} dk_2 \int_{-\infty}^{\infty} \frac{dk_3}{2k_1} R \exp(i[k_1^R(x_1 + h) + k_2 x_2 + k_3 x_3]),$$

$$u_T(\mathbf{x}; \omega) = \frac{i}{(2\pi)^2} \int_{-\infty}^{\infty} dk_2 \int_{-\infty}^{\infty} \frac{dk_3}{2k_1} T \exp(i[k_1^T x_1 + k_1 h + k_2 x_2 + k_3 x_3]).$$

In these equations, k_1 is defined in (6.5.5), k_1^R and k_1^T are defined by (6.5.11), and R and T are defined by (6.5.13).

References

Abramowitz, M., and Stegun, I. [1965]. "Handbook of Mathematical Functions." Dover, New York.

Bleistein, N., and Cohen, J. K. [1977]. Nonuniqueness in the inverse source problem in acoustics and electromagnetics, *J. Math. Phys.* **18**, 194–201.

Bowman, J. J., Senior, T. B. A., and Uslenghi, P. L. E. [1969]. "Electromagnetic and Acoustic Scattering by Simple Shapes." North-Holland, Amsterdam, and Wiley (Interscience), New York.

Coddington, E. A., and Levinson, N. [1955]. "Theory of Ordinary Differential Equations." McGraw-Hill, New York.

Courant, R., and Hilbert, D. [1962]. "Methods of Mathematical Physics," Vol. 2. Wiley (Interscience), New York.

Erdélyi, A., Magnus, W., Oberhettinger, F., and Tricomi, F. G. [1953]. "Higher Transcendental Functions," Vol. 2. McGraw-Hill, New York.

Garabedian, P. R. [1964]. "Partial Differential Equations." Wiley, New York.

Gradshteyn, I. S., and Ryzhik, I. M. [1965]. "Table of Integrals, Series and Products." Academic Press, New York. [Corrected and enlarged edition, 1980.]

Hildebrand, F. B. [1962]. "Advanced Calculus for Engineers." Prentice-Hall, Englewood Cliffs, New Jersey.

Jackson, J. D. [1975]. "Classical Electrodynamics," 2nd ed. Wiley, New York.

John F. [1982]. "Partial Differential Equations," 4th ed. Springer-Verlag, New York.

Lebedev, N. N. [1972]. "Special Functions and Their Applications." Dover, New York.

Stratton, J. A. [1941]. "Electromagnetic Theory." McGraw-Hill, New York.

Watson, G. N. [1966]. "A Treatise on the Theory of Bessel Functions," 2nd ed. Cambridge Univ. Press, Cambridge.

Whittaker, E. T., and Watson, G. N. [1958]. "A Course in Modern Analysis." Cambridge Univ. Press, Cambridge.

7 MORE ON ASYMPTOTICS

The purpose of this chapter is to develop the *method of steepest descents*, which has great utility for the high-frequency analysis of wave problems. The method of steepest descents depends crucially on *Watson's lemma*, which will be developed in the first section, and on certain properties of analytic functions to be developed in the second section.

7.1 WATSON'S LEMMA

We shall be concerned here with the asymptotic expansion of integrals of the type

$$I(\lambda) = \int_0^\infty f(t)e^{-\lambda t}\, dt, \tag{7.1.1}$$

in the limit $\lambda \to \infty$. We remind the reader that in practice we shall implement the results of this analysis for λ finite but large.

We shall show that for a class of functions that includes many of the type that arise in practice, the asymptotic expansion of $I(\lambda)$ as $\lambda \to \infty$ is determined by the asymptotic expansion of $f(t)$ as $t \to 0+$. We shall assume that

(i) $f(t)$ is locally absolutely integrable on any finite subinterval of $(0, \infty)$;

(ii) $f(t)$ grows no worse that a linear exponential at infinity; that is, for some real number a,

$$f(t) = O(e^{at}), \qquad t \to \infty; \tag{7.1.2}$$

(iii)

$$f(t) \sim \sum_{k=0}^\infty c_k t^{a_k}, \qquad t \to 0+, \tag{7.1.3}$$

with Re $a_0 > -1$ and the real parts of the set of numbers $\{a_k\}$ monotonically increasing to infinity. We remark that this last condition guarantees that t^{a_k} is an *asymptotic sequence* necessary if (7.1.3) is to be an asymptotic expansion of the function $f(t)$.

Under these assumptions, the asymptotic expansion of $I(\lambda)$ is obtained from term-by-term integration of the asymptotic expansion of $f(t)$; that is,

$$I(\lambda) \sim \sum_{k=0}^{\infty} c_k \int_0^{\infty} t^{a_k} e^{-\lambda t}\, dt = \sum_{k=0}^{\infty} \frac{c_k \Gamma(a_k + 1)}{\lambda^{a_k + 1}}, \qquad \lambda \to \infty. \qquad (7.1.4)$$

This is an asymptotic expansion with respect to the asymptotic sequence $\{\lambda^{-a_k-1}\}$. This result is *Watson's lemma*. The proof can be found in Bleistein and Handelsman [1975], Copson [1965], and Olver [1974].

The exponential integral (2.5.8) can be recast in a form to which Watson's lemma may be applied. To do so, set

$$t = \lambda(s + 1) \qquad (7.1.5)$$

and then replace s by t again to obtain

$$I(\lambda) = \lambda \int_0^{\infty} \frac{e^{-\lambda t}}{1 + t}\, dt. \qquad (7.1.6)$$

Except for the multiplicative factor of λ, this is exactly of the form (7.1.1) with

$$f(t) = \frac{1}{1 + t} \sim \sum_{k=0}^{\infty} (-1)^k t^k. \qquad (7.1.7)$$

In fact, this asymptotic expansion is convergent in the unit disk in the complex t plane, but that is not relevant to the asymptotic expansion of $I(\lambda)$ under consideration here.

By applying (7.1.4), we conclude that

$$I(\lambda) \sim \lambda \sum_{k=0}^{\infty} \frac{(-1)^k \Gamma(k + 1)}{\lambda^{k+1}} = \sum_{k=0}^{\infty} \frac{(-1)^k k!}{\lambda^k}, \qquad (7.1.8)$$

which is the result (2.5.11).

As a second example, we consider

$$f(t) = \sin 2t^{1/2} \sim \sum_{k=0}^{\infty} \frac{(-1)^k 2^{2k+1} t^{(k+1/2)}}{(2k + 1)!}. \qquad (7.1.9)$$

Substitution of this function into (7.1.4) yields the asymptotic expansion

$$I(\lambda) \sim \sum_{k=0}^{\infty} \frac{(-1)^k 2^{2k+1} \Gamma(k + \frac{3}{2})}{(2k + 1)!\, \lambda^{k+3/2}}. \qquad (7.1.10)$$

This result is actually convergent for all $\lambda > 0$. An alternative derivation of this asymptotic expansion is carried out in Exercise 7.6.

In either of the examples considered here, the interval of integration $(0, \infty)$ could as easily have been a *ray* directed into the right half t plane. We leave the verification of this observation to the exercises. However, we note that this observation is more generally true. That is, suppose that we are given an integral in the complex t plane from the origin to infinity somewhere in the right half plane. Suppose further that by application of Cauchy's theorem, we can show that the integral is equal to an integral along the real axis plus, perhaps, other contributions. Then Watson's lemma may, at least, be applicable to the integral along the real axis. The replacement of one contour by another may often involve other contributions, such as residues and branch-cut integrals. However, as long as these singularities reside in the right half plane, such contributions will contain a factor $\exp\{\lambda t_0\}$, where $\operatorname{Re} t_0 > 0$. Hence, such contributions will be exponentially smaller than (asymptotically zero when compared to) the asymptotic series provided by Watson's lemma. We remark that the two cases treated above, $f(t) = 1/(1 + t)$ and $f(t) = \sin 2t^{1/2}$, were such that the integral along a ray directed into the right half plane is equal to the integral along the real line. Thus, all of these integrals, being equal, have the same asymptotic expansion.

Similarly, for an integral on the real line with λ complex but approaching infinity along a ray in the right half λ plane, the asymptotic expansion is given by Watson's lemma. As above, the character of the integrand as $\lambda \to \infty$ is dominated by the decay of $\exp\{-\lambda t\}$ for large enough values of the product λt.

Bleistein and Handelsman [1975] and Olver [1974] provide extensions of Watson's lemma. We mention two of the extensions here. First, we remark that if the asymptotic expansion (7.1.3) contains terms with integer powers of $\log t$, with only a finite set of such powers associated with each power of t, then the asymptotic expansion is again obtained by term-by-term integration. The actual formula for this case is left to the exercises.

A second case of interest arises when the interval of integration in (7.1.1) is replaced by a "loop" contour starting from infinity, enclosing the real axis in a counterclockwise manner, and returning to infinity again, as in Fig. 7.1. We shall denote this contour by $0+$. In this case, if (7.1.3) holds in a sector $\delta \leq \arg t \leq 2\pi - \delta$, with $\delta < \pi/2$, then the asymptotic expansion of $I(\lambda)$ is again obtained by termwise integration of the asymptotic expansion of $f(t)$.

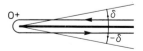

Fig. 7.1. The contour $0+$ and domain of validity for the asymptotic expansion of $f(t)$.

The resulting asymptotic expansion depends on the fact that

$$\int_{0+} t^z e^{-t}\, dt = -\frac{2\pi i e^{i\pi z}}{\Gamma(-z)} = 2i e^{i\pi z}\Gamma(1+z)\sin\pi z. \qquad (7.1.11)$$

(The second form is convenient for real z negative except when z is a negative integer.) Consequently, we find that

$$\int_{0+} f(t)e^{-\lambda t}\, dt \sim \sum_{k=0}^{\infty} c_k \int_{0+} t^{a_k} e^{-\lambda t}\, dt = \sum_{k=0}^{\infty} -\frac{2\pi i c_k e^{i\pi a_k}}{\Gamma(-a_k)\lambda^{a_k+1}}. \qquad (7.1.12)$$

As an example of this type, let us consider

$$I(\lambda) = \int_{-\infty}^{\infty} \frac{e^{-\lambda\tau^2}}{1+\tau^2}\, d\tau, \qquad (7.1.13)$$

which can be recast in the form (7.1.12) by introducing the change of variable of integration $\tau^2 = t$. Here we take $\arg\tau = 0$ on the positive real axis and $\arg\tau = \pi$ on the negative real axis and interpret the original interval of integration as a path of integration in the complex τ plane passing *above* the origin. We remark that this interpretation is not unique. However, it is valid and allows us to obtain the desired representation in the most straightforward manner. Alternative definitions of $\arg\tau$ must, of necessity, lead to the same fianl result.

The image of the path of integration under the prescribed transformation is, indeed, the contour $0+$ in Fig. 7.1. Thus, we find that

$$I(\lambda) = -\int_{0+} \frac{e^{-\lambda t}}{1+t} t^{-1/2}\, dt. \qquad (7.1.14)$$

We leave it as an exercise for the reader to perform the necessary substitution into (7.1.12) and to use the appropriate relationship between $\Gamma(z)$ and $\Gamma(1-z)$ to obtain

$$I(\lambda) \sim \sum_{k=0}^{\infty} \frac{(-1)^k \Gamma(k+\tfrac{1}{2})}{\lambda^{k+1/2}}. \qquad (7.1.15)$$

Exercises

7.1 Calculate the asymptotic expansion of the integral (7.1.1) with $f(t)$ given by:
  (a) $f(t) = \log(1+t)$.
  (b) $f(t) = J_0(t)$.
  (c) $f(t) = \begin{cases} (b-t)^p,\ 0 \le t < b,\ b > 0,\ p > -1, \\ \text{anything satisfying conditions (i) and (ii) of} \\ \text{Watson's lemma,}\ t \ge b. \end{cases}$

7.2   (a)   Find two terms of the asymptotic expansion of (7.1.1) for the function

$$f(t) = (t^2 - it)^{p-1/2}, \qquad \text{Re } p > -\tfrac{1}{2}.$$

For this function, we define

$$-\pi/2 \le \arg(t - i) < 0, \qquad \text{for} \quad 0 \le t < \infty.$$

  (b)   From a table of Laplace transforms, it can be verified that for this function f,

$$I(\lambda) = \tfrac{1}{2}i\sqrt{\pi}\,\Gamma(p + \tfrac{1}{2})\lambda^{-p}e^{-i\lambda/2}H_p^{(1)}(\lambda/2), \qquad \lambda > 0.$$

Use this result to verify that the leading term of the asymptotic expansion agrees with the first of Eq. (6.3.18).

7.3   By formal differentiation of the integral representation of the gamma function (2.7.10), we find that

$$\left[\frac{d}{dz}\right]^m \Gamma(z) = \int_0^\infty (\log z)^m t^{z-1} e^{-t}\, dt,$$

with the case $m = 1$ defining $\Psi(z)$, the *digamma function*, or logarithmic derivative of the gamma function,

$$\frac{d}{dz}\Gamma(z) = \Gamma'(z) = \Gamma(z)\Psi(z).$$

Suppose that (7.13) is replaced by the asymptotic expansion

$$f(t) \sim \sum_{k=0}^{\infty} \sum_{m=0}^{M(k)} c_{mk} t^{a_k} (\log t)^m,$$

with $M(k)$ finite for each k. Then, assuming that formal substitution is valid for this case, derive the asymptotic expansion

$$I(\lambda) \sim \sum_{k=0}^{\infty} \sum_{m=0}^{M(k)} c_{mk} \left[\frac{d}{da_k}\right]^m \left[\frac{\Gamma(a_k + 1)}{\lambda^{a_k+1}}\right].$$

Here the asymptotic sequence is

$$\{\lambda^{-a_0}(\log \lambda)^{M(0)}, \lambda^{-a_0}(\log \lambda)^{M(0)-1}, \ldots, \lambda^{-a_0}, \lambda^{-a_1}(\log \lambda)^{M(1)}, \ldots\}.$$

7.4   (a)   In (7.1.1), let $f(t) = Y_0(t)$, the Bessel function of the second kind. Use the result of the preceding exercise to show that

$$I(\lambda) \sim -\frac{2}{\pi\lambda}\log(2\lambda) + O(\lambda^{-3}\log \lambda).$$

(b) Find the exact Laplace transform of the Bessel function of the second kind and obtain the same result by expanding that Laplace transform for large positive argument.

7.5 (a) Verify that $\{t^{a_k}\}$ is an asymptotic sequence in the limit $t \to 0+$ when $\{\operatorname{Re} a_k\}$ is a monotonically increasing sequence with limit $+\infty$.

(b) Repeat (a) for the sequence $\{\lambda^{-a_k - 1}\}$ in the limit $\lambda \to \infty$.

7.6 (a) Calculate a closed-form expression for the integral (7.1.1) with f given by (7.1.9), proceeding as follows.

(i) Replace the sine function by a sum of exponentials.

(ii) Set $\sigma = \pm\sqrt{t}$ in the two integrals, and rewrite the sum as an integral on the interval $(-\infty, \infty)$.

(iii) Complete the square in the exponent and calculate the resulting integral. Obtain the result

$$I(\lambda) = \frac{e^{-1/\lambda}}{\lambda^{3/2}}.$$

(b) Use the power series representation of the exponential function to obtain an asymptotic expansion of $I(\lambda)$.

(c) Verify that this result agrees with (7.1.10). Here the identity

$$\sqrt{2\pi}\, \Gamma(2z) = 2^{2z-1/2}\Gamma(z)\Gamma(z + \tfrac{1}{2})$$

will prove useful.

7.7 In (7.1.12), suppose that

$$f(t) = t^{-1/2}/(1 + t^{1/2}).$$

Show that

$$I(\lambda) \sim \sum_{k=0}^{\infty} 2\, \frac{\Gamma(k + \tfrac{1}{2})}{\lambda^{k+1/2}}.$$

7.8 Suppose that the conditions (i), (ii), and (iii) of this section hold in some domain that includes the ray in the t plane at angle θ, $-\pi/2 < \theta < \pi/2$, and also contains the sector between this ray and the positive real axis. Suppose also that $f(t)$ is analytic in this same domain. Connect the ray and the real axis with two arcs, one of radius ε, which will ultimately approach zero, and the other of radius R, which will ultimately approach infinity.

(a) On the smaller arc, estimate the integrand by an upper bound dependent on a power of t and estimate the entire integral by multiplying this estimate by a bound on the length of path to show that the integral on this arc approaches zero as the radius approaches zero.

(b) On the larger arc, choose λ large enough to obtain a bound on the integrand that decays exponentially to zero with increasing radius. Show that the integral on this path also decays to zero, in this case, as the radius approaches infinity.

(c)   Finally, conclude that the integral on the ray is equal to the integral on the real line.

7.9   Let

$$I(\lambda) = \int_0^\infty \frac{e^{i\lambda t}}{1 + t^4} \, dt.$$

(a)   Use Cauchy's theorem to justify replacing this integral by one along the imaginary axis in a complex t plane. Then, with $t = i\sigma$, show that

$$I(\lambda) = i \int_0^\infty \frac{e^{-\lambda\sigma}}{1 + \sigma^4} \, d\sigma + \frac{\pi}{2} \exp\left[-\frac{\lambda}{\sqrt{2}}(1 - i) - i\pi/4 \right].$$

(b)   Use Watson's lemma and conclude that

$$I(\lambda) \sim i \sum_{k=0}^\infty (-1)^k \frac{\Gamma(4k + 1)}{\lambda^{4k+1}}.$$

7.10   Let

$$I(\lambda) = \int_0^\infty \frac{\exp[-\lambda e^{i\pi/4} t]}{1 + t^2} \, dt.$$

Use Cauchy's theorem to justify replacing the integral along the real axis by an integral along the ray at angle $-\pi/4$ in the complex t plane. Then introduce the variable of integration σ by $t = \sigma e^{-i\pi/4}$ and obtain the asymptotic expansion

$$I(\lambda) \sim e^{-i\pi/4} \sum_{k=0}^\infty i^k \frac{\Gamma(2k + 1)}{\lambda^{2k+1}}.$$

7.11   In the preceding exercise, replace the angle $\pi/4$ in the definition of $I(\lambda)$ by θ, $-\pi/2 < \theta < \pi/2$. Show that

$$I(\lambda) \sim e^{-i\theta} \sum_{k=0}^\infty (-1)^k e^{-2ik\theta} \frac{\Gamma(2k + 1)}{\lambda^{2k+1}}.$$

7.12   Let

$$I(\lambda) = \int_0^\infty f(t)e^{-\lambda t^2} \, dt,$$

with f satisfying conditions (i), (ii), and (iii) of this section. (In (ii), at could be replaced by at^2.) Introduce the change of variable of integration $t^2 = \sigma$, and conclude that

$$I(\lambda) \sim \frac{1}{2} \sum_{k=0}^\infty \frac{c_k \Gamma((a_k + 1)/2)}{\lambda^{(a_k + 1)/2}}.$$

7.13 In the preceding exercise, replace t^2 in the definition of $I(\lambda)$ by t^p and conclude that

$$I(\lambda) \sim \frac{1}{p} \sum_{k=0}^{\infty} \frac{c_k \Gamma((a_k + 1)/p)}{\lambda^{(a_k + 1)/p}}.$$

7.14 (a) Verify (7.1.15).
(b) Calculate the asymptotic expansion of the integral (7.1.12) for f given by (7.1.9) and $0 < \arg t \leq 2\pi$.

7.2 THE METHOD OF STEEPEST DESCENTS: PRELIMINARY RESULTS

For the remainder of this chapter, we shall consider integrals of the form

$$I(\lambda) = \int_C g(z) e^{\lambda w(z)} \, dz. \tag{7.2.1}$$

In this equation, C is a contour in the complex z plane with

$$z = x + iy, \qquad w(z) = u(x, y) + iv(x, y). \tag{7.2.2}$$

Either the inverse Fourier transform, with $w = i\Phi$, or the inverse Laplace transform is an integral of this type. The choice of traditional notation of complex function theory anticipates our exploitation of that theory in the development of a technique for the asymptotic expansion of $I(\lambda)$ as $\lambda \to \infty$.

In the preceding section, we have seen that an integral with the real part of the exponent linearly (in fact, *monotonically* is good enough) decreasing toward $-\infty$ has a particularly simple asymptotic expansion derived totally in terms of the local behavior of the integrand near the endpoint of integration. In Section 2.6, in the context of Fourier integrals, we introduced the idea of critical points—points that were important to the development of the asymptotic expansion of an integral. It is these two features that provide a motivation for the development of the method of steepest descents. In this method, we exploit Cauchy's theorem to deform the given contour of integration onto a sum of contours on each of which the exponent is monotonically decreasing toward $-\infty$ and then calculate the asymptotic expansion of each integral by using the results of the preceding section. Implicit here is the assumption that the integral on the sum of contours is related to $I(\lambda)$ in a known manner, usually equality up to a sum of residues and/or a sum of contour integrals that are known to be asymptotically negligible when compared to $I(\lambda)$ itself.

We focus our attention on the function $w(z)$. We shall assume that $w(z)$ is analytic and nonconstant in some domain D of the complex z plane. For any point in D, say, $z_0 = x_0 + iy_0$, we define a *direction of descent* from z_0 as a direction in which the real part of w, $u(x, y)$, *decreases* from its value at that point. Analogously, a *direction of ascent* is one in which $u(x, y)$ *increases* from its value at (x_0, y_0).

A directed curve starting at z_0 along which the tangent direction is always a direction of descent will be called a *curve* or *path of descent*. Analogously, if the tangent is always a direction of ascent, the curve is called a *curve* or *path of ascent*.

There are many (one or more continua) of directions of descent and ascent at each point in the z plane. More precisely, there are sectors of descent and ascent bounded by directions in which u neither increases nor decreases. Of special interest are the directions in which the *rate* of descent or ascent is maximal (say, with respect to arc length). These are called the *directions of steepest descent and ascent*, respectively. At a point at which $\mathbf{V}u \neq \mathbf{0}$, there is only one direction of steepest descent, namely, the direction of $-\mathbf{V}u$, and one direction of steepest ascent, namely, the direction of $\mathbf{V}u$. Figure 7.2

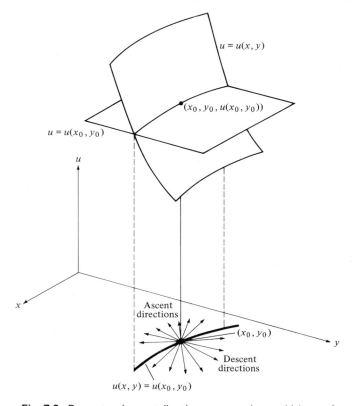

Fig. 7.2. Descent and ascent directions near a point at which $u \neq \mathbf{0}$.

depicts a surface $u(x, y)$ in the neighborhood of a point at which $\nabla u \neq 0$.

A *path or curve of steepest descent (ascent)* is a curve along which the tangent direction at each point is a direction of steepest descent (ascent). We remark that given a path of steepest descent, the same curve with opposite orientation is a path of steepest ascent (and vice versa).

The words *descent* and *ascent* are prompted by consideration of the surface $u(x, y)$, which is suggestive of a rolling countryside. To carry that analogy further, given two points z_0 and z_1, if $u(x_1, y_1) < u(x_0, y_0)$, then we will say that z_1 is in a *valley* of $w(z)$ with respect to the point z_0. Similarly, if $u(x_1, y_1) > u(x_0, y_0)$, we will say that z_1 is on a *hill* of $w(z)$ with respect to z_0.

Thus, paths of steepest descent proceed progressively deeper into a valley of w with respect to the initial point on the path, while paths of ascent proceed progressively higher onto hills. Of course, this progression is most rapid on paths of steepest descent and ascent.

Curves along which u remains equal to its value at z_0 are called *boundary curves* between hills and valleys.

Motivated by the preceding section, let us consider the exponent $w(z) = -z$, for which $u = -x$, and choose as reference point the origin $z_0 = 0$. All rays directed into the right half plane are paths of descent with respect to their left endpoint, while all rays directed into the left half plane are paths of ascent with respect to their right endpoint. Any horizontal line directed toward the right is a path of steepest *descent* with respect to its left endpoint; a horizontal line directed to the left is a path of steepest *ascent* with respect to its right endpoint. The open right half plane, excluding the imaginary axis, is in the valley of $w(z)$ with respect to the origin, while the open left half plane is on a hill. The imaginary axis is the boundary between the hill and the valley.

For more general exponents, the identification of paths of steepest descent (and ascent) is facilitated by the following observation:

The directions of steepest descent and steepest ascent from any point $z_0 = x_0 + iy_0$ are the directions tangent to $\operatorname{Im} w(z) = \mathrm{const}$, that is, tangent to the curves

$$v(x, y) = v(x_0, y_0). \tag{7.2.3}$$

To see why this is so, let us consider δw, the differential variation in w near z_0,

$$\delta w = w(z) - w(z_0) = \delta u + i\,\delta v. \tag{7.2.4}$$

In the direction of interest, in the differential limit,

$$\delta v = 0 \quad \text{and} \quad \delta w = \delta u. \tag{7.2.5}$$

For any direction away from z_0,

$$|\delta w|^2 = |\delta u|^2 + |\delta v|^2, \tag{7.2.6}$$

so that

$$|\delta u| \le |\delta w|, \tag{7.2.7}$$

with equality holding *only* when $\delta v = 0$.

Thus, the variation in u is maximal in the directions tangent to the curves of constant v. Consequently, the paths of steepest ascent and descent are the curves of constant v.

We remark that by avoiding use of the gradient in the preceding argument, we have included directions and paths of steepest descent away from points such as the origin for the functions

$$w(z) = z^\gamma, \quad \gamma > 0, \quad \text{but not an integer.} \tag{7.2.8}$$

We leave the determination of the directions of steepest descent for this class of functions to the exercises.

We focus our attention now on the question of directions of steepest ascent and descent at a point of analyticity of the function $w(z)$. The answer is given in the following theorem.

Theorem   Suppose that at the point z_0, all of the derivatives up to order $n - 1$ vanish; that is,

$$\frac{d^q w}{dz^q} = 0, \quad q = 1, 2, \ldots, n - 1, \quad \frac{1}{n!}\frac{d^n w}{dz^n} = ae^{i\alpha}, \quad a > 0, \quad z = z_0. \tag{7.2.9}$$

If $z - z_0 = \rho e^{i\theta}$, then the directions of steepest descent, steepest ascent, and constant u at $z = z_0$ are as given in Table 7.1.

Proof   We introduce

$$\delta w = w(z) - w(z_0) = ae^{i\alpha}\rho^n e^{in\theta}[1 + O(\rho)], \quad \rho \to 0, \tag{7.2.10}$$

which follows from (7.2.9) and the polar representation introduced after that equation. The directions of steepest descent here are those directions in which δw is negative, that is, for those directions that satisfy

$$\alpha + n\theta = (2p + 1)\pi, \tag{7.2.11}$$

where p is an integer. The first row of Table 7.1 is the solution of this equation for θ. The range on p in the table provides the distinct choices for θ.

Table 7.1

Directions of	θ	p
Steepest descent	$(2p + 1)\pi/n - \alpha/n$	$p = 0, 1, \ldots, n - 1$
Steepest ascent	$2p\pi/n - \alpha/n$	$p = 0, 1, \ldots, n - 1$
Constant u	$(p + \frac{1}{2})\pi/n - \alpha/n$	$p = 0, 1, \ldots, 2n - 1$

The directions of steepest ascent are determined in a similar manner by requiring that δw be positive:

$$\alpha + n\theta = 2p\pi. \tag{7.2.12}$$

Again, the table provides the distinct choices for θ.

Finally, in the directions of constant u,

$$\alpha + n\theta = (p + \tfrac{1}{2})\pi, \tag{7.2.13}$$

and the table again provides the distinct choices for θ.

This completes the proof.

Let us consider the angles in the table starting from $-\alpha/n$ and moving counterclockwise about the point z_0. The direction defined by $\theta = -\alpha/n$ is a direction of steepest ascent with $p = 0$. The next-distinguished direction provided by the table is a direction of constant u at $\theta = -\alpha/n + \pi/2n$, again with $p = 0$. This is followed by a direction of steepest descent for $\theta = -\alpha/n + \pi/n$ for $p = 0$. This is followed then, in order, by directions of constant u, ascent, constant u, descent, ..., continuing around the circle until all distinguished directions are exhausted.

Let us consider two boundary curves (curves of constant u) emanating from $z = z_0$, with initial directions

$$\theta = -(\alpha/n) + (2k + 1 \pm \tfrac{1}{2})(\pi/n), \qquad k \quad \text{an integer.} \tag{7.2.14}$$

For θ in the range between these boundary directions,

$$(2k + \tfrac{1}{2})\pi < \alpha + n\theta = \arg \delta w < (2k + \tfrac{3}{2})\pi \tag{7.2.15}$$

and

$$\delta u = a\rho^n \cos(\alpha + n\theta) < 0. \tag{7.2.16}$$

That is, the region between these two boundary curves is a valley of $w(z)$ with respect to the point z_0. Note that this valley contains a direction of *steepest* descent (as indeed it should!) with $p = k$ in the first row of Table 7.1.

In a similar fashion, the boundary curve in (7.2.14) with the larger value of θ and the next boundary curve in the sequence will bound a *hill* of the exponent and contain in their range a direction of steepest *ascent*.

Let us now consider the result of the theorem for the case $n = 2$. The table then fills in as

$$\theta = -\frac{\alpha}{2} + \frac{\pi}{2}, \quad -\frac{\alpha}{2} + \frac{3\pi}{2}, \qquad \text{(descent)},$$

$$\theta = -\frac{\alpha}{2}, \quad -\frac{\alpha}{2} + \pi, \qquad \text{(ascent)}, \tag{7.2.17}$$

$$\theta = -\frac{\alpha}{2} + \frac{\pi}{4}, \quad -\frac{\alpha}{2} + \frac{3\pi}{4}, \quad -\frac{\alpha}{2} + \frac{5\pi}{4}, \quad -\frac{\alpha}{2} + \frac{7\pi}{4}, \qquad \text{(constant } u\text{)}.$$

Thus, we find that the two steepest descent directions are opposite to one another, as are the two steepest ascent directions, while these two pairs are perpendicular to one another. The four directions of constant u bisect the pairs of directions of ascent and descent. Figure 7.3 depicts the local structure of the function u in the neighborhood of z_0. We can see that the surface is locally a saddle. Consequently, *any point* at which at least one derivative of $w(z)$ vanishes is called a *saddle point*. When it is necessary to distinguish a saddle point where only the first derivative vanishes and the second does not, we identify this case as a *simple* saddle point. Consistent with this identification, we define the order of the saddle point as the order of the last vanishing derivative. In (7.2.9), then, the order of the saddle point is $n - 1$.

When $n = 3$, there will be three hills and three valleys symmetrically placed about the point z_0. The surface for this case is called a *monkey saddle*.

While it is a relatively straightforward matter to determine directions of steepest descent at a particular point, it is for all but a few cases rather difficult to determine paths of steepest descent in the large by analytical means. Fortunately, it is not really necessary to determine these paths in the large. The reason for this will be discussed later. However, the development of the theory requires that we gain insight into the identification of such paths in order to develop the asymptotic theory. To this end, we shall proceed to consider some examples.

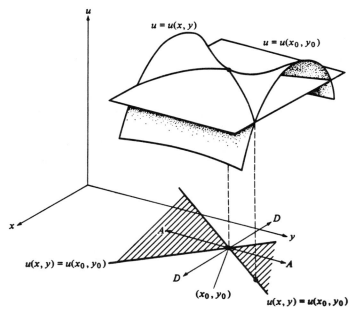

Fig. 7.3. The surface $u(x, y)$ in the neighborhood of a (simple) saddle point.

We consider first, the exponent

$$w(z) = z^2 - 2(1 + i)z, \tag{7.2.18}$$

with derivatives

$$w'(z) = 2z - 2(1 + i), \qquad w''(z) = 2. \tag{7.2.19}$$

We can readily see that this function has a simple saddle point at $z_0 = 1 + i$, with

$$w(z_0) = -2i, \qquad w''(z_0) = 2. \tag{7.2.20}$$

Focusing on this saddle point for the moment, we can see, first, that $\alpha = 0$; that is, the second derivative is (real and) positive at the saddle point. Thus, we can readily fill in the table as listed in (7.2.17) for simple saddle points as

$$\theta = \frac{\pi}{2}, \quad \frac{3\pi}{2}, \qquad \text{(descent)},$$

$$\theta = 0, \quad \pi, \qquad \text{(ascent)}, \tag{7.2.21}$$

$$\theta = \frac{\pi}{4}, \quad \frac{3\pi}{4}, \quad \frac{5\pi}{4}, \quad \frac{7\pi}{4}, \qquad \text{(constant } u\text{)}.$$

The choice of α is nonunique. However, other choices will yield the same set of directions in a different order. This is verified in the exercises.

To study other points in the complex plane, we might proceed to use Table 7.1 with $n = 1$. However, this is somewhat cumbersome. It is easier to study u, v, and ∇u, as we shall quickly see. From (7.2.19), we find for this example that

$$u = x^2 - y^2 - 2(x - y), \qquad \nabla u = 2[x - 1, y - 1],$$
$$v(x, y) = 2[xy - x - y] = 2(x - 1)(y - 1) - 2. \tag{7.2.22}$$

From the second line here, we see that the curves of constant v—the paths of steepest ascent and descent—are sections of the family of hyperboli centered at the point $(x_0, y_0) = (1, 1)$, the saddle point. In the special case $v(x, y) = v(x_0, y_0) = -2$, the hyperbola degenerates into the horizontal and vertical lines through $(1, 1)$. Speaking in terms of the semi-infinite lines directed *away* from this point, the vertical lines are paths of steepest descent, while the horizontal lines are paths of steepest ascent.

For this particular example, it is possible to determine analytically the path of steepest descent from any point in the complex plane. Nonetheless, in anticipation of examples in which this is not the case, we demonstrate in Fig. 7.4 a computer aid to this analysis implemented on this example. The figure shows an array of steepest descent directions for the function of this

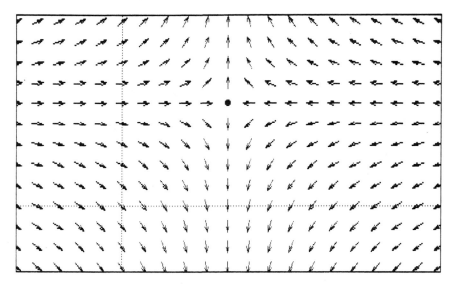

Fig. 7.4. Computer plot of the directions of steepest descent for the function of the example, $w(z) = z^2 - 2(1 + i)z$, saddle point at $z = 1 + i$, $-1 \leq x \leq 3$, $0.65 \leq y \leq 1.85$, x and y increments $= 0.2$.

example. The figure was generated on a Northstar Advantage using Northstar GBASIC and printed on an Epson printer. The program was given to me by A. Nuttall, who developed it on a Hewlett-Packard HP-85. The program simply plots arrows in the direction of $-\nabla u$ on a user-specified array of points. Examination of this figure will quickly confirm agreement with the preceding analytical discussion.

For the saddle point $1 + i$, the boundary curves between hill and valley are the curves $u = 0$, since this is the value of u at the saddle point. These are merely the four rays making angles $\pm \pi/4$ with the horizontal and emanating from the point $(1, 1)$:

$$(x - y)(x + y - 2) = 0; \qquad x - y = 0; \qquad x + y - 2 = 0. \quad (7.2.23)$$

We remark that for any other reference point, the boundary curves must have these two straight lines as asymptotes. That is, at infinity all valleys are the same!

As a second example, let us consider the function

$$w(z) = i[z^{1/2} - z/2], \qquad -\pi/2 < \arg z \leq 3\pi/2, \qquad (7.2.24)$$

for which the first two derivatives are

$$w'(z) = (i/2)[z^{-1/2} - 1], \qquad w''(z) = -(i/4)z^{-3/2}. \quad (7.2.25)$$

Our choice of arg z introduces a branch cut on the negative imaginary axis and yields a positive square root on the positive axis. Thus, we see that there is a saddle point at $z = 1$, with

$$w(1) = i/2, \qquad w''(1) = -i/4. \tag{7.2.26}$$

It is now straightforward to fill in the values in Table 7.1 or (7.2.17) for $z = 1$. First, $n = 2$, $a = \frac{1}{4}$, and $\alpha = -\pi/2$. Then

$$\theta = 3\pi/4, \quad 7\pi/4, \qquad \text{(descent)},$$
$$\theta = \pi/4, \quad 5\pi/4, \qquad \text{(ascent)}, \tag{7.2.27}$$
$$\theta = 0, \quad \pi/2, \quad \pi, \quad 3\pi/2, \qquad \text{(constant } u\text{)}.$$

Another point of interest for this exponent is the origin. In order to study the neighborhood of this point more closely, observe that $w(0) = 0$ and that δw is then approximated by the first term in (7.2.24). We write this approximation in polar representation as

$$\delta w \sim \rho^{1/2} e^{i(\phi + \pi)/2}, \qquad z = \rho e^{i\phi}. \tag{7.2.28}$$

We can see here that δw will be real and negative when $\phi = \pi$; this makes the total argument of δw equal to π.

Although it is possible to determine the paths of steepest descent exactly for both the origin and the saddle point, we will first proceed in a qualitative manner to analyze these paths. From (7.2.24) we see that if $|z| \to \infty$, it is necessary that $y \to -\infty$ or remain finite since the second term dominates at infinity. The latter choice is quickly eliminated, because the first term cannot approach $-\infty$ as $x \to \infty$ with y finite. As $y \to -\infty$ with x positive in order that the imaginary part of $w(z)$ remain finite (as it must on a path of steepest descent), y must approach infinity faster than x so that the y dependence in the square root can balance the linear second term. Under this assumption, $w(z)$ on the path of steepest descent can be approximated by

$$w(z) \sim i[e^{-i\pi/4}|y|^{1/2} - z/2]. \tag{7.2.29}$$

Thus, for $v(x, y)$ to remain finite, it is necessary that

$$(|y|^{1/2}/\sqrt{2}) - (x/2) \sim 0; \qquad |y| \sim x^2/2. \tag{7.2.30}$$

Thus, all steepest descent paths in this quadrant must be parabolic near infinity. We can perform similar analysis in the left half plane and discover that the same is true there, the parabolas again turning downward.

The analysis just described is typical. It is not often that one can analytically determine the paths of steepest descent. Indeed, for this example, except for some special points, anything more than such qualitative analysis is impractical. However, such qualitative information about the descent paths will almost always suffice. When it does not, we can always resort to computer analysis for fields of descent directions, as described earlier, or for actual construction of the steepest descent paths. Bleistein and Handelsman [1975] have described an analytical technique for finding fields of descent directions. However, even in that technique, it is necessary at least to find the curves on which u_x, u_y, v_x, and v_y are equal to zero.

For the two distinguished points, the saddle point and the origin, the equations of the paths of steepest descent are determined by setting Im w in (7.2.24) equal to $\frac{1}{2}$ or 0, respectively. After some algebra, this leads to the equations

$$2y = 1 - x^2, \qquad 4y^2 = x^3(x - 4), \tag{7.2.31}$$

respectively. In the former equation, $-\infty < x < \infty$; in the latter, $-\infty < x \leq 0$. In Fig. 7.5, we display a computer plot of the directions of steepest descent for a neighborhood of the origin. The features already discussed can be seen to be consistent with this figure.

In Exercise 2.11, we outlined a proof of *Jordan's lemma*. Because of the significance of that result in the method of steepest descent, we repeat that discussion here. Thus, let us consider the integral

$$I = \int_C e^{iz} \, dz. \tag{7.2.32}$$

The contour C is the counterclockwise semicircular path of radius R on which $0 \leq \arg z \leq \pi$. Jordan's lemma states that

$$|I| \leq \int_C |e^{iz}| |dz| \leq \pi. \tag{7.2.33}$$

The result is independent of R and hence remains true in the limit $R \to \infty$.

Let us consider the integral

$$I = \int_C f(z) e^{iz} \, dz, \tag{7.2.34}$$

with C as for (7.2.33). By using standard estimation techniques, we obtain the estimate

$$|I| \leq \max_{z \text{ on } C} |f(z)| \int_C |e^{iz}| |dz| \leq \pi \max_{z \text{ on } C} |f(z)|. \tag{7.2.35}$$

Now we can see that a sufficient condition for the integral in (7.2.34) to decay to zero with increasing radius is that

$$\max_{z \text{ on } C} |f(z)| \to 0 \qquad \text{as} \qquad |z| \to \infty. \tag{7.2.36}$$

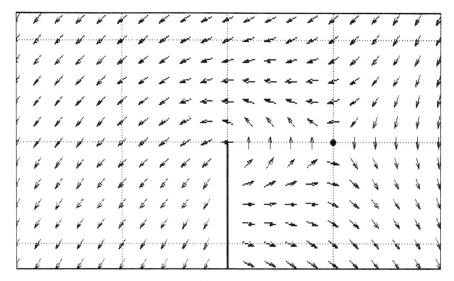

Fig. 7.5. Computer plot of the directions of steepest descent for a neighborhood of the origin. $w(z) = i[z^{1/2} - z/2]$, saddle point at $z = 1$, branch point at $z = 0$, $-z \leq x \leq 2$, $-1.25 \leq y \leq 1.25$, x and y increments $= 0.2$.

This result has important consequences for replacement of contours on boundary curves by contours of descent. For the integrand in (7.2.34)—despite the lack of a large parameter λ—suppose that we were contemplating replacement of the integral along the positive real axis with an integral along a ray in the valley of the exponent, that is, a ray in the upper half z plane. Then the integral (7.2.34) would be the path connecting those two rays. We see here that this integral on an arc makes no contribution to the final replacement so long as (7.2.36) is true.

In most applications, the exponent will not be iz and the path of integration will not be an arc, but some connecting path between a contour integral on the boundary of a valley and a descent contour in the adjacent valley. However, if we can determine that a transformation to a linear exponent yields an integral with the essential features of (7.2.34), then the result obtained here must hold. That is, as the connecting contour recedes toward a region where the exponent approaches infinity in absolute value, the integral on the connecting curve must approach zero.

To complete this discussion, we shall now prove Jordan's lemma.

Proof The first part of (7.2.33) is merely the result that the absolute value of an integral is less than or equal to the integral of the absolute value. Thus, we consider only the second part. We denote by R and ϕ the polar co-ordinates of z. Then

$$z = R^{i\phi}, \qquad dz = iRe^{i\phi}\, d\phi, \qquad |dz| = R\, d\phi, \qquad |e^{iz}| = e^{-R\sin\phi}. \quad (7.2.37)$$

We can now use these results in (7.2.33) to find that

$$\int_C |e^{iz}|\,|dz| = \int_0^\pi e^{-R\sin\phi} R\, d\phi = 2\int_0^{\pi/2} e^{-R\sin\phi} R\, d\phi. \quad (7.2.38)$$

To obtain the second part here, we use the fact that the integrand is even about $\pi/2$.

In order to estimate this integral, we shall first simplify the exponent by obtaining a linear bound on the function $\sin\phi$. By considering the graph of $\sin\phi$ on the interval $(0, \pi/2)$, it becomes apparent that the straight line connecting the endpoints $(0, 0)$ and $(\pi/2, 1)$ always lies below $\sin\phi$. That is,

$$0 \le \sin\phi \le (2/\pi)\phi, \qquad 0 \le \phi \le \pi/2. \quad (7.2.39)$$

Consequently,

$$\int_C |e^{iz}|\,|dz| \le 2\int_0^{\pi/2} e^{-2R\phi/\pi} R\, d\phi = \pi[1 - e^{-R}] \le \pi. \quad (7.2.40)$$

This completes the proof.

Exercises

7.15 Suppose that in (7.2.9), α is replaced by $\alpha + 2k\pi$, with k an integer. Show that this does not change the set of directions in Table 7.1 but only shifts p by $-k$.

7.16 Suppose that

$$w(z) - w(z_0) \sim ae^{i\alpha}(z - z_0)^\gamma, \qquad \gamma > 0,$$

in some sector of the z plane (or multisheeted Riemann surface) with apex at z_0. Show that the directions of steepest descent at z_0 are given by

$$\theta_p = -(\alpha/\gamma) + [(2p + 1)\pi/\gamma], \qquad p \quad \text{an integer},$$

and that the associated valleys are bounded by $\theta_p \pm (\pi/2\gamma)$.

7.17 Suppose now that

$$w(z) \sim a^{i\alpha}(z - z_0)^\gamma, \qquad \gamma < 0,$$

again in some sector of the z plane or multisheeted Riemann surface with apex at z_0. Show that now $w(z) \to -\infty$ as $\to z_0$ at angle θ_p of the preceding exercise. Furthermore, in the sector with boundaries $\theta_p \pm \pi/2$, $\operatorname{Re} w(z) \to -\infty$

as $z \to z_0$. We remark that a path of steepest descent might actually terminate at such a point in the finite part of the z plane with valleys having boundaries that "pinch" together at such a singularity.

7.18   Let

$$w(z) = \log z - z, \qquad -\pi < \arg z \le \pi.$$

Show that $w(z)$ has a saddle point at $z = 1$, with paths of steepest descent being the semi-infinite x axis, $1 \le x < \infty$ and the line segment $0 < x \le 1$.

7.19   Let

$$w(z) = z - z^3/3.$$

(a)   Show that $w(z)$ has two saddle points at $z = \pm 1$, with the directions of steepest descent being vertical at -1 and horizontal at $+1$.

(b)   Show that the paths of steepest ascent and descent from either of these saddle point are sections of the curves

$$y[x^2 - (y^2/3) - 1] = 0, \qquad y = 0, \quad x^2 - (y^2/3) - 1 = 0.$$

(c)   Conclude that the paths of steepest descent from the point -1 are the two sections of hyperbola that leave that point vertically and have as asymptote at infinity the rays from the origin making angles $\pm 2\pi/3$ with the positive x axis. The paths of steepest descent from $+1$ are along the x axis. One path extends to plus infinity, while the other terminates at the saddle point at -1.

(d)   Show that the valleys with respect to the point $z = 1$ have as asymptotes the sectors $\pi/2 \le |\arg z| \le 5\pi/6$. A more detailed depiction of the hills and valleys of this exponent with respect to the two saddle points is shown in Fig. 7.6.

7.20   Let

$$w(z) = i \cos z = i \cos x \cosh y + \sin x \sinh y, \qquad -\pi/2 < x \le 3\pi/2.$$

(a)   Show that $w(z)$ has saddle points at $z = 0, \pi$, with the steepest descent directions at π being $\pi/4, -3\pi/4$, and the steepest descent directions at 0 being $-\pi/4, 3\pi/4$.

(b)   By considering $u(x, y)$, show that on the path of steepest descent, y must approach $+\infty$ on such a path when $\sin x$ is positive and that y must approach $-\infty$ on such a path when $\sin x$ is negative.

7.21   Let

$$I(\lambda) = \int_C e^{i\lambda z^2} \, dz,$$

with C the counterclockwise quarter circle of radius R in the first quadrant

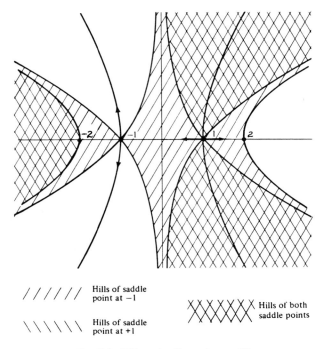

////// Hills of saddle
        point at −1

\\\\\\ Hills of saddle
        point at +1

XXXXXX Hills of both
XXXXXX saddle points

Fig. 7.6. Hills and valleys of $z - z^3/3$.

of the z plane, $0 \le \phi \le \pi/2$. Show that

$$\lim_{R \to \infty} I(\lambda) = 0.$$

7.22 (a) The purpose of this exercise is to develop Jordan's lemma and its corollary for integrals in the lower half z plane. Let

$$I = \int_C e^{-iz} \, dz,$$

with C now a sector of a circle of radius R in the lower half z plane, $-\pi \le \arg z \le 0$ on C. Show that, again, as in Jordan's lemma, $|I| \le \pi$, independent of R.

(b)   Now consider the integral

$$I = \int_C f(z) e^{-iz} \, dz,$$

on the same contour C. Show that if (7.2.36) is true, then

$$\lim_{R \to \infty} I = 0.$$

7.3   FORMULAS FOR THE METHOD OF STEEPEST DESCENTS

Let us suppose now that we are given an integral of the form (7.2.1) in which the contour C is replaced by a path D, which is a path of steepest descent away from some point z_0. Furthermore, let us suppose that z_0 is a saddle point of order $n - 1$ as defined by (7.2.9) and (7.2.10). Then, the path of steepest descent must initially have one of the directions of Table 7.1; let us denote that direction by θ_p,

$$\theta_p = (2p + 1)\pi/n - \alpha/n, \tag{7.3.1}$$

for some fixed integer p.

On the path D, $w(z) - w(z_0)$ must be real and decrease monotonically to $-\infty$. Thus, we may introduce the new variable of integration

$$t = -[w(z) - w(z_0)] \tag{7.3.2}$$

and rewrite (7.2.1) as an integral of the form (7.1.1)

$$I(\lambda) = e^{\lambda w(z_0)} \int_0^\infty f(t) e^{-\lambda t} \, dt. \tag{7.3.3}$$

In this equation, $f(t)$ is defined by

$$f(t) = g(z) \frac{dz}{dt} = -\frac{g(z)}{w'(z)} \bigg|_{z = w^{-1}(w(z_0) - t)} \tag{7.3.4}$$

When $f(t)$ has an asymptotic expansion of the form (7.1.3), $I(\lambda)$ has an expansion of the form (7.1.4) multiplied by an exponential factor as in (7.3.4); that is,

$$I(\lambda) \sim e^{\lambda w(z_0)} \sum_{k=0}^\infty \frac{c_k \Gamma(a_k + 1)}{\lambda^{a_k + 1}}. \tag{7.3.5}$$

It is, admittedly, extremely tedious to find even two terms of the asymptotic series for $f(t)$ in terms of the given functions $g(z)$ and $w(z)$. However, determination of the first term is fairly straightforward, as we shall now demonstrate for a few examples. We remark that for this first term, we need only know the leading term in the expansion of $g(z)$ itself and the leading term in the expansion of dz/dt.

Example: Saddle Point at a Regular Point of $g(z)$ First, let us consider the case in which $g(z)$ is analytic at z_0, with $g(z_0) \neq 0$. Then, this constant is the leading term in the asymptotic expansion of $g(z)$. Now let us consider dz/dt, which is not as transparent. From (7.2.9) and (7.3.2), we may write the approximation for t as

$$t \sim -[w(z) - w(z_0)] \sim -ae^{i\alpha}\rho^n e^{in\theta_p} = -a\rho^n e^{(2p+1)\pi i}. \tag{7.3.6}$$

We shall now choose the factor of -1 on the right side of this equation so as to make $\arg t = 0$. That is, we set

$$t = e^{-(2p+1)\pi i}[w(z) - w(z_0)] \sim a e^{-in\theta_p}(z - z_0)^n. \qquad (7.3.7)$$

Now we solve this equation approximately for $z - z_0$ and thence for dz/dt:

$$z - z_0 \sim \left[\frac{t}{a}\right]^{1/n} e^{i\theta_p}; \qquad \frac{dz}{dt} \sim \frac{t^{1/n-1} e^{i\theta_p}}{na^{1/n}}. \qquad (7.3.8)$$

Now from (7.3.4) and (7.3.8), it follows that

$$c_0 = \frac{g(z_0)}{n}\left[\frac{n!\, e^{i\theta_p}}{|w^{(n)}(z_0)|}\right]^{1/n}, \qquad a_0 = 1/n - 1. \qquad (7.3.9)$$

We now substitute these results into (7.3.5) to obtain the leading term of the asymptotic expansion

$$I(\lambda) \sim \frac{g(z_0)}{n}\left[\frac{n!}{\lambda|w^{(n)}(z_0)|}\right]^{1/n} \Gamma\left[\frac{1}{n}\right] e^{[\lambda w(z_0) + i\theta_p]}. \qquad (7.3.10)$$

We see in this result that the leading term of the asymptotic expansion is expressed in terms of the location of the saddle point z_0; the direction of steepest descent θ_p, which, in turn, depends on the order n of the first non-vanishing derivative at the saddle point and the argument α of that derivative; and the values of the exponent $w(z_0)$, amplitude function $g(z_0)$, and, finally, the value of the first nonvanishing derivative $|w^{(n)}(z_0)|$.

Special Case: Simple Saddle Point A special case of this result that arises most often is the case $n = 2$, namely, the simple saddle point. Now (7.3.10) becomes

$$I(\lambda) \sim g(z_0)\sqrt{\frac{\pi}{2\lambda|w''(z_0)|}}\, e^{[\lambda w(z_0) + i\theta_p]}. \qquad (7.3.11)$$

There are two distinct choices of p and, hence, θ_p for this example. They yield leading terms of opposite sign for the two steepest descent paths from a simple saddle point. The two paths leave the saddle point in opposite directions. Quite often, the deformation of the contour C of a given integral will produce a contour passing *through* the saddle point from one valley to the other. In such a case, the replacement is given in terms of the difference of integrals on the two steepest descent paths. That difference is just twice the result (7.3.11) with one of the choices of p, namely, the choice for which the descent path D and the original contour C have the same orientation. We remark that for analytic functions w and g, the next term $[O(\lambda^{-1})]$ in each of the two expansions is of the *same* sign and identical. Consequently, these terms will subtract when the two integrals are combined. The following terms $[O(\lambda^{-3/2})]$ are of *opposite* signs again but otherwise identical. Thus, these terms will again add to yield twice one of them, the choice again depending on the orientation of the original contour. This is the same phenomenon as was discussed in Section 2.7 in the context of the method of stationary phase.

Example: Branch Point of $g(z)$ Let us suppose now that z_0 is not only a saddle point of $w(z)$, but also a branch point of $g(z)$; that is,

$$g(z) \sim g_0(z - z_0)^{\beta - 1}, \qquad z \to z_0, \tag{7.3.12}$$

in some sector containing the direction of steepest descent. Then, we use the first equation in (7.3.8) as a leading order estimate of $z - z_0$ and use this result in (7.3.12) to obtain

$$g(z) \sim g_0 \left[\frac{te^{i\theta_p}}{a} \right]^{(\beta - 1)/n}, \qquad c_0 = \frac{g_0}{n} \left[\frac{n! \, e^{i\theta_p}}{|w^{(n)}(z_0)|} \right]^{\beta/n}, \qquad a_0 = \beta/n - 1. \tag{7.3.13}$$

We can now use these results in (7.3.5) to write the leading term of the asymptotic expansion of $I(\lambda)$ for this case. The result is

$$I(\lambda) \sim \frac{g_0}{n} \left[\frac{n!}{\lambda |w^{(n)}(z_0)|} \right]^{\beta/n} \Gamma \left[\frac{\beta}{n} \right] e^{[\lambda w(z_0) + i\beta\theta_p]}. \tag{7.3.14}$$

We remark that when we leave the coefficient a in its original form from Eq. (7.2.9), the formula remains valid even when n is not an integer. That is, in this case, we set

$$w(z) \sim w(z_0) + ae^{i\alpha}(z - z_0)^n \tag{7.3.15}$$

and obtain the result

$$I(\lambda) \sim \frac{g_0 \Gamma(\beta/n)}{n(\lambda a)^{\beta/n}} e^{[\lambda w(z_0) + i\beta\theta_p]}. \tag{7.3.16}$$

It is interesting in (7.3.14) to examine the dependence of the algebraic order in λ on the order of vanishing of the function $g(z)$ at z_0, $\beta - 1$, and on the order of the saddle point or the order of vanishing of $\delta w(z)$, n. We see here that increasing β—increasing order of vanishing of $g(z)$ at z_0—causes an increasing order of λ in the denominator, that is, a *decreasing* algebraic order in λ of $I(\lambda)$. On the other hand, increasing n—increasing the order of vanishing of $\delta w(z)$ at z_0—causes a *decrease* in the order of λ in the denominator, that is, an *increasing* algebraic order in λ of $I(\lambda)$.

Qualitatively, larger n characterizes an exponent that decays less rapidly in the neighborhood of the saddle point along the path of integration; hence, a larger integral. On the other hand, larger β characterizes more rapid decay toward zero of the integrand as $z \to z_0$.

Special Case: $n = 1$ The case in which there is a branch point at z_0 but no saddle point is of special interest. Therefore, we write down the result (7.3.14) for this case as

$$I(\lambda) \sim \frac{g_0 \Gamma(\beta)}{[\lambda |w'(z_0)|]^\beta} e^{[\lambda w(z_0) + i\beta\theta_1]}, \qquad \theta_1 = \pi - \alpha. \qquad (7.3.17)$$

From the results presented here, we can see that detailed information about the path of steepest descent is unnecessary. We need only know about this path at the saddle point itself. Indeed, let us suppose that D_1 is a directed path from z_0 that is the same as the steepest descent path D for some finite length but then differs from D although remaining a path of descent. Then, the asymptotic expansion of the integral along D_1 would, of necessity, differ from the integral along D at worst by a quantity that is exponentially smaller in λ than the asymptotic series arising from the critical point z_0. Thus, we may say that the paths D and D_1 are *asymptotically equivalent*. We remark that this observation is the basis for a method known as the *saddle point method*. Also, there are examples in which those exponentially small differences between the integrals on these two paths have a physical interpretation. Thus, it behooves the user to recognize these differences.

Suppose now that D_2 is *any* descent path from z_0 and that all of the approximations for $w(z)$ and $g(z)$ are valid in some sector with apex z_0 containing both D and D_2 for some finite length. Then, again, the integrals can only differ by terms that are exponentially smaller in λ than the contribution from z_0 itself, these contributions arising from poles or other critical points in the region between the two paths, which, of necessity, lie wholly in the valley of $w(z)$ with respect to z_0.

In summary, then, we require the following detailed knowledge about the path of steepest descent: (i) the direction of the path at the saddle point and (ii) the series expansion of $w(z)$ and $g(z)$ at z_0 to a sufficient number of terms to write down the series for $f(t)$ to the desired order. Except for this information, qualitative knowledge of the path of steepest descent or only a path of descent will suffice for application of the method of steepest descents.

Exercises

7.23   (a)   Let

$$I(\lambda) = \int_{-\infty}^{\infty} \frac{e^{i\lambda z^2}}{1 + z^6} dz.$$

Show that the exponent has a simple saddle point at $z = 0$, with steepest descent paths from that point being the rays at angles $\pi/4$ and $-3\pi/4$.

(b)   With those two contours denoted by D_1 and D_2, respectively, show that

$$I(\lambda) = \int_{D_1 - D_2} \frac{e^{i\lambda z^2}}{1 + z^6}\, dz + \frac{2\pi}{3} \exp[-\lambda(\sqrt{3} - i)/2 - i\pi/3].$$

(c)   Use the steepest descent formula (7.3.11) to conclude that

$$I(\lambda) \sim \sqrt{\pi/\lambda}\; e^{i\pi/4}.$$

(d)   In the integral over the path D_1, introduce the new variable of integration σ, $z = \sigma e^{\pi i/4}$, and apply Watson's lemma to the resulting integral to obtain the asymptotic series

$$\int_{D_1} \frac{e^{i\lambda z^2}}{1 + z^6}\, dz \sim \frac{e^{i\pi/4}}{2} \sum_{k=0}^{\infty} \frac{i^k \Gamma(3k + \frac{1}{2})}{\lambda^{3k + 1/2}}.$$

(e)   On the path D_2, introduce the new variable of integration σ, $z = \sigma e^{5\pi i/4}$, and conclude that this integral is indeed the negative of the integral on the path D_1.

7.24   Let

$$I(\lambda) = \int_0^\infty \frac{e^{i\lambda z^2}}{1 + z^4}\, dz.$$

Use the method of steepest descents to determine that

$$I(\lambda) \sim \tfrac{1}{2}\sqrt{\pi/\lambda}\; e^{i\pi/4}.$$

7.25   Let

$$I(\lambda) = \int_0^\infty \frac{e^{i\lambda z^3}}{1 + z^2}\, dz.$$

Show that

$$I(\lambda) = \tfrac{1}{3}[2/\lambda]^{1/3}\Gamma(\tfrac{1}{3})e^{\pi i/6}.$$

7.4   THE METHOD OF STEEPEST DESCENTS: IMPLEMENTATION

We shall now develop the *method of steepest descents* for the asymptotic analysis as $\lambda \to \infty$ of integrals of the type (7.2.1). The method consists of five basic steps.

(1) Identify the possible critical points of the integrand. These consist of (i) the endpoints of integration, (ii) points at which $g(z)$ or $w(z)$ fails to be analytic, and (iii) points at which $w'(z) = 0$, that is, *saddle points*.

(2) Determine the paths of (steepest) descent from each of the critical points except for poles, and

(3) justify via Cauchy's integral theorem, perhaps with the aid of Jordan's lemma, the deformation of the original contour of integration C onto one or more of the paths of (steepest) descent, possibly with the addition of residues, to account for poles in the region enclosed between contours.

(4) Determine the asymptotic expansion of the integrals on the descent paths that arise as a result of the deformation in (3) by using the formulas of Section 7.3.

(5) Sum the asymptotic expansions obtained to determine the asymptotic expansion of $I(\lambda)$.

Twice we have set the word *steepest* in parentheses to remind the reader that complete detail about the path of steepest descent is necessary only in the neighborhood of the critical point; away from that point, a path of descent, that is, qualitative information, will suffice. Step (1) is a straightforward application of complex function theory. Step (2) was discussed in Section 7.2 and step (4) in Section 7.3. Step (3) is the heart of the method of steepest descents; once done, the method is reduced to calculation.

As implied by the sentence continuation, steps (2) and (3) are not carried out in strict sequential order. There is no need to analyze paths of steepest descent from critical points that in advance can be seen for certainty not to contribute to the asymptotic expansion of the original integral. With practice, then, steps (2) and (3) become progressively more integrated into a single step.

Suppose that we have a set of candidate critical points in step (1), say, $z_0, z_1, \ldots, z_n$. Then, as can be seen from (7.3.5), the asymptotic expansions from these critical points have the asymptotic order $O(e^{\lambda w(z_j)}\lambda^{-a_{0j}})$, $j = 1, \ldots, n$. We might argue, then, that only the terms with the largest exponential order in λ ought to be retained, with all other contributions viewed as being asymptotically zero in comparison. In applications, however, it is often true that each critical point is associated with a different physical phenomena. Thus, we may keep these *subdominant contributions* because of their physical significance. Also, as a function of other parameters in the integrand, such as observation point in space and/or time, the value of the exponent may change in such manner as to interchange the roles of dominant and subdominant critical points. Such an interchange is a manifestation of the *Stokes phenomenon*, which was mentioned in Section 4.7. This relationship between interchange of dominance of critical points and the Stokes phenomenon has been discussed extensively by Bleistein and Handelsman [1975].

We shall now demonstrate the method of steepest descent through the application to some specific integrals. To begin, let us consider the integral

$$I(\lambda) = \int_C \frac{e^{i\lambda[(z^2/2) - \gamma z]}}{z^2 - 1} \, dz. \tag{7.4.1}$$

The contour C is a path from $-\infty$ to ∞ passing above the poles of the integrand at $z = \pm 1$. In addition to these poles, the only possible critical points are saddle points of the exponent:

$$w(z) = i[(z^2/2) - \gamma z], \qquad w'(z) = i[z - \gamma], \qquad w''(z) = i. \tag{7.4.2}$$

From the second equation here, we see that $w(z)$ has a saddle point at $z = z_0 = \gamma$; from the third equation, we see that $n = 2$ and $\alpha = \pi/2$ at the saddle point. Thus, the directions of steepest descent at the saddle point, as determined from Table 7.1 in Section 7.2 or from (7.2.17), are $\pi/4$ and $7\pi/4$. By setting $v(x, y) = v(\mathrm{Re}\ \gamma, \mathrm{Im}\ \gamma)$, we find that the paths of steepest descent and ascent through the saddle point are the straight lines (degenerate hyperbolas)

$$x - \mathrm{Re}\ \gamma \mp [y - \mathrm{Im}\ \gamma] = 0. \tag{7.4.3}$$

The determination of descent directions at the saddle point leads us to conclude that the upper sign in this equation yields the paths of steepest descent while the lower sign yields the paths of steepest ascent.

Let us restrict our attention to real values of γ. There are then five cases of interest: $\gamma < -1$, $\gamma = -1$, $-1 < \gamma < 1$, $\gamma = 1$, and $1 < \gamma$. We shall discuss the first, third, and fifth of these cases now. We shall discuss the case $\gamma = -1$ at the end of the section because this case requires that we extend the theory. That extension would apply to the case $\gamma = 1$ as well.

Figure 7.7 depicts the paths of steepest descent through the saddle point. It also depicts two contours C_1 and C_2 that together replace a finite section

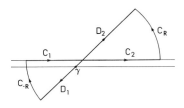

Fig. 7.7. Contours for the discussion of the integral (7.4.1).

of the original contour C. Two connecting paths C_{-R} and C_R are also shown. By using the extension of Jordan's lemma developed in Exercise 7.22, we can conclude that in the limit as $R \to \infty$ the integral along each of the last pair of contours approaches zero while in the same limit, the sum of the integrals along C_1 and C_2 approaches the integral along C itself.

We conclude then that

$$I(\lambda) = \int_{D_2} - \int_{D_1} \frac{e^{i\lambda[(z^2/2) - \gamma z]}}{z^2 - 1} \, dz + 2\pi i \sum \text{residues.} \tag{7.4.4}$$

Whether or not there are residue contributions depends on whether or not the closed paths $C_1 + D'_1 + C_{-R}$ and $C_2 + C_R - D'_2$ enclose either of the poles at $z = \pm 1$. Here by D'_1 we mean D_1 plus the small piece of D_2 connecting the saddle point with C. Similarly, D'_2 excludes this small connector.[†] The sector $C_2 + C_R - D'_2$ will not enclose either of these poles for any choice of γ. The first contour will enclose

(i)   no poles when $\gamma < -1$;
(ii) the pole at -1 when $-1 < \gamma < 1$;
(iii) both poles when $1 < \gamma$.

Note that the closed path on the left is traversed in the clockwise manner, and, hence, that residue must be taken with a *minus sign*:

$$\pm \text{ residue}(z = \pm 1) = \tfrac{1}{2} e^{i\lambda[1/2 \mp \gamma]}. \tag{7.4.5}$$

Now we shall calculate the leading term of the asymptotic expansion of the integrals along the paths of steepest descent. The result is given by (7.3.11) with $g(z) = 1/(z^2 - 1)$, $\theta_p = \pi/4, 7\pi/4$, and $w(z)$ given by (7.4.2). The result for the integral on D_2 is

$$\int_{D_2} \frac{e^{i\lambda[(z^2/2) - \gamma z]}}{z^2 - 1} \, dz \sim \frac{1}{\gamma^2 - 1} \sqrt{\frac{\pi}{2\lambda}} e^{-i\lambda\gamma^2/2 + i\pi/4}, \tag{7.4.6}$$

with the integral on the contour D_1 being just the negative of this one. Thus, as can be seen from (7.4.4), the integral along C will have as saddle point contribution just twice the result in (7.4.6). This phenomenon for simple saddle points was noted in Section 7.3.

[†] This is a minor detail. Were we to replace C by a contour on the axis except for semicircles over the poles, these "tails" used to obtain closed contours would be unnecessary. However, it would then be necessary to depict each case of saddle point relative to poles in separate figures.

By properly summing the residue and saddle point contributions, we obtain

$$
I(\lambda) \sim
\begin{cases}
\dfrac{1}{\gamma^2 - 1}\sqrt{\dfrac{2\pi}{\lambda}}\, e^{-i\lambda\gamma^2/2 + i\pi/4}, & \gamma < -1, \\[3ex]
\pi i e^{i\lambda[1/2 + \gamma]} + \dfrac{1}{\gamma^2 - 1}\sqrt{\dfrac{2\pi}{\lambda}}\, e^{-i\lambda\gamma^2/2 + i\pi/4}, & -1 < \gamma < 1, \quad (7.4.7) \\[3ex]
2\pi i e^{i\lambda/2}\cos\gamma + \dfrac{1}{\gamma^2 - 1}\sqrt{\dfrac{2\pi}{\lambda}}\, e^{-i\lambda\gamma^2/2 + i\pi/4}, & 1 < \gamma.
\end{cases}
$$

We turn now to another example,

$$
I(\lambda) = \int_0^\infty z^{-3/4}\exp(i\lambda[z^{1/2} - z/2])\,dz. \tag{7.4.8}
$$

We shall use the same range of the argument z as was used in Section 7.2 and defined by (7.2.4). The integrand has only two critical points, the saddle point at $z = 1$ and the endpoint at the origin, which point is also a branch point of the integrand. We have already determined the paths of steepest descent for these two critical points in Section 7.2; see (7.2.31). The paths are shown in Fig. 7.8.

The original path of integration, denoted by C in Fig. 7.8, can be replaced by the integral on the path $D_1 - D_2 + D_3$. The integrals along the path connecting C to D_1 and along the path connecting D_2 and D_3 can be shown to decay to zero as the radius of each such path increases beyond all bounds. This, again, is an application of Jordan's lemma, in this case, with a reversal

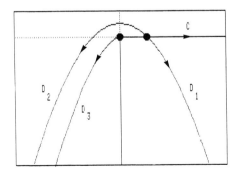

Fig. 7.8. Descent paths for the integral in (7.4.8).

of sign in the exponent and some preliminary analysis to obtain exactly a linear exponent. The details are carried out in the exercises.

We denote by I_3 the integral along the path D_3. The leading term of the asymptotic expansion of this integral is obtained by using (7.3.16), with

$$a = 1, \qquad \alpha = \pi/2, \qquad \theta_p = \pi, \qquad w(z_0) = 0, \qquad n = \tfrac{1}{2}, \qquad g_0 = 1, \qquad \beta = \tfrac{1}{4}.$$

The first two values were determined from (7.3.15) and (7.2.24). The descent direction θ_p was determined from (7.2.28). The values of n and $w(z_0)$ are determined from (7.2.24) and g_0 and β from comparison of (7.4.8) with (7.3.12).

By using these values in (7.3.16), we obtain the result

$$I_3(\lambda) \sim 2\sqrt{\pi/\lambda}\, e^{i\pi/4}. \tag{7.4.9}$$

The asymptotic expansion of the contribution from the saddle point can be determined by using the formula (7.3.11) along with (7.2.26) and (7.2.27). The result for $I(\lambda)$ then becomes

$$I(\lambda) \sim 2\sqrt{\pi/\lambda}\, e^{i\pi/4} + \sqrt{2\pi/\lambda}\, e^{i\lambda/2 - i\pi/4}. \tag{7.4.10}$$

We remark that both of these terms are of the same order in λ, with one term having arisen from a saddle point at which both $w(z)$ and $g(z)$ are analytic and the other having arisen from an endpoint at which both $w(z)$ and $g(z)$ have a branch point.

As a third example, we consider the Sommerfeld integral representations of the Hankel functions $H_0^{(j)}$, $j = 1, 2$, given by

$$\pi H_0^{(j)}(kr) = \int_{C_j} e^{ikr\cos z}\, dz, \qquad j = 1, 2. \tag{7.4.11}$$

The contours C_1 and C_2 are shown in Fig. 7.9. We have used kr rather than λ for the large parameter because this is the form of independent variable that most often arises in practice, with r the cylindrical radius and k the wave number.

The exponent

$$w(z) = \cos z = \sin x \sinh y + i \cos x \cosh y \tag{7.4.12}$$

is 2π periodic in $x = \mathrm{Re}\, z$. Furthermore, $\mathrm{Re}\, w(z) \to +\infty$ whenever $\sin x$ is positive and $y \to +\infty$ or whenever $\sin x$ is negative and $y \to -\infty$. These regions are stippled in the figure. The alternate unstippled regions at infinity are the regions where $\mathrm{Re}\, w(z) \to -\infty$; that is, they are the valleys at infinity of the exponent. The contours C_1 and C_2 each pass from one such valley at infinity to another.

If there are any saddle points in the region of the figure, these saddle

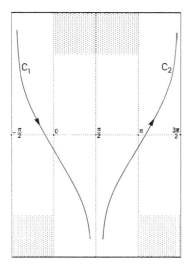

Fig. 7.9. Sommerfeld contours for the Hankel functions.

points must repeat with period 2π in the entire z plane. Let us therefore focus our attention on the analysis of this one region. This is an example of combining steps (2) and (3) in the outline at the beginning of this section. By differentiating (7.4.12), we find that

$$w'(z) = -i \sin z, \qquad w''(z) = -i \cos z. \qquad (7.4.13)$$

There are two saddle points in the region of interest,

$$z_+ = 0, \quad z_- = \pi; \qquad w(z_\pm) = \pm i, \quad w'(z_\pm) = \mp i. \qquad (7.4.14)$$

At z_+, the directions of steepest descent are $-\pi/4$ and $3\pi/4$, while at z_-, they are $\pi/4$ and $7\pi/4$. The reader should verify these facts by using (7.2.17).

The paths of steepest descent are given by

$$\cos x \cosh y = \pm 1. \qquad (7.4.15)$$

A pointwise determination of these contours without the aid of a computer would be quite tedious. However, a qualitative determination of the contours is not too difficult. We already know that $y \to \pm\infty$ on the paths of steepest descent with $\mp \sin x > 0$ on these paths. To the left of z_+, say, $-\pi < x < 0$, $\sin x < 0$ and, therefore, $y \to +\infty$. In (7.4.15), if the left side is to remain finite in this limit, then $\cos x$ must approach zero as y approaches infinity. Thus, we conclude that $x \to -\pi/2$ along the path of steepest descent.

Arguing in similar fashion, one can see that the paths of steepest descent in the region $0 < x < \pi$ must have the line $x = \pi/2$ as asymptote, and for

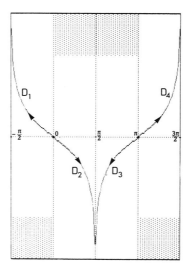

Fig. 7.10. Descent contours for $w(z) = i \cos z$.

$\pi < x < 2\pi$, the descent contour must have the line $x = 3\pi/2$ as asymptote. These results, along with the initial directions of descent determined earlier, lead us to conclude that the descent contours for this exponent are as shown in Fig. 7.10. We leave as an exercise to confirm that the leading terms of the asymptotic expansions of the integrals (7.4.11) are

$$H_0^{(1)}(kr) \sim \sqrt{2/\pi kr}\, e^{ikr - i\pi/4}, \qquad H_0^{(2)}(kr) \sim \sqrt{2/\pi kr}\, e^{-ikr + i\pi/4}. \qquad (7.4.16)$$

As a last example for this section, we return to consideration of the case $\gamma = -1$ for the integral (7.4.1),

$$I(\lambda) = \int_C \frac{e^{i\lambda[(z^2/2) + z]}}{z^2 - 1}\, dz, \qquad (7.4.17)$$

with C as described following (7.4.1). We cannot replace the contour C by contours of steepest descent emanating from the pole at $z = -1$. Instead, let us consider the contour D of Fig. 7.11, which consists of the path of constant v connecting the two valleys of the exponent except for the avoidance of the pole by a small semicircle. As we have in each of the preceding examples, we can confirm that the integrals on C and D are equal. Let us now introduce the new variable of integration σ:

$$w(z) - w(-1) = (i/2)[z + 1]^2 = t, \qquad (7.4.18)$$

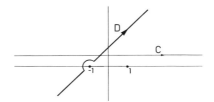

Fig. 7.11. The contour D.

which has as solution

$$z + 1 = \sqrt{2}\, e^{i\pi/4} t^{1/2}, \qquad 0 \le \arg t < 2\pi. \tag{7.4.19}$$

We have chosen the square root in this equation such that when $\arg t = 0$, $\arg[z + 1] = \pi/4$, so that the image of the descent path at angle $\pi/4$ is the positive real axis in the σ plane. See Fig. 7.11. Clockwise traversal of the semicircular section of the path in this figure corresponds to clockwise traversal of a full circle in the t plane, and then the image of the descent path at angle $5\pi/4$ is the real axis at angle 2π. Thus, the entire path D has as image the path denoted by $0+$ in Fig. 7.1, but with *opposite* orientation. For this example, by using (7.4.19), it is possible to make all substitutions explicitly in (7.4.17) and obtain the result

$$I(\lambda) = \frac{e^{-i\lambda/2}}{2} \int_{0+} \frac{e^{-\lambda t}}{2 - \sqrt{2}\, e^{i\pi/4} t^{1/2}} \frac{dt}{t}. \tag{7.4.20}$$

This integral is now in the form of the first equation in (7.1.12). Therefore,

$$
\begin{aligned}
I(\lambda) &\sim \frac{e^{-i\lambda/2}}{4} \sum_{k=0}^{\infty} \frac{e^{ik\pi/4}}{2^{k/2}} \int_{0+} t^{k/2-1}\, e^{-\lambda t}\, dt \\
&\sim \frac{e^{-i\lambda/2}}{4} \int_{0+} e^{-\lambda t} \frac{dt}{t} + \frac{e^{-i\lambda/2}}{4} \sum_{p=0}^{\infty} \frac{e^{i[p+1/2]\pi/2}}{2^{p+1/2}} \int_{0+} t^{p-1/2} e^{-\lambda t}\, dt \\
&\sim \frac{\pi i e^{-i\lambda/2}}{2} - \frac{e^{-i\lambda/2}}{2} \sum_{p=0}^{\infty} \frac{e^{[p+1/2]\pi i/2}}{(2\lambda)^{p+1/2}} \Gamma\!\left(p + \frac{1}{2}\right).
\end{aligned} \tag{7.4.21}
$$

In the second line, we have observed that all even values of k except $k = 0$ lead to functions that are analytic inside of $0+$ and, hence, have integral zero. In the third line, we have used the second definition in (7.1.11) to evaluate the integrals of the preceding line.

The essential features of the analysis of the case when the saddle point is also a pole can be seen in this example. The transformation (7.4.18) will always map z into a t plane with a branch cut. For the simple saddle point, the image contour is always one that encircles the branch point at the origin.

Then, the leading term of the asymptotic expansion is at least tractable with the aid of Watson's lemma for loop integrals. The case of higher-order saddle points requires that we first write the given integral as an integral over a sum of contours connecting adjacent (in the sense of a clockwise or counterclockwise manner) valleys. Then, we apply this method to each integral over adjacent valleys. Each asymptotic series for $f(t)$ will be slightly different, because a saddle point whose first nonvanishing derivative has order n will produce n inversions from z to t, one for each mapping of an adjacent set— valley, hill, valley—onto a t plane with a branch cut.

Exercises

7.26 (a) The purpose of this exercise is to verify the replacement of contours in the second example of this section. Therefore, let us define

$$ I_R = \int_{C_R} z^{-3/4} e^{i\lambda[z^{1/2} - z/2]} \, dz. $$

The contour C_R is a sector of a circle of radius R connecting the paths D_1 and C in Fig. 7.8 in a counterclockwise direction. Set $z^{1/2} - z/2 = -t$ and confirm the estimate in the neighborhood of C_R $t = (z/2)[1 + o(1)]$. From this estimate and the observation that t is real and positive for z real and positive, conclude that the image of C_R is a contour in the lower half t plane on which $|t|$ is large and becomes unbounded with increasing R.

(b) Now conclude that

$$ z = 2t[1 + o(1)], \qquad dz/dt = 2[1 + o(1)], \qquad z^{-3/4} = t^{-3/4}[1 + o(1)] $$

for $|t|$ large.

(c) Conclude that

$$ I_R = \int_{C_{R'}} f(t) e^{-i\lambda t} \, dt. $$

Here $C_{R'}$ is a contour in the lower half t plane on which $|t| \geq R'$ and R' approaches infinity with R. Furthermore, $f(t)$ is analytic in the neighborhood of $C_{R'}$ and approaches zero with increasing R'.

(d) Thus, use the extension of Jordan's lemma in Exercise 7.25 to conclude that I_R must decay to zero with increasing R.

(e) Introduce a path connecting D_2 and D_3 in Fig. 7.8. Show that the integral along this path must also decay to zero as the minimum value of y on this path approaches negative infinity.

7.27 Use the formula (7.3.11) to verify (7.4.16).

7.28 (a)   Consider now the Sommerfeld integral representations for the Hankel functions

$$\pi H_{ka}^{(j)}(kr) = \int_{C_j} e^{\{ikr\cos z + ika(z - \pi/2)\}} \, dz, \qquad j = 1, 2.$$

In this equation, the contours are again those of Fig. 7.9. Proceed formally with large parameter k and

$$w(z; a, r) = ir \cos z + ia(z - \pi/2), \qquad a < r.$$

Show that w has two saddle points $z_\pm$,

$$0 < z_+ = \sin^{-1}(a/r) < \pi/2, \qquad z_- = \pi - z_+,$$

but that the descent directions are as in the case $a = 0$ discussed in this section.

(b)   Verify that the paths of steepest descent are qualitatively as they were in the case $a = 0$ except for the shift in the location of the saddle points.

(c)   Obtain the asymptotic expansions

$$H_{ka}^{(1)}(kr) \sim \sqrt{2/\pi k} \, \exp(ik[\sqrt{r^2 - a^2} - a \cos^{-1}[a/r] - (\pi/4)]),$$

with the asymptotic expansion of $H_{ka}^{(2)}(kr)$ being the complex conjugate of this result.

7.29 (a)   Consider again the Hankel functions of the preceding exercise for the case $a = r$. Show that the exponent has a second-order saddle point at $z = \pi/2$ and that the directions of descent at the saddle point make angles $\pi/6$, $5\pi/6$, and $3\pi/2$ with positive x axis.

(b)   Show that one path of steepest descent is the semi-infinite vertical line extending downward from the saddle point and that the other two descent paths are qualitatively the same as the paths D_1 and D_2 of Fig. 7.10, except for their behavior near the single second-order saddle point.

(c)   Obtain the asymptotic expansions

$$H_{ka}^{(1)}(ka) \sim -\frac{\Gamma(\tfrac{1}{3})}{\pi(ka)^{1/3}} \left[\frac{4}{3} \right]^{1/6} e^{2\pi i/3},$$

with the asymptotic expansion of $H_{ka}^{(2)}(ka)$ being the complex conjugate of this result.

7.30 (a)   Let

$$I(\lambda) = \int_C e^{\lambda w(z)} \, dz, \qquad w(z) = z - z^3/3,$$

and C is the contour on which $-\infty < y < \infty$ and $x = \text{const} \leq 0$. This exponent was studied in Exercise 7.20. Show that

$$I(\lambda) \sim \sqrt{\pi/\lambda}\, e^{-2\lambda/3 + i\pi/2}.$$

(b)   Now suppose that

$$I(\lambda) = \int_C \frac{e^{\lambda w(z)}}{z + 1}\, dz,$$

with w and C as in (a). Deform the path of integration onto the paths of steepest descent except for a clockwise semicircle around the branch point. Then show that the transformation $w(z) + \frac{2}{3} = te^{i\pi}$ leads to a mapping of this contour onto the contour $0+$ of Fig. 7.1 but directed oppositely. We can write this result in the form (7.1.12) but for a multiplier of $\exp(-2\lambda/3)$ and account for the orientation with an adjustment of the sign of $f(t)$.

(c)   Show that $f(t) = -(1/z + 1)(dz/dt)$.

(d)   In order that we obtain a two-term asymptotic expansion of $I(\lambda)$, we require a two-term asymptotic expansion of $f(t)$ for t near zero or z near -1. Obtain the two-term expansion $f(t) \sim -(1/2t) - (i/4t^{1/2})$. One way to obtain this result is to write the two-term Taylor expansion

$$te^{i\pi} = (z + 1)^2 - \tfrac{1}{3}(z + 1)^3$$

and solve for $z + 1$:

$$z + 1 = t^{1/2}e^{i\pi/2}[1 - \tfrac{1}{3}(z + 1)]^{1/2}.$$

Now use the leading order approximation for $z + 1$ on the right side and the binomial expansion of the square root to obtain a two-term expansion

$$z + 1 = t^{1/2}e^{i\pi/2}[1 + \tfrac{1}{6}t^{1/2}e^{i\pi/2}].$$

Now proceed to determine the two-term expansion of $f(t)$ as given above.

(e)   Obtain the two-term asymptotic expansion

$$I(\lambda) \sim -e^{-2\lambda/3}[\pi i + \tfrac{1}{2}\sqrt{\pi/\lambda}].$$

References

Bleistein, N., and Handelsman, R. A. [1975]. "Asymptotic Expansions of Integrals." Holt, Rinehart and Winston, New York. [Also to be published by Dover, New York.]

Copson, E. T. [1965]. "Asymptotic Expansions." Cambridge Univ. Press, Cambridge.

Nuttall, A. [1981]. An Aid in Steepest Descent Evaluation of Integrals. NUSC Tech. Rep. 6433. Naval Undersea Systems Center, New London, Connecticut.

Olver, F. W. J. [1974]. "Asymptotics and Special Functions." Academic Press, New York.

Sommerfeld, A. [1949]. "Partial Differential Equations in Physics." Academic Press, New York.

8 ASYMPTOTIC TECHNIQUES FOR DIRECT SCATTERING PROBLEMS

In this chapter, we shall discuss the use of asymptotic methods to analyze direct scattering problems. This is a vast area of application and ongoing research, and so our discussion can serve at best as a brief introduction. We shall describe three methods used for analysis of wave problems. The first of these is the method of steepest descent, discussed in Chapter 7. In the first section, we shall describe in detail the implementation of this method on a problem for the Helmholtz equation. In the next two sections, we shall describe ray methods, again in the context of the Helmholtz equation. In the last section of this chapter, we shall discuss the Kirchhoff approximation.

8.1 SCATTERING BY A HALF-SPACE: ANALYSIS BY STEEPEST DESCENTS

Let us suppose that $u(x, z, \omega)$ satisfies the differential equation

$$\frac{\partial^2 u}{\partial x^2} + \frac{\partial^2 u}{\partial z^2} + \frac{\omega^2}{v^2} u = -\delta(x)\delta(z). \tag{8.1.1}$$

The function v is defined by

$$v = \begin{cases} c_0, & z < H, \\ c_1, & z > H. \end{cases} \tag{8.1.2}$$

We require that u and u_z be continuous across $z = H$ and that either u be analytic in the upper half ω plane or that, for ω real, u satisfy the Sommerfeld radiation condition stated in Exercise 6.11. We shall follow the geophysical convention here of taking z positive as *downward.* Furthermore, we shall consider the case $c_1 > c_0$, which is more typical and provides a problem with the mathematical structure that serves our purpose as regards exposition.

This problem models the propagation from an acoustic point source in two dimensions—a line source in three dimensions—in a medium with a planar boundary across which the velocity is discontinuous but the density is not. We might view this problem then as propagation and scattering by the interface in a two-layered acoustic earth of infinite extent in both directions. The analysis of this problem will introduce features that arise in more physically realistic problems. Therefore, it is worth studying as a prototype.

We shall be concerned with the high-frequency asymptotic solution. That is, we shall assume that $\omega L / v$ is large, with L a "typical length scale" of the problem. The only explicit length scale in the problem is H. Thus, this criterion should certainly hold for H. We shall proceed formally using ω as the large parameter and discover along the way other length scales that must satisfy this criterion. We remark that Exercise 1.11 dealt with the propagation of wave fronts for just such a medium. Thus, we can expect that the rays and wave fronts of that exercise will arise here.

To begin, we introduce the transverse (x) Fourier transform

$$\tilde{u}(k, z, \omega) = \int_{-\infty}^{\infty} u(x, z, \omega) e^{-ikx} \, dx. \tag{8.1.3}$$

By applying this Fourier transform to (8.1.1), we obtain the ordinary differential equation

$$\tilde{u}'' + [(\omega^2/c^2) - k^2]\tilde{u} = -\delta(z), \tag{8.1.4}$$

with prime meaning d/dz. The functions $\tilde{u}$ and $\tilde{u}'$ must be continuous across $z = H$, and $\tilde{u}$ must also be analytic in an upper half ω plane or satisfy a one-dimensional Sommerfeld condition (Exercise 6.11); that is, it must be "outgoing at infinity" and bounded.

Let us define

$$k_3 = \sqrt{(\omega^2/c_0^2) - k^2}, \qquad k_4 = \sqrt{(\omega^2/c_1^2) - k^2}. \tag{8.1.5}$$

We shall define these square roots for the imaginary part of ω positive. We shall also restrict real ω to be positive for the present and discuss real ω negative later. Our definition of k_3 is as shown in Fig. 8.1. In the limit, as $\text{Im } \omega \to 0$, k_3 becomes purely real or purely imaginary for k real, with the indicated inequalities still holding. The same definition relative to the branch points at $\pm \omega/c_1$ is taken for k_4. For this choice, the functions $e^{-ik_3 z}$, $e^{ik_4 z}$ decay to zero, respectively, for k real as $z \to -\infty$ for the first function or as $z \to \infty$ for the second function. Similarly, for z in its appropriate half-space and $\text{Im } \omega \to \infty$, these solutions decay to zero. Furthermore, in the ω plane

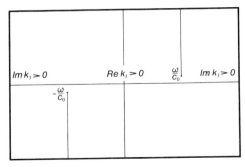

Fig. 8.1. Definition of k_3.

with branch cuts at $\pm k$ excluded from the upper half plane, these same two functions are seen to be analytic in the upper half plane and also to satisfy the Sommerfeld radiation condition for real ω. Thus, either the condition of analyticity in some upper half ω plane or the imposition of the Sommerfeld radiation condition picks the same distinguished solutions to the ordinary differential equation. This analysis is completely equivalent to the discussion in Section 4.4, in which the free-space Green's function was calculated [Eq. (4.4.25)].

We think of the total solution in $z < 0$ as being comprised of that free-space Green's function—a primary or incident wave—plus a reflected wave that must be outgoing; in $z > 0$, the solution should consist of an outgoing transmitted wave. Therefore, we can write the result

$$u(k, z, \omega) = -\frac{e^{ik_3|z|}}{2ik_3} - R\frac{e^{-ik_3[z-2H]}}{2ik_3}, \qquad z < H,$$

$$u(k, z, \omega) = -T\frac{e^{\{ik_4 z + i[k_3 - k_4]H\}}}{2ik_3}, \qquad z > H. \qquad (8.1.6)$$

In this equation, we have anticipated evaluation of the functions and their derivatives at $z = H$ to satisfy the continuity conditions and have therefore added a constant to the phases of the reflected and transmitted waves. Also, we have introduced the same scale factor in the amplitude of these terms as appears in the free-space Green's function. These extra multipliers simplify the determination of the coefficients R and T and will reduce them to classical reflection and transmission coefficients, independent of the coordinate system but dependent only on the local properties of the medium in the neighborhood of $z = H$.

The conditions that u and its first derivative be continuous across $z = H$ now leads to two algebraic equations for R and T, namely,

$$-1 - R = -T, \qquad -k_3 + k_3 R = -k_4 T. \qquad (8.1.7)$$

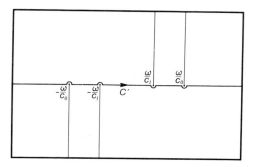

Fig. 8.2. Integration path in k plane for Im $\omega = 0$.

We solve for R and T and find that

$$R = \frac{k_3 - k_4}{k_3 + k_4}, \qquad T = \frac{2k_3}{k_3 + k_4}. \tag{8.1.8}$$

Thus, we obtain the solution

$$u(x, z, \omega) = \begin{cases} u_P(x, z, \omega) + u_S(x, z, \omega), & z < H, \\ u_T(x, z, \omega), & z > H. \end{cases} \tag{8.1.9}$$

Each of these functions is expressed as in inverse Fourier transform:

$$u_P(x, z, \omega) = -\frac{1}{4\pi i} \int_{C'} \frac{dk}{k_3} \exp\{ikx + ik_3|z|\}, \tag{8.1.10}$$

$$u_S(x, z, \omega) = -\frac{1}{4\pi i} \int_{C'} \frac{k_3 - k_4}{k_3 + k_4} \frac{dk}{k_3} \exp\{ikx - ik_3[z - 2H]\}, \tag{8.1.11}$$

$$u_T(x, z, \omega) = -\frac{1}{2\pi i} \int_{C'} \frac{dk}{k_3 + k_4} \exp\{ikx + ik_4 z + [k_3 - k_4]H\}. \tag{8.1.12}$$

For Im $\omega > 0$, the contour C' is just the real k axis. If we are to allow Im ω to approach zero, then this contour must be deformed away from the branch points at $\pm\omega/c_0$, $\pm\omega/c_1$, as shown in Fig. 8.2. For u_P, there are no branch points at $\pm\omega/c_1$ and the contour C' could be further straightened near that point.

We take the limit Im $\omega = 0$ and exploit the fact that ω is real and positive to rescale the variable of integration and obtain ω/c_0 as a multiplier of the exponent in each of the representations (8.1.10)–(8.1.12). Therefore, let us set

$$k = \lambda p, \qquad \lambda = \omega/c_0 \tag{8.1.13}$$

and rewrite the three integrals (8.1.10)–(8.1.12) as

$$u_\text{P}(x, z, \omega) = -\frac{1}{4\pi i} \int_C \frac{dp}{p_3} \exp(i\lambda\{px + p_3|z|\}), \tag{8.1.14}$$

$$u_\text{S}(x, z, \omega) = -\frac{1}{4\pi i} \int_C \frac{p_3 - p_4}{p_3 + p_4} \frac{dp}{p_3} \exp(i\lambda\{px - p_3[z - 2H]\}), \tag{8.1.15}$$

$$u_\text{T}(x, z, \omega) = -\frac{1}{2\pi i} \int_C \frac{dp}{p_3 + p_4} \exp(i\lambda\{px + p_4 z + [p_3 - p_4]H\}). \tag{8.1.16}$$

In these equations,

$$p_3 = \sqrt{1 - p^2}, \qquad p_4 = \sqrt{n^2 - p^2}, \qquad n^2 = c_0^2/c_1^2. \tag{8.1.17}$$

The branch points have now been moved to ± 1 and $\pm n$. Otherwise, the contour C is just like the contour C'. Each of these functions is positive for p real and between the appropriate pair of branch points. Outside of the branch points, these functions are purely imaginary with argument $\pi/2$.

We shall use the method of steepest descent (with large dimensional parameter λ) to analyze the integral representations (8.1.15)–(8.1.17).

ANALYSIS OF u_P

We consider the exponent

$$w(p) = i\{px + p_3|z|\}, \tag{8.1.18}$$

with derivatives

$$w'(p) = i\{x - (p/p_3)|z|\}, \qquad w''(p) = -i(|z|/p_3^3). \tag{8.1.19}$$

We find the saddle points by setting w' equal to zero. By rewriting that equation as $x = \cdots$ and squaring both sides, we obtain a quadratic equation for p. Only one root has the proper sign to make w' equal zero; thus,

$$w'(p_0) = 0 \quad \Rightarrow \quad x = (p_0/p_3)|z|, \qquad p_0 = x/\rho, \qquad \rho = \sqrt{x^2 + z^2}. \tag{8.1.20}$$

This root lies between the branch points at ± 1. At the saddle point,

$$w(p_0) = i\rho, \qquad w''(p_0) = -i\rho. \tag{8.1.21}$$

The directions of steepest descent at the saddle point are $-\pi/4$ and $3\pi/4$. For this problem, we shall only find the descent paths in the large qualitatively. Let us examine the possibility that the path of steepest descent extends to large values of $|p|$ in the right half plane. In that region, $p_3 \sim ip$ and

$$w(p) \sim p[ix - |z|] + O(|p|^{-1})$$
$$= -[x \operatorname{Im} p + |z| \operatorname{Re} p] + i[x \operatorname{Re} p - |z| \operatorname{Im} p] + O(|p|^{-1}). \tag{8.1.22}$$

On the path of steepest descent, the imaginary part of $w(p)$ must be constant and, hence, remain finite. Therefore, the path of steepest descent would have to have as asymptote the line

$$\text{Im } p = (x/|z|) \text{ Re } p - \rho, \tag{8.1.23}$$

with the constant $-\rho$ determined from (8.1.21).

The slope of this line has the same sign as x. When x is negative, the path of steepest descent is directed downward for large $|p|$ just as it is at the saddle point. Thus, it is reasonable to expect that the path of steepest descent simply continues on its trajectory downward to the right from the saddle point.

When x is positive, the path of steepest descent must be directed upward in the right half plane. In order for this to happen, the path of steepest descent must turn upward to the right of the saddle point and cross the real p axis. We leave it to the reader to check that the steepest descent path crosses the axis between the branch points only at the saddle point. For p to the right of the branch point at $+1$, the steepest descent path will cross the axis if

$$px = \rho; \qquad p = \rho/x, \tag{8.1.24}$$

which root is indeed to the right of the branch point when x is positive. For $x = 0$, from (8.1.22), we see that for Re w to approach negative infinity, it is necessary that Re $p \to \infty$. Furthermore, for Im $w = \rho$ in that limit, it is necessary that the path of steepest descent have as asymptote the line Im $p = -\rho/|z|$. The paths of steepest descent for saddle points at ± 0.6 are shown in Fig. 8.3. We leave it to the exercises to show that the paths of steepest descent in the left half plane are as shown in the figure. We remark

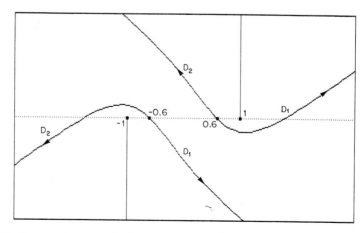

Fig. 8.3. Steepest descent paths for saddle points at 0.6 and -0.6; $0.6 = (1 \times 1)/\rho < n$.

that the slopes on these paths are just the *negative* of what they are for the corresponding sign of x and for p in the right half plane. This happens because in this half plane, $p_3 \sim -ip$. Similarly, the asymptote for $x = 0$ is now the line $\text{Im } p = \rho/|z|$.

Replacing the path C by a sum of descent paths now requires only that we check that the regions between them are indeed completely in the valley of the exponent. We leave this as an exercise. Now we can use the formula (7.3.11) and the discussion following that equation for simple saddle points to write the leading term of the asymptotic expansion of u_p. The result is

$$u_p(x, z, \omega) \sim \tfrac{1}{2}\sqrt{c_0/2\pi\rho\omega} \, \exp(i\omega\rho/c_0 + i\pi/4). \qquad (8.1.25)$$

In order to relate this result to the discussion of the eikonal equation in Chapter 1, we note that the phase of this primary wave is just the conoidal solution of the eikonal equation (1.4.1) and that the condition of stationarity of the exponent can be viewed as parametrically defining the rays. In fact, if we make the identifications by using (8.1.20),

$$p_0 = \sin \tau, \qquad z = 2\sigma \cos \tau, \qquad x = 2\sigma \sin \tau, \qquad \rho = 2\sigma, \qquad (8.1.26)$$

then further identification of the results here with those of Exercise 1.11a requires only a change of coordinate system; in particular, we remark that the z coordinate of this section corresponds to x in Exercise 1.11 up to a shift and that x in this discussion corresponds to y in Exercise 1.11.

We leave it to an exercise to verify that the effect of replacing ω by $-\omega$ is to transform the solutions (8.1.10)–(8.1.12) into their complex conjugates.

In order that the approximation (8.1.25) be valid, it is necessary that the product $\lambda|w''(p_0)| = \omega\rho/c_0$, which will appear to progressively higher powers in the terms of the complete asymptotic expansion of u_p, should be large. Thus, we would expect this asymptotic expansion to be valid at distances that are large compared to the wavelengths at the frequencies of interest.

ANALYSIS OF u_S

Let us now turn to the analysis of u_S as defined by (8.1.15). For this integral, we introduce the exponent

$$w(p) = i\{px + p_3[2H - z]\}. \qquad (8.1.27)$$

The analysis of the saddle points and steepest descent paths from them is exactly the same as for the preceding case if only we replace $|z|$ by $2H - z$, which is positive in the region of interest. A new feature that arises in this case is that the paths of steepest descent may cross the branch cuts from the branch points at $\pm n$. This will occur when the saddle point of the exponent lies between the branch points at n and 1 or between the branch points at

$-n$ and -1. We shall avoid this case for the moment by assuming that

$$|x|/\rho_1 < n, \qquad \rho_1 = \sqrt{x^2 + [2H - z]^2}. \tag{8.1.28}$$

For this case, a diagram depicting the steepest descent paths would look exactly like Fig. 8.3 except for the inclusion of a branch cut from n to $n + i\infty$ and from $-n$ to $-n - i\infty$—lines parallel to the existing branch cuts located above the pair of steepest descent paths on the right and below the steepest descent paths on the left.

For $w(p)$ defined by (8.1.27), the saddle point is located at

$$p_0 = x/\rho_1, \tag{8.1.29}$$

with

$$w(p_0) = i\rho_1. \tag{8.1.30}$$

Again, the condition of stationarity (8.1.29) can be interpreted as a parametric representation of a family of rays. However, this time it is the family of *reflected rays* of Exercise 1.11, with ρ_1 being the phase of the reflected waves. Thus, we shall denote the saddle point contribution of u_S as the *reflected wave* and denote it by u_R; that is, we set

$$u_S(x, z, \omega) = u_R(x, z, \omega), \qquad |x|/\rho_1 < n, \tag{8.1.31}$$

with u_R having the asymptotic expansion

$$u_R(x, z, \omega) \sim R(p_0) \frac{1}{2} \sqrt{\frac{c_0}{2\pi\rho_1\omega}} \, e^{i\omega\rho_1/c_0 + i\pi/4};$$

$$R(p_0) = \frac{p_3(p_0) - p_4(p_0)}{p_3(p_0) + p_4(p_0)}, \qquad p_0 = \frac{x}{\rho_1}. \tag{8.1.32}$$

We expect that this asymptotic expansion will remain valid as long as the product $\lambda|w''(k_0)| = \omega\rho_1/c_0$ is large. The minimum value of ρ_1 is H. Therefore, as long as the interface is "many" wavelengths from the source for the frequencies of interest, this asymptotic expansion will be valid and useful.

Let us now consider the domain complementary to that defined by (8.1.28); that is, let us suppose that

$$n < |x|/\rho_1 < 1. \tag{8.1.33}$$

Before we proceed with the asymptotic analysis, it will prove useful to interpret this condition geometrically in the (x, z) domain. We introduce the polar angle θ measured from the vertical at the point $(0, 2H)$. Then

$$x/\rho_1 = p_0 = \sin\theta, \qquad (2H - z)/\rho_1 = \cos\theta. \tag{8.1.34}$$

Equality in (8.1.33) then defines a *critical angle* θ_C, $\sin \theta_C = n$, such that for angles nearer the vertical $\theta < \theta_C$, u_S is asymptotically a reflected wave only. For angles nearer to the horizontal than θ_C, $\theta > \theta_C$, there will be an additional contribution to u_S. In Exercise 1.11, this additional wave was called the *lateral wave* or *head wave*. Thus, we anticipate that whatever additional contribution arises owing to the interaction between the saddle point and the branch point, it is likely to provide the asymptotic expansion of this head wave.

Let us return to the asymptotic analysis for the case defined by (8.1.33). Now we cannot deform the contour C onto paths of steepest descent without crossing one or the other of the branch cuts. For x positive, we would cross the branch cut from $+n$; for x negative, we would cross the branch cut from $-n$. Thus, for $(x, 2H - z)$ in the range (8.1.33), we must analyze the branch points $\pm n$ as critical points, respectively, for $\pm x > 0$. Let us first investigate the directions of descent at the branch points. For the range defined by (8.1.33), neither of these points is a saddle point. We consider first the case x positive and focus our attention on the point $p = n$. In this case, from (8.1.19),

$$w'(n) = i\{x - (n/\sqrt{1 - n^2})[2H - z]\} = i[2H - z][\tan \theta - \tan \theta_C]. \quad (8.1.35)$$

For the range (8.1.33), $\tan \theta > \tan \theta_C$, and the direction of steepest descent is $\pi/2$. In a completely analogous manner, we can show that for x negative and in the range (8.1.33), the direction of steepest descent at $-n$ is $-\pi/2$. That is, the directions of the branch cuts are the directions of steepest descent, respectively, in either case of interest. We can further show that away from the branch points, the branch cuts remain paths of descent, although not paths of steepest descent, up to the intersection of the steepest descent path from the saddle point with the branch cut. This suffices for our further analysis.

For the interested reader, we point out that the paths of steepest descent away from the branch points can be determined in the same detail as were the steepest descent paths from the saddle points. However, these paths would extend onto a *second Riemann sheet* of the function $p_4(p)$.

In Fig. 8.4, we depict a replacement of the original contour C for either choice of the sign of x; again, we have used saddle points at ± 0.6 for our sample saddle points with $\pm x$ positive. In each case, the contours D_1 and D_2 represent steepest descent paths from the saddle points. The contour D_3 consists of a loop or keyhole contour around the branch point and then a descent path away from the branch cut. Actually, we have used the steepest descent path from the saddle point for this last piece of contour, but that is not relevant.

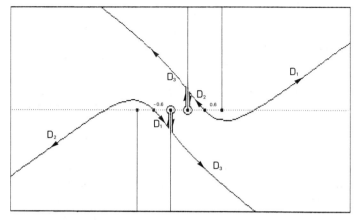

Fig. 8.4. Steepest descent paths for saddle points at 0.6 and -0.6; $0.6 = (1 \times 1)/\rho > n$.

We can see that C can be replaced by $D_1 - D_2 \mp D_3$, with the integral on the first pair of paths again producing u_R as defined by (8.1.32). An important difference in this result is that now R is complex. Let us redefine R as a function of the polar angle θ introduced earlier. Then, we find that

$$R = \frac{\cos \theta - \sqrt{\sin^2 \theta_C - \sin^2 \theta}}{\cos \theta + \sqrt{\sin^2 \theta_C - \sin^2 \theta}}, \qquad |\sin \theta| < n = \sin \theta_C,$$

(8.1.36)

$$R = \frac{\cos \theta - i\sqrt{\sin^2 \theta - \sin^2 \theta_C}}{\cos \theta + i\sqrt{\sin^2 \theta - \sin^2 \theta_C}}, \qquad |\sin \theta| > n = \sin \theta_C.$$

We note that

$$\frac{c_1 - c_0}{c_1 + c_0} \le R < 1, \quad 0 \le \theta_C; \qquad |R| = 1, \quad \theta_C \le \theta < \frac{\pi}{2}. \quad (8.1.37)$$

That is, the reflection coefficient varies from a minimum value at *normal incidence*, given by the *normal reflection coefficient*, to unity at the critical angle. Beyond the critical angle, the reflection coefficient is complex with modulus equal to unity.

We now turn to consideration of the integral on D_3. In particular, we shall consider the case in which $n < x/\rho_1$, so that the upper set of contours in Fig. 8.4 is of interest. Let us define then, in this case,

$$u_H(x, z, \omega) = \frac{1}{4\pi i} \int_{D_3} \frac{p_3 - p_4}{p_3 + p_4} \frac{dp}{p_3} \exp(i\lambda \{px - p_3[z - 2H]\}). \quad (8.1.38)$$

We remark that the integrand is just the negative of the integrand in (8.1.15).

This is consistent with the replacement of C by $D_1 - D_2 - D_3$ for this case. With this definition,

$$u_S(x, z, \omega) = u_R(x, z, \omega) + u_H(x, z, \omega), \qquad |x|/\rho_1 > \sin \theta_C = n. \quad (8.1.39)$$

On D_3, we can disregard all but the keyhole contour of finite extent with only exponentially small error. This is so because the error in such an approximation is an integral on a contour completely in the valley of the exponent with respect to either the branch point or the saddle point. On the small circle around the branch point, the integrand is bounded and the length of the path is proportional to the radius of the path. Thus, the integral on this circle approaches zero as the radius of the path shrinks to zero. Therefore, in all further considerations, we shall proceed under the assumption that this radius will be allowed to approach zero.

Let us define D as the contour along the right side of the branch cut extending upward from the branch point as far as the finite segment of keyhole contour but oppositely directed when compared to the same segment of D_3. The negative of the integral along D is asymptotically equal to the right segment of D_3. The integral along D, with p_4 replaced by its negative, is also asymptotically equal to the portion of D_3 to the right of the branch cut. That is,

$$u_H(x, z, \omega) \sim \frac{1}{4\pi i} \int_D \left[\frac{p_3 + p_4}{p_3 - p_4} - \frac{p_3 - p_4}{p_3 + p_4} \right] \frac{dp}{p_3} \exp(i\lambda\{px - p_3(z - 2H)\})$$

$$= \frac{1}{\pi i(1 - n^2)} \int_D p_4 \, dp \, \exp(i\lambda\{px - p_3[z - 2H]\}). \quad (8.1.40)$$

This is an example of the standard contour integral (7.2.1) in which there is a branch point of the amplitude as in (7.3.12) at a point that is not a saddle point; that is, the index n of the order of the saddle point is equal to unity. To make the identification between (8.1.40) and (7.3.12) easier, we first write $p_4 = i\sqrt{p - n}\sqrt{p + n}$. The asymptotic expansion of the integral in (8.1.40) is given by (7.3.14) with

$$g_0 = \frac{\sqrt{1 + n}}{\pi(1 - n^2)}, \qquad |w'| = \frac{\rho}{\sqrt{1 - n^2}} \sin(\theta - \theta_C), \qquad \theta_p = \frac{\pi}{2}, \qquad \beta = \frac{3}{2}. \quad (8.1.41)$$

By using these results in (8.1.40), we obtain the result

$$u_H(x, z, \omega) \sim \frac{\sqrt{\pi}}{2} \left[\frac{1 + \sin \theta_C}{1 - \sin \theta_C} \right]^{1/4} \left[\frac{c_0}{\omega\rho \sin(\theta - \theta_C)} \right]^{3/2}$$

$$\cdot \exp\left(\frac{i\omega\{x \sin \theta_C + [2H - z] \cos \theta_C\}}{c_0} + \frac{3\pi i}{4} \right). \quad (8.1.42)$$

We leave as an exercise the verification that the phase $\{x \sin \theta_c + [2H - z] \cos \theta_c\}/c_0$ is just the head wave solution of the eikonal equation of Exercise 1.11.

THE TRANSMITTED WAVE u_T

We consider now the solution in the region $z > H$, the transmitted wave, defined by (8.1.16). For this integral, the exponent to be studied is

$$w(p) = i\{px + p_4 z + [p_3 - p_4]H\}, \tag{8.1.43}$$

with derivatives

$$w'(p) = i\left\{x - \frac{p}{p_3}z - \left[\frac{p}{p_3} - \frac{p}{p_4}\right]H\right\},$$

$$w''(p) = -i\left\{\frac{z}{p_3^3} + \left[\frac{1}{p_3^3} - \frac{1}{p_4^3}\right]H\right\}. \tag{8.1.44}$$

The location of the saddle points for this exponent is more difficult than for the previous cases. The condition that w have a saddle point is

$$x - \frac{p}{p_3}z - \left[\frac{p}{p_3} - \frac{p}{p_4}\right]H = 0. \tag{8.1.45}$$

Let us first consider the possibility of saddle points on the real axis in the region $-n < p < n$. In this interval, both p_3 and p_4 are real, and an explicit solution of (8.1.45) is not available (or useful). We shall also assume that x is nonnegative, since we have seen that the results are symmetric in x. To specialize even further, let us consider this condition for $z = H$, that is, right on the interface. Now the equation for the saddle points becomes

$$x - (p/p_3)H = 0. \tag{8.1.46}$$

This is exactly the equation that determines the saddle points for the exponent of the integrand for u_P [Eq. (8.1.19)] for points (x, z) at the interface. Thus, for $z > H$, we view (8.1.46) as providing a continuation of the incident rays into the lower medium. From (8.1.45), we can see that the direction of that ray is given by $(p_0, p_4(p_0))$ in this second medium; this ray continues the incident ray with direction $(p_0, p_3(p_0))$.

Let us define θ_i and θ_r as the angles that a given ray makes with the normal to the interface. By taking the dot product of the two direction vectors with the normal $(0, 1)$ and dividing by 1 and n, respectively, to normalize, we obtain the expressions for the cosines of these two angles as

$$\cos \theta_i = p_3, \qquad \cos \theta_r = p_4/n. \tag{8.1.47}$$

By solving for the sines of these angles and taking the quotient, we obtain
Snell's law,

$$\sin \theta_i / \sin \theta_r = n, \qquad (8.1.48)$$

which confirms that the rays defined by setting w' equal to zero are indeed
the *refracted rays* and that the contribution from this saddle point is the refracted wave. Furthermore, since $n < 1$, we see that $\theta_r = \pi/2$ when $\sin \theta_i = n$.
That is, when the incident angle at the interface is the *critical angle*, the
refracted ray propagates along the interface. As the incident angle varies
between zero, at normal incidence, and the critical angle, the refracted rays
cover the right half of the lower region. Of course, the left half region is
covered by the negative angles corresponding to incident rays to the left
of the z axis.

We shall denote the contribution due to this saddle point by u_r and content
ourselves with a parametric representation in terms of the angle τ. We leave
to the exercises the verification of the result

$$u_r(x, z, \omega)$$

$$\sim \sqrt{\frac{c_0}{2\pi\omega}} \; \frac{\exp\left(\dfrac{i\omega}{c_0}\{x \sin \tau + (z - H)\sqrt{n^2 - \sin^2 \tau} + H \cos \tau\} + \dfrac{i\pi}{4}\right)}{\left[\sqrt{n^2 - \sin^2 \tau} + \cos \tau\right]\sqrt{\dfrac{z - H}{(n^2 - \sin^2 \tau)^{3/2}} + \dfrac{H}{\cos^3 \tau}}}; \qquad (8.1.49)$$

$$z - H = x \frac{\sqrt{n^2 - \sin^2 \tau}}{\sin \tau} - H \frac{\sqrt{n^2 - \sin^2 \tau}}{\cos \tau}.$$

The parameter τ labels a refracted ray, which is the continuation into the
lower medium of the incident ray from the source point with same parameter
τ. For each choice of τ, the first equation in (8.1.49) gives the asymptotic
value of the refracted field. The rays for this diffracted field are initiated on
the interface between the boundary points $x = \pm H \tan \tau_C$ defined by the
critical angle τ_C, $\sin \tau_C = n$ and its negative. We can check further that on
this finite segment of the interface, the refracted wave and its normal derivative asymptotically balance the sum of the incident field and the reflected field.

On the interface outside of the critical points, reflected and head rays are
initiated into the upper region. The head wave in (8.1.42) can be seen to be
of lower order in ω than the incident wave (8.1.25) and the reflected wave
(8.1.32). Therefore, there must be another wave in the lower medium to
balance the sum of incident and reflected waves at the interface outside the
critical points already defined.

We remark that the condition of stationarity (8.1.46) is satisfied for x
values in this region by the same values of p that yield the saddle points for

the incident wave (8.1.19), with $z = H$. Thus, there is another simple saddle point contribution, at least at the interface. However, for $z > H$, the full condition of stationarity (8.1.45) is a complex valued equation, since p_4 is complex and now has a nonzero coefficient. Finding this complex root in the large is extremely difficult. However, for small values of $z - H$, we might think to search for a root with a small imaginary part. Indeed, let us set $p = p_1 + i\varepsilon$, with p_1 and ε real. We leave it to an exercise to confirm that to leading order

$$p_1 \sim \frac{x}{\sqrt{x^2 + H^2}}, \qquad \varepsilon \sim (z - H)\frac{H}{2x}\frac{\sqrt{x^2 + H^2}}{\sqrt{x^2(1 - n^2) + H^2}}. \qquad (8.1.50)$$

At this saddle point, w has a negative real part, so that, in fact, the corresponding field contribution decays exponentially with increasing $z - H$. However, at $z = H$, there is no exponential decay, and this field balances the incident plus reflected field. This field contribution is called the *evanescent field*.

In Fig. 8.5, we show the source point and its image below the interface. In addition, the critical angles θ_C and τ_C are shown along with a set of rays. A typical incident ray at angle less than critical is labeled by P, the corresponding reflected ray by R, and the refracted ray by r. The critically reflected ray is labeled by C, the critically refracted ray by t, and a typical head or lateral ray by L.

Exercises

8.1   Consider $w(p)$ defined by (8.1.18). Set $\text{Im } p = \rho$ defined by (8.1.20). Show that the only solution p on the interval $(-1, 1)$ is the saddle point. That is, the path of steepest descent can only cross the real axis in this interval at the saddle point itself.

8.2   (a)   The purpose of this exercise is to confirm that for the integral representation of u_p in (8.1.14), the path of integration C can be replaced by

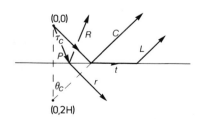

Fig. 8.5. Geometry of rays and critical angles.

the difference of paths of steepest descent in Fig. 8.3. Thus, let x be positive and z be nonzero, and define C_R as an arc of radius R connecting C and D_1 on the right. Confirm that C_R is in the valley of the exponent and that the integrand decays exponentially to zero with increasing R everywhere on C_R. Show that the integral on C_R must approach zero with increasing R.

(b)   Now suppose that $z = 0$. Verify that the integrand satisfied the conditions of the corollary to Jordan's lemma to conclude that, again, the integral on C_R decays to zero. All other cases follow in the same manner.

8.3   (a)   Suppose that $\omega < 0$. Show that the solutions to (8.1.4) that satisfy the Sommerfeld radiation condition in Exercise 6.11 are

$$e^{ik_3 z}, \quad z \to -\infty; \qquad e^{-ik_4 z}, \quad z \to \infty.$$

Note that these solutions have the opposite signs of the choices in the discussion of this section. Conclude that we can account for negative ω in the solutions (8.1.6) by replacing p_3 and p_4 by sgn ωp_3 and sgn ωp_4, respectively.

(b)   In (8.1.10)–(8.1.12), take ω to be real, and let the semicircular arcs around the branch points shrink to zero radius. Show that replacing ω by its negative and then making the change of variable of integration from k to $-k$ yields the integral with complex conjugate integrand. Thus, the transformation from ω to $-\omega$ is achieved by taking the complex conjugate of the results developed in this section.

8.4   Verify that for $x < 0$ and $|x|/\rho_1 > n$, the direction of steepest descent in (8.1.27) at $p = -n$ is $-\pi/2$.

8.5   (a)   Consider the steepest descent path from $p = n$ for the exponent in (8.1.27). We have shown that the direction of this path is $\pi/2$. Set $p = n + i\sigma$ and show that for σ small, the path of steepest descent curves to the left of the vertical.

(b)   Consider this steepest descent path for large $|p|$ in the left half plane. Show that the analysis of the asymptote to this path is exactly as in the discussion in this section except that ρ must be replaced by $nx + \sqrt{1 - n^2}\,(2H - z)$.

8.6   In Fig. 8.5, choose (x, z) on the ray labeled L. Show that the phase of u_H in (8.1.42) is given by the length of the ray P plus the length of the ray L, the sum multiplied by $1/c_0$, plus the length of the ray t multiplied by $1/c_1$. That is, the phase at this point is the product of the frequency ω multiplied by the travel time on a ray path that propagates to the boundary at the critical angle, propagates in the second medium along the boundary, and then propagates at the critical angle again in the first medium to the point (x, z).

8.7   Consider the exponent $w(p)$ defined by (8.1.43). Show that the analysis of the steepest descent paths for large values of $|p|$ is exactly as it was for the discussion of the exponent in (8.1.18), except that $|z|$ is replaced by z (which is positive anyway).

8.8 Carry out the saddle point evaluation to confirm (8.1.49).

8.9 Suppose that $z = H$ and that $|x| < Hn/\sqrt{1 - n^2} = H \tan \tau_C$. Show that

$$u_P + u_R \sim u_r, \qquad \frac{\partial u_P}{\partial z} + \frac{\partial u_R}{\partial z} \sim \frac{\partial u_r}{\partial z},$$

to leading order in ω.

8.10 Verify (8.1.50).

8.11 Consider the same problem as in this section except in three space dimensions, with x replaced by $\mathbf{x} = (x_1, x_2)$ and the source now replaced by $\delta(x_1)\delta(x_2)\delta(z)$. Introduce the two-dimensional Fourier transform

$$\tilde{u}(k_1, k_2, z) = \int_{-\infty}^{\infty} u(\mathbf{x}, z)e^{-i\mathbf{k}\cdot\mathbf{x}} \, dx_1 \, dx_2, \qquad \mathbf{k} = (k_1, k_2).$$

(a) Show that the solution to this problem is given by (8.1.10)–(8.1.12), with k and x replaced by $\mathbf{k}$ and $\mathbf{x}$, respectively, kx replaced by $\mathbf{k}\cdot\mathbf{x}$, k replaced by the magnitude of $\mathbf{k}$, and dk replaced by $dk_1 \, dk_2$.

(b) Consider u_P with ω positive. Scale by ω/c_0, and write the integral in polar coordinates. Formally use the method of stationary phase on the angular integral and obtain the result

$$u_P \sim \frac{i}{2}\sqrt{\frac{\omega}{2\pi x c_0}} \sum_{\pm} \int_0^{\infty} \frac{\sqrt{p} \, e^{\mp i\pi/4}}{p_3} \, dp \, \exp(i\lambda\{px + p_3|z|\}).$$

In this equation, $x = \sqrt{\mathbf{x}\cdot\mathbf{x}}$ and $\lambda = \omega/c_0$.

(c) Replace the lower term in the sum by an integral over negative values of p, and conclude that except for the avoidance of the branch point at the origin by passing over it, this result is equivalent to an integral on $-\infty < p < \infty$ exactly like the integral considered in this section. The only new feature is a branch cut from the origin extending downward to $-i\infty$. Explain why this branch cut does not affect the leading order asymptotic analysis of this section.

(d) Compare the result of (c) to (8.1.14) to conclude that the asymptotic solution for u_P can be obtained from the discussion of this section if we only multiply the result (8.1.25) by

$$\sqrt{\omega p/2\pi x c_0} \, e^{-i\pi/4},$$

with p evaluated at the stationary point (8.1.20). Thus, obtain the result

$$u_P \sim \exp(i\omega x/c_0)/4\pi x.$$

which is the exact solution in three dimensions.

(e)   Apply the same scaling to the results for u_S and u_T to obtain the asymptotic solutions for these cases.

8.2   INTRODUCTION TO RAY METHODS

Ray methods provide a natural synthesis of mathematical and physical insights into wave propagation. Mathematically, ray methods are an extension to partial differential equations of the WKBJ method for ordinary differential equations. Physically, ray methods extend the basic concepts of geometrical optics to a large class of optical wave phenomena and then extend these results to other wave phenomena as well.

We shall present in this and the next section an introduction to the application of the ray method to the Helmholtz equation. As in the preceding section, this discussion is by no means complete or definitive. A complete discussion of the ray method for the Helmholtz equation would require a volume in itself. Much has been written about the numerical implementation of this method in complex geometries and about the extension of the method to the systems of equations of elastic and electromagnetic wave propagation. Implementation on other equations and systems also abounds. As in Section 8.1, however, we consider the features introduced here as being fundamental and prototypical.

The ray method is a formal asymptotic method with rigorous justification available only under limitations that do not account for applications for which the method has been demonstrated to be valid. Furthermore, the method remains valid even in cases in which the dimensionless large parameters of interest are not really large at all, even smaller than our rule-of-thumb value of three. As in earlier sections, the dimensionless parameter is a "natural" length scale of the problem at hand measured in units of wavelengths. Equivalently, it is the ratio of some natural length scale of the problem, other than a wavelength, to a typical wavelength. See the discussion in Section 4.7 for ordinary differential equations.

The use of numerical techniques in support of ray methods greatly enhances their utility. There is also a great practical advantage in applying numerical techniques to the ray solution over applying numerical techniques to the original equation(s). This is a matter of length scales again. A numerical scheme applied to the original equation or system would require, of necessity, discretization on a length scale that is a fraction of a wavelength, whereas the discretization of the ray solution uses the other longer-length scales of the problem at hand. Thus, there are savings in computer capacity that allow us to obtain ray method solutions to more complicated problems than might be obtained otherwise. Furthermore, the ray method reduces problems for partial differential equations to the solution of a system of first-order ordinary differential equations that are linear in the derivatives of the unknown constituents of the ray solution.

FORMAL SERIES SOLUTION

We begin our development of the ray method by considering a function $u(\mathbf{x}, \omega)$, $\mathbf{x} = (x_1, x_2, x_3)$, which satisfies the homogeneous Helmholtz equation

$$\nabla^2 u + (\omega^2/c^2(\mathbf{x}))u = 0. \tag{8.2.1}$$

As in the discussion in Section 4.7 for the case of one independent variable, we shall assume a solution in the form (4.7.4), now in three independent variables,

$$u(\mathbf{x}; \omega) \sim \omega^\beta e^{i\omega\tau(\mathbf{x})} \sum_{j=0}^{\infty} \frac{A_j(\mathbf{x})}{(i\omega)^j}. \tag{8.2.2}$$

We remark that β, the power of ω, cannot be determined from the homogeneous equation (8.2.1). It will be determined in matching the solution u to prescribed data. This will be seen in examples in the next section. By substituting this series into (8.2.1), we obtain the equation

$$\sum_{j=0}^{\infty} \{(i\omega)^{2-j}A_j[(\nabla\tau)^2 - c^{-2}] + (i\omega)^{1-j}[2\,\nabla\tau\cdot\nabla A_j + A_j\,\nabla^2\tau]$$

$$+ (i\omega)^{-j}\,\nabla^2 A_j\} = 0. \tag{8.2.3}$$

We shall determine the series solution by setting the coefficient of each power of ω separately equal to zero. The highest power is two, and for any nonzero choice of A_0 this coefficient will be zero if we choose τ to satisfy the equation

$$(\nabla\tau)^2 = c^{-2}; \tag{8.2.4}$$

that is, τ must be a solution of the *eikonal equation*, which was discussed in Chapter 1, but with n^2 replaced by c^{-2}. Had we assumed a series in the parameter ω/c_0, with c_0 some reference velocity, then the phase here would have been exactly as in the first chapter. Our choice here is motivated by seismological applications in which the *travel time* τ is taken as the distinguished phase.

When τ is chosen to satisfy (8.2.4), not only the highest-order term in ω in (8.2.3) is eliminated, but the entire first series is eliminated. Setting the coefficient of ω equal to zero yields the equation

$$2\,\nabla\tau\cdot\nabla A_0 + A_0\,\nabla^2\tau = 0, \tag{8.2.5}$$

while equating the coefficients of the subsequent powers of ω equal to zero yields the system of equations

$$2\,\mathbf{V}\tau \cdot \mathbf{V}A_j\,\nabla^2\tau = -\nabla^2 A_{j-1}, \qquad j = 1, 2, \ldots. \tag{8.2.6}$$

We introduce new variables defined by

$$p = 1/c, \qquad p_j = \partial\tau/\partial x_j, \qquad j = 1, 2, 3. \tag{8.2.7}$$

The variable p, inverse velocity, is called *slowness*. The vector $\mathbf{p} = (p_1, p_2, p_3)$ has the dimension of inverse velocity as well and is called the *slowness vector*. In terms of these variables, we rewrite the eikonal equation as

$$\sum_{j=1}^{3} p_j p_j = p^2. \tag{8.2.8}$$

Thus, the eikonal equation states that the slowness vector has the slowness p as its magnitude.

The characteristic equations for (8.2.8), known as the *ray equations*, are

$$\frac{dx_j}{d\sigma} = \lambda p_j, \qquad \frac{dp_j}{d\sigma} = \lambda\,\frac{\partial p}{\partial x_j}, \qquad j = 1, 2, 3; \qquad \frac{\partial\tau}{\partial\sigma} = \lambda p^2. \tag{8.2.9}$$

The first six equations here define the characteristics that are the rays of the ray method. The seventh equation governs the propagation of τ along the rays. We shall discuss choices of σ later. We remind the reader that the rays are directed along the gradient to τ, that is, orthogonal to the phase fronts.

Let us consider now (8.2.5), the equation for the leading order amplitude, known as the (first) *transport equation*. We shall show that this equation is also an ordinary differential equation with respect to σ. Before doing so, however, we shall present an analytical solution formula that provides some physical insight into the nature of the propagation modeled by the function $u(\mathbf{x}; \omega)$. This solution formula results from the observation that multiplication by A_0 in (8.2.5) produces an exact divergence, namely, $\mathbf{V} \cdot [A_0^2\,\mathbf{V}\tau]$. Thus, for any volume D,

$$0 = \int_D \mathbf{V} \cdot [A_0^2\,\mathbf{V}\tau]\,dV = \int_{\partial D} A_0^2\,\mathbf{V}\tau \cdot \hat{\mathbf{n}}\,dS. \tag{8.2.10}$$

In the second line, $\hat{\mathbf{n}}$ denotes the outward unit normal to the boundary ∂D of the domain D.

We choose the domain D as follows. First, introduce a differential element of a surface of constant τ. Call this surface dS_1. Then, let D consist of the tube of rays between this surface element and the intersection of this family of rays with another surface of constant value of τ; call this surface dS_2. See Fig. 8.6. The part of D labeled by B in the figure has the ray direction as a

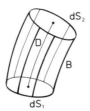

Fig. 8.6. Domain of integration D.

tangent at each point. Thus, the dot product of the normal with the gradient of τ is zero on this part of ∂D, and the integral over the boundary reduces to the integral over the two ends of the ray tube on which τ is constant.

Since the ends were of small cross-sectional area, we approximate the integral on each end by the area of the end multiplied by the integrand evaluated at some point of the surface, say, on some central ray on which we take the value of σ to be σ_1 or σ_2 corresponding to the labeling on dS. We remark further that the normal on dS is directed along $\nabla\tau$ at the larger value of τ and is oppositely directed at the smaller value of τ. Furthermore, the magnitude of the dot product $\nabla\tau \cdot \hat{\mathbf{n}}$ must be p. Thus, we conclude that

$$A_0^2(\sigma_2) = A_0^2(\sigma_1)\frac{p(\sigma_1)\,dS_1}{p(\sigma_2)\,dS_2}. \tag{8.2.11}$$

This equation has the interpretation that $A_0^2 p = A_0^2/c$ (and hence $|A_0^2|/c$) is *preserved in ray tubes*. If we think of this quantity as defining an energy density, then we conclude that the flux of energy through a ray tube is zero for this leading order approximation in ω.

The differential surface elements can be expressed in terms of elementary quantities on the rays. Let us label the rays on dS_1 by the pair of parameters (γ_1, γ_2). Then, the integration over the end surface can be carried out in γ_1 and γ_2 if we set

$$dS = \left|\hat{\mathbf{n}} \cdot \frac{d\mathbf{x}}{d\gamma_1} \times \frac{d\mathbf{x}}{d\gamma_2}\right| d\gamma_1\,d\gamma_2, \tag{8.2.12}$$

where $\times$ denotes the vector cross product. The vector

$$\frac{d\mathbf{x}}{d\gamma_1} \times \frac{d\mathbf{x}}{d\gamma_2} d\gamma_1\,d\gamma_2$$

has as magnitude the area of a differential parallelogram in the surface of constant σ (not necessarily the same as constant τ). The dot product then projects this area onto a surface of constant τ. The domains dS_1 and dS_2 extend over exactly the same range of parameters (γ_1, γ_2). Thus, the difference between the surface areas on the two end caps can only arise through differences in the *Jacobian of mapping via rays*:

$$J(\sigma) = \left| \hat{\mathbf{n}} \cdot \frac{d\mathbf{x}}{d\gamma_1} \times \frac{d\mathbf{x}}{\gamma_2} \right|. \tag{8.2.13}$$

Therefore, we rewrite (8.2.11) as

$$A_0^2(\sigma) = A_0^2(\sigma_1) \frac{p(\sigma_1)J(\sigma_1)}{p(\sigma)J(\sigma)}. \tag{8.2.14}$$

We have replaced the *specific value* σ_2 by the *generic value* σ in this equation.

TRANSPORT EQUATIONS AS ORDINARY DIFFERENTIAL EQUATIONS

We return now to the development of a system of ordinary differential equations for the amplitude coefficients in the solution series. We use (8.2.9) in (8.2.5) to obtain the equation

$$\frac{2}{\lambda} \frac{d\mathbf{x}}{d\sigma} \cdot \nabla A_0 + A_0 \nabla^2 \tau = 0, \tag{8.2.15}$$

which after multiplication by λA_0 can be rewritten as

$$\frac{dA_0^2}{d\sigma} = -A_0^2 \lambda \nabla^2 \tau. \tag{8.2.16}$$

When a ray solution for τ is known, the product $\lambda \nabla^2 \tau$ is a known function of σ, and this is an ordinary differential equation in σ for A_0^2.

A comparison of this differential equation with the result of differentiation of the solution formula (8.2.14) suggests a relationship between $\nabla^2 \tau$ and the Jacobian J. To obtain this relationship, we differentiate (8.2.14) and obtain the result

$$\frac{dA_0^2(\sigma)}{d\sigma} = A_0^2(\sigma_1)p(\sigma_1)J(\sigma_1) \frac{d}{d\sigma} \left[\frac{1}{p(\sigma)J(\sigma)} \right]$$

$$= -A_0^2 \frac{d}{d\sigma} \{ \log[p(\sigma)J(\sigma)] \}. \tag{8.2.17}$$

A comparison of this differential equation with (8.1.16) suggests that

$$\lambda \, \nabla^2 \tau = \frac{d}{d\sigma} \{\log[p(\sigma)J(\sigma)]\}. \tag{8.2.18}$$

We shall verify this result at the end of this section.

We leave it to an exercise to verify that the higher-order transport equations (8.2.6) have solutions given by

$$A_j(\sigma) = \frac{A_j(\sigma)\sqrt{p(\sigma_1)J(\sigma_1)}}{\sqrt{p(\sigma)J(\sigma)}}$$

$$- \frac{1}{p(\sigma)J(\sigma)} \int_{\sigma_1}^{\sigma} \lambda(\sigma')\sqrt{p(\sigma')J(\sigma')} \, \nabla^2 A_{j-1}(\sigma') \, d\sigma'. \tag{8.2.19}$$

This result is valid for all integers $j > 0$. If we define $A_{-1} = 0$, then the result is valid for all integers $j \geq 0$. In all functions here we have omitted dependence on other variables, namely, the ray parameters γ_1 and γ_2. Furthermore, the dependence of λ and p on σ is more accurately written as $\lambda(\mathbf{x}(\sigma, \gamma_1, \gamma_2))$ and $p(\mathbf{x}(\sigma, \gamma_1, \gamma_2))$. We omit these details for the sake of brevity in the solution formula.

We remark, also, that a solution in terms of the Jacobian J as stated is not always satisfactory. The reason is that except in cases in which analytic solutions of the ray equations are available, it may be difficult to calculate the indicated derivatives required for the determination of J. Either we must compute the Laplacian of τ determined as a function of $(\gamma_1, \gamma_2, \sigma)$, or we must differentiate the solution $\mathbf{x}(\gamma_1, \gamma_2, \sigma)$ with respect to the parameters (γ_1, γ_2).

An alternative approach to the determination of J is provided by deriving additional ordinary differential equations along the rays for the quantities that make up J. These quantities can then be determined along with τ as solutions that propagate along the rays. We remark that the normal vector appearing in the definition of J in (8.2.13) is just $\mathbf{p}/p$, which is determined as part of the solution of (8.2.9). Thus, we need only determine the other elements in J. To do this, we define

$$y_{jk}(\gamma_1, \gamma_2, \sigma) = \frac{\partial x_j(\gamma_1, \gamma_2, \sigma)}{\partial \gamma_k},$$

$$\quad\quad\quad\quad\quad\quad\quad\quad\quad\quad\quad j = 1, 2, 3, \quad k = 1, 2. \tag{8.2.20}$$

$$z_{jk}(\gamma_1, \gamma_2, \sigma) = \frac{\partial p_j(\gamma_1, \gamma_2, \sigma)}{\partial \gamma_k},$$

We have introduced 12 new unknowns that are the partial derivatives of $\mathbf{x}$ and $\mathbf{p}$ with respect to the ray parameters γ_1 and γ_2. We leave it to an exercise to verify that these 12 unknowns satisfy the following system of ordinary differential equations—derivable from the ray equations—with respect to σ:

$$\frac{dy_{jk}}{d\sigma} = \lambda z_{jk} + p_j \sum_{i=1}^{3} \frac{\partial \lambda}{\partial x_i} y_{ik},$$

$$\frac{dz_{jk}}{d\sigma} = \sum_{i=1}^{3} \frac{\partial}{\partial x_i} \left[\lambda p \frac{\partial p}{\partial x_j} \right] y_{ik}, \qquad j = 1, 2, 3, \quad k = 1, 2. \qquad (8.2.21)$$

These 12 equations can be appended to the original 7 for determination of the rays and τ to yield a system of 19 (!) *first-order ordinary differential equations* governing the propagation of the leading order approximation of the solution u. Of course, each correction term in the amplitude series requires the addition of only one first-order ordinary differential equation. Complete determination of a solution to this system would require initial data described parametrically in terms of the parameters γ_1 and γ_2.

SPECIAL CHOICES OF RAY PARAMETER σ

For reference purposes, we shall write the ray equations for three choices of the parameter λ, which, in turn, provide three choices of σ with different interpretations.

Case 1 $\lambda = 1$ This would likely be the parameterization of choice in a case in which analytical solutions were accessible, such as in homogeneous media. The ray equations (8.2.9) and (8.2.21) now take on the form

$$\frac{dx_j}{d\sigma} = p_j, \qquad \frac{dp_j}{d\sigma} = p \frac{dp}{dx_j}, \qquad \frac{d\tau}{d\sigma} = p^2,$$

$$\frac{dy_{jk}}{d\sigma} = z_{jk}, \qquad \frac{dz_{jk}}{d\sigma} = \sum_{i=1}^{3} \frac{\partial}{\partial x_i} \left[p \frac{\partial p}{\partial x_j} \right] y_{ik}, \qquad j = 1, 2, 3, \quad k = 1, 2. \qquad (8.2.22)$$

Case 2 $\lambda = c = 1/p$ For this choice, the derivative of the position vector has magnitude equal to unity and the parameter σ is just arc length along the ray. Therefore, we shall distinguish this choice of scale by replacing σ by s:

$$\frac{dx_j}{ds} = \frac{p_j}{p}, \qquad \frac{dp_j}{ds} = \frac{dp}{dx_j}, \qquad \frac{d\tau}{ds} = p,$$

$$\frac{dy_{jk}}{ds} = \frac{z_{jk}}{p} - \frac{p_j}{p^2} \sum_{i=1}^{3} \frac{\partial p}{\partial x_i} y_{ik}, \qquad (8.2.23)$$

$$\frac{dz_{jk}}{ds} = \sum_{i=1}^{3} \frac{\partial}{\partial x_i} \left[\frac{1}{p} \frac{\partial p}{\partial x_j} \right] y_{ik}, \qquad j = 1, 2, 3, \quad k = 1, 2.$$

Case 3 $\lambda = c^2 = 1/p^2$ This last case is distinguished by the fact that the derivative of τ with respect to σ is equal to unity. Thus, the parameter along the ray can be taken to be τ itself so long as we remember to set its initial value—not arbitrarily equal to zero as we did in Chapter 1, but— equal to the correct value of τ for each ray as defined as a function of the other parameters γ_1 and γ_2. The ray equations now become

$$\frac{dx_j}{d\tau} = \frac{p_j}{p^2}, \qquad \frac{dp_j}{d\tau} = \frac{1}{p}\frac{dp}{dx_j},$$

$$\frac{dy_{jk}}{d\tau} = \frac{z_{jk}}{p^2} - \frac{2p_j}{p^3}\sum_{i=1}^{3}\frac{\partial p}{\partial x_i}y_{ik}, \tag{8.2.24}$$

$$\frac{dz_{jk}}{d\tau} = \sum_{i=1}^{3}\frac{\partial}{\partial x_i}\left[p\frac{\partial p}{\partial x_j}\right]y_{ik},$$

$$j = 1, 2, 3, \quad k = 1, 2.$$

This last parameterization seems to be the one of choice for numerical computation of rays, amplitude, and phase. However, I find a rescaling by some reference velocity as an attractive alternative, since the velocity as a raw number is often extremely large in dimensional variables. For example, in seismology, with units of meters and seconds, the magnitude of the velocity is $O(10^3)$. We leave to Exercise 8.16 the development of the system of ray equations for a ray parameter scaled by a reference velocity.

In the next section, we shall discuss the identification of initial data for this system of equations. We have already seen how this is done for τ in Chapter 1. Therefore, it is necessary now to determine the initial values of the constituents of A_j to use the formulas (8.2.14) and (8.2.19).

TWO DIMENSIONS

There are some simplifications that occur with the reduction of the number of dimensions. We consider, again, Eq. (8.2.1), with the Laplace operator and **x** now two-dimensional. We proceed with the formal solution (8.2.2) and obtain (8.2.8) with the upper limit on j being 2 instead of 3. In the ray equations (8.2.9), the upper limit on j is again 2. The discussion of the use of the divergence theorem to solve for A_0 goes as before except that the domain D is now two-dimensional and the boundary ∂D is now made up of curves that are either along the rays or orthogonal to them.

Thus, the first substantive change occurs in the result (8.2.11), where we must interpret dS as a differential arc length rather than as a differential surface area. We replace (8.2.12) then by the statement

$$dS = \left|\hat{\mathbf{n}} \cdot \frac{d\mathbf{x}}{dy_1}\right| dy_1, \tag{8.2.25}$$

with γ_1 labeling the rays and $\hat{n}$ a unit normal to the curve of constant τ. We now define J by

$$J = \left| \hat{n} \cdot \frac{d\mathbf{x}}{d\gamma_1} \right|. \qquad (8.2.26)$$

The identity (8.2.18) still holds. Verification of this two-dimensional result is left as an exercise to be modeled on the three-dimensional result presented in the verification of (8.2.18) below. The solution formulas (8.2.19) still hold as well. The set of definitions in (8.2.20) and the system of equations (8.2.21) is reduced in number because the range on j is 1–2, while k only takes on the value of 1 and can be eliminated altogether. Thus, in place of those two equations, we write

$$y_j(\gamma_1, \sigma) = \frac{\partial x_j(\gamma_1, \sigma)}{\partial \gamma_1}, \qquad z_j(\gamma_1, \sigma) = \frac{\partial p_j(\gamma_1, \sigma)}{\partial \gamma_1}, \qquad j = 1, 2; \quad (8.2.27)$$

$$\frac{dy_j}{d\sigma} = \lambda z_j + p_j \sum_{i=1}^{2} \frac{\partial \lambda}{\partial x_i} y_i, \qquad \frac{dz_j}{d\sigma} = \sum_{i=1}^{2} \frac{\partial}{\partial x_i}\left[\lambda p \frac{\partial p}{\partial x_j} \right] y_i, \qquad j = 1, 2. \quad (8.2.28)$$

VERIFICATION OF (8.2.18)

We close this section with a proof of the result (8.2.18). To do so, we note first that the normal to the surface of constant τ appearing in the definition of J (8.2.13) is proportional to the gradient of τ, which is also proportional to the ray tangent from (8.2.9). That is, using the ray equation (8.2.9), we can write

$$J = \frac{1}{\lambda p} \left| \frac{d\mathbf{x}}{d\sigma} \cdot \frac{d\mathbf{x}}{d\gamma_1} \times \frac{d\mathbf{x}}{d\gamma_2} \right|. \qquad (8.2.29)$$

Let us introduce the alternative notation γ_3 for σ. We shall use either of these as is convenient.

The triple scalar product appearing in the last equation can also be interpreted as a determinant. Thus, we can write

$$\lambda p J = |K|, \qquad K = \det(K_{ij}), \qquad K_{ij} = \partial x_i / \partial \sigma_j, \qquad i, j = 1, 2, 3. \quad (8.2.30)$$

We now use the rule for differentiation of a determinant in order to write

$$\frac{dK}{d\sigma} = \sum_{i,j=1}^{3} \frac{\partial}{\partial \sigma}\left[\frac{\partial x_i}{\partial \gamma_j} \right] \text{cof}\left[\frac{\partial x_i}{\partial \gamma_j} \right]. \qquad (8.2.31)$$

We have used the term cof to denote the cofactor of an element in a determinant. In the first factor, we shall interchange the order of differentiation; σ is, after all, one of the three independent parameters γ_1, γ_2, γ_3, and we assume enough smoothness to allow the interchange. The second factor is, within a scale of K itself, just the element of another matrix L_{ij}. The transpose of that matrix is the inverse of the matrix K_{ij}. That is,

$$L_{ij} = \text{cof}\left[\frac{\partial x_i}{\partial \gamma_j}\right] = K \frac{\partial \gamma_i}{\partial x_i}, \qquad j = 1, 2, 3, \qquad (8.2.32)$$

and

$$\frac{dK}{d\sigma} = \sum_{i,j=1}^{3} \frac{\partial}{\partial \gamma_j}\left[\frac{\partial x_i}{\partial \sigma}\right] K \frac{\partial \gamma_j}{\partial x_i}. \qquad (8.2.33)$$

The sum on j is now recognized as just the chain rule differentiation of $\partial x_i/\partial \sigma$ with respect to each x_i. Furthermore, we can use the ray equations (8.2.9) themselves to replace $\partial x_i/\partial \sigma$ by λp_i. The result of these two operations is

$$\frac{dK}{d\sigma} = K \sum_{i=1}^{3} \frac{\partial}{\partial x_i}[\lambda p_i] = \lambda K \nabla^2 \tau + K \sum_{i=1}^{3} p_i \frac{\partial \lambda}{\partial x_i}. \qquad (8.2.34)$$

In the last expression, we have used the definition $p_i = \partial \tau/\partial x_i$ to identify one term of the derivative of the product in the middle expression; in the last term, we again use (8.2.9) to identify that sum as a derivative with respect to σ. The result is

$$\lambda \nabla^2 \tau = \frac{1}{K}\frac{dK}{d\sigma} - \frac{1}{\lambda}\frac{d\lambda}{d\sigma} = \frac{d}{d\sigma}\{\log[|K|/\lambda]\} = \frac{d}{d\sigma}\{\log[pJ]\}. \qquad (8.2.35)$$

The last line is just the result (8.2.18) and the verification is complete.

Exercises

8.12 (a) Use the results of this section to rewrite (8.2.6) in the form

$$d[A_j\sqrt{pJ}]/d\sigma = -\tfrac{1}{2}\lambda\sqrt{pJ}\,\nabla^2 A_{j-1}.$$

(b) Integrate the result of (a) to obtain (8.2.19).

8.13 Verify that the functions y_{jk} and z_{jk}, defined by (8.2.20), satisfy the system of differential equations (8.2.21). To do this, differentiate in (8.2.20), interchange the order of differentiation of σ and γ_k, and use the ray equations (8.2.9).

8.14 When the surfaces of constant σ are also the surfaces of constant τ, the triple scalar product in (8.2.12) reduces to the magnitude of the cross product multiplied by the two differentials. Show that

$$\left|\frac{d\mathbf{x}}{d\gamma_1} \times \frac{d\mathbf{x}}{d\gamma_2}\right|^2 = g = |\det(g_{ij})|; \qquad g_{ij} = \frac{d\mathbf{x}}{d\gamma_i}\cdot\frac{d\mathbf{x}}{d\gamma_j}, \qquad i, j = 1, 2.$$

8.15 The purpose of this exercise is to provide some practice with carrying out analytically for a particular example the computations indicated by the results of this section. The field generated will be a reflected wave for a point source over a half-space in three dimensions under the assumption of a reflection coefficient of unity and a constant velocity c in the region $z \leq H$. Thus, the geometry of this problem is as in Exercise 8.11. We shall use as parameters γ_1 and γ_2, respectively, the polar angles θ, measured from the z axis and ϕ measured in the x–y plane from the x axis. Thus, let us suppose that on the surface $z = H$, the following data are given.

$$x(\theta, \phi, 0) = \tan \theta \cos \phi, \qquad y(\theta, \phi, 0) = \tan \theta \sin \phi, \qquad z(\theta, \phi, 0) = H;$$

$$p_1(\theta, \phi, 0) = p \sin \theta \cos \phi, \qquad p_2(\theta, \phi, 0) = p \sin \theta \sin \phi, \qquad p_3(\theta, \phi, 0) = p \cos \theta;$$

$$\tau(\theta, \phi, 0) = Hp \sec \theta, \qquad A_0(\theta, \phi, 0) = (\cos \theta)/4\pi H;$$

$$y_{11}(\theta, \phi, 0) = H \sec^2 \theta \cos \phi, \qquad y_{21}(\theta, \phi, 0) = H \sec^2 \theta \sin \phi, \qquad y_{31}(\theta, \phi, 0) = 0;$$

$$y_{12}(\theta, \phi, 0) = -H \tan \theta \sin \phi, \qquad y_{22}(\theta, \phi, 0) = H \tan \theta \cos \phi, \qquad y_{32}(\theta, \phi, 0) = 0;$$

$$z_{11}(\theta, \phi, 0) = p \cos \theta \cos \phi, \qquad z_{21}(\theta, \phi, 0) = p \cos^2 \theta \sin \phi, \qquad z_{31}(\theta, \phi, 0) = p \sin \theta;$$

$$z_{12}(\theta, \phi, 0) = -p \sin \theta \sin \phi, \qquad z_{22}(\theta, \phi, 0) = p \sin \theta \cos \phi, \qquad z_{32}(\theta, \phi, 0) = 0.$$

(a) Show that the solutions of the ray equations (8.2.22) ($\lambda = 1$) are as follows:

$$\mathbf{z}_1 = \mathbf{z}_1(\theta, \phi, 0), \qquad \mathbf{z}_2 = \mathbf{z}_2(\theta, \phi, 0), \qquad \mathbf{y}_1 = \mathbf{y}_1(\theta, \phi, 0) + \sigma \mathbf{z}_1(\theta, \phi, 0),$$

$$\mathbf{y}_2 = \mathbf{y}_2(\theta, \phi, 0) + \sigma \mathbf{z}_2(\theta, \phi, 0), \qquad \mathbf{p} = \mathbf{p}(\theta, \phi, 0),$$

$$\mathbf{x} = \mathbf{x}(\theta, \phi, 0) + \sigma \mathbf{p}(\theta, \phi, 0), \qquad \tau = \tau(\theta, \phi, 0) + \sigma p.$$

(b) Verify by differentiation of the preceding results that $\mathbf{y}_1 = d\mathbf{x}/d\gamma_1$, $\mathbf{y}_2 = d\mathbf{x}/d\gamma_2$.

(c) Use the definition (8.2.29) to determine J. Show that $J = c \sin \theta$ $\cdot [H \sec \theta + \sigma p]^2$ and thus that $A_0 = (4\pi[H \sec \theta + \sigma p])^{-1}$.

(d) Show that the vector cross product $\mathbf{y}_1 \times \mathbf{y}_2$ is not normal to the vector $\mathbf{p}$. Explain why.

(e) Introduce r as the distance between $\mathbf{x}$ and the point $(0, 0, 2H)$, which is the image of the origin with respect to the plane $z = H$. Show that $r = H \sec \theta + \sigma p$ and that, therefore, $\tau = pr$, $A_0 = (4\pi r)^{-1}$.

8.16 The purpose of this exercise is to develop the system of ray equations for variables scaled by a reference speed.

(a) Consider the ray equations (8.2.9), with the scale $\lambda = c_0$, a constant reference speed. Also, introduce the new functions ϕ, $\mathbf{q}$, and n defined by

$$\tau = \phi/c_0, \qquad \mathbf{q} = \nabla \phi, \qquad n = c_0/c.$$

Show that the system (8.2.9) now becomes

$$dx/d\sigma = \mathbf{q}, \qquad d\mathbf{q}/d\sigma = n\,\nabla n, \qquad d\phi/d\sigma = n^2.$$

(b)   Now define

$$B_j = c_0^j A_j, \qquad j = 0, 1, \ldots,$$

and show from (8.2.6) and (8.2.18) that

$$2\,\nabla\phi\cdot\nabla B_j + B_j\,\nabla^2\phi = -\nabla^2 B_{j-1}, \qquad j = 0, 1, \ldots, \qquad \nabla^2\phi = d[\log nJ]/d\sigma.$$

In the first equation, we take $B_{-1} = 0$. [*Remark*: The scaling on the B_j's as compared to the A_j's changes the formal large parameter $i\omega$ in (8.2.2) $i\omega/c_0$.]

(c)   Introduce $w_{jk} = \lambda z_{jk}, j = 1, 2, 3, k = 1, 2$. Show that the system (8.2.21) now becomes

$$\frac{dy_{jk}}{d\sigma} = w_{jk}, \qquad \frac{dw_{jk}}{d\sigma} = \sum_{i=1}^{3} y_{ik}\frac{\partial}{\partial x_i}\left[n\frac{\partial n}{\partial x_j}\right], \qquad j = 1, 2, 3, \quad k = 1, 2.$$

8.3   DETERMINATION OF RAY DATA

In the preceding section, we derived the system of ray equations (8.2.9), (8.2.16), or (8.2.28) for the determination of the amplitude and phase of an asymptotic solution to the homogeneous Helmholtz equation (8.2.1). We shall now address the question of determining *initial data* for this system of equations, also known as *ray data*. We take the point of view that a source or an incidence of a wave of the type (8.2.2) on a scattering surface gives rise to a wave, again of the type (8.2.2). The subsequent propagation of this wave is then governed by the *homogeneous* Helmholtz equation. Thus, our objective here will be to show how these ray data are determined for waves initiated by a source or waves initiated in response to a scattering mechanism. We shall do this for a number of prototypical examples.

RAY DATA FOR A POINT SOURCE

Let us consider first the case of a point source

$$f(\mathbf{x}, \omega) = -\delta(\mathbf{x} - \mathbf{x}_0)e^{i\omega\tau_0}. \tag{8.3.1}$$

We should expect that a point source of this type will give rise to a wave emanating from the source point. Furthermore, the $\mathbf{x}$-independent phase of the source would have to be a phase contribution to the solution as well. Therefore, we obtain directly from the nature of the source the ray data

$$\mathbf{x} = \mathbf{x}_0, \qquad \tau = \tau_0, \qquad \sigma = 0. \tag{8.3.2}$$

If all of the rays are to emanate from this point, then the solution of the eikonal equation we seek is the conoidal solution for which $\mathbf{p}$ is not determined initially, but only its magnitude p is given by the eikonal equation (8.2.8) itself. The two ray parameters, then, prescribe the initial direction of the vector $\mathbf{p}$. For example, we may use the polar angles of the initial value of $\mathbf{p}$; see the discussion in Section 1.4. Thus, we set

$$\mathbf{p} = p(\mathbf{x}_0)\hat{\mathbf{p}}; \qquad \hat{\mathbf{p}} = (\sin \gamma_1 \cos \gamma_2, \sin \gamma_1 \sin \gamma_2, \cos \gamma_1). \qquad (8.3.3)$$

From the definitions (8.2.20), we can now determine the initial values of y_{jk} and z_{jk}. These values are

$$z_{11} = p(\mathbf{x}_0) \cos \gamma_1 \cos \gamma_2, \qquad z_{21} = p(\mathbf{x}_0) \cos \gamma_1 \sin \gamma_2,$$

$$z_{31} = -p(\mathbf{x}_0) \sin \gamma_1,$$

$$z_{12} = -p(\mathbf{x}_0) \sin \gamma_1 \cos \gamma_2, \qquad z_{22} = p(\mathbf{x}_0) \sin \gamma_1 \sin \gamma_2, \qquad z_{32} = 0; \qquad (8.3.4)$$

$$y_{jk} = 0, \qquad \sigma = 0, \qquad j = 1, 2, 3, \quad k = 1, 2.$$

From the definition of J in (8.2.29), we see that for these data, $J(0) = 0$ and that the solution formula (8.2.14), with $\sigma_1 = 0$, is not valid. Let us rewrite (8.2.14) in the form

$$A_0^2(\sigma)p(\sigma)J(\sigma) = A_0^2(\sigma_1)p(\sigma_1)J(\sigma_1) = C^2(\gamma_1, \gamma_2). \qquad (8.3.5)$$

Here C^2 is a constant with respect to σ and can depend only on the parameters γ_1 and γ_2.

Let us consider $J(\sigma)$ for σ near zero. If we approximate $\mathbf{p}$ and z_{jk} by their initial values, then

$$y_{jk} \approx z_{jk}(0)\sigma, \qquad j = 1, 2, 3, \quad k = 1, 2, \qquad (8.3.6)$$

and we use this result in (8.2.9) to find that

$$J(\sigma) \approx [\sigma \lambda p(\mathbf{x}_0)]^2 \sin \gamma_1. \qquad (8.3.7)$$

We see from this equation that $J(\sigma)$ vanishes quadratically in σ as $\sigma \to 0$. Thus, if $C^2(\gamma_1, \gamma_2)$ in (8.3.5) is to be finite, we must expect that $A_0^2(\sigma)$ must be singular to the same order as $J(\sigma)$. Indeed, we can check this quite readily by arguing that in the limit, as $\sigma \to 0$, the solution to the problem with variable $p(\mathbf{x})$ should be the same as the solution to the problem with constant value $p(\mathbf{x}) \equiv p(\mathbf{x}_0)$. For the latter case, the solution to the Helmholtz equation with source (8.3.1) is just the Green's function (5.3.24), multiplied by the phase factor $\exp\{i\omega\tau_0\}$. This *exact solution* is of the form of our asymptotic solution (8.2.2), with $\beta = 0$, and we conclude that

$$A_0 \approx (4\pi r)^{-1}, \qquad r = |\mathbf{x} - \mathbf{x}_0|. \qquad (8.3.8)$$

Now we need only express r in terms of σ to obtain an approximate solution of (8.3.5), from which we can estimate $C(\gamma_1, \gamma_2)$ by letting $\sigma \to 0$.

The solution for $\mathbf{x}$ for small σ is readily obtained from the ray data for $\mathbf{x}$ and $\mathbf{p}$ given in (8.3.2) and (8.3.3) and the ray equations (8.2.9). The result is

$$\mathbf{x} \approx \mathbf{x}_0 + \lambda p(\mathbf{x}_0)\hat{\mathbf{p}}\sigma, \tag{8.3.9}$$

from which we conclude that

$$r \approx \lambda p(\mathbf{x}_0)\sigma. \tag{8.3.10}$$

Now we use (8.3.7), (8.3.8), and (8.3.10) in (8.3.5) to conclude that

$$C(\gamma_1, \gamma_2) = \sqrt{p(\mathbf{x}_0)} \sin \gamma_1/4\pi. \tag{8.3.11}$$

Thus, for this point source problem, we do not directly find an initial value for A_0 because this amplitude is singular at the source point but, instead, find a constant $C^2(\gamma_1, \gamma_2)$ that allows us to use (8.3.5) to write the solution for A_0, despite its singular behavior,

$$A_0(\sigma) = \sqrt{p(\mathbf{x}_0)} \sin \gamma_1/4\pi\sqrt{p(\mathbf{x}(\sigma))J(\sigma)}. \tag{8.3.12}$$

Analysis of the solution formula (8.2.19) for $j > 0$ reveals that each A_j is progressively more singular than the preceding coefficient. Determination of the ray data for these coefficients is quite involved and will not be discussed here.

RAY DATA FOR A LINE SOURCE

We consider now the case of a line source of the form

$$f(\mathbf{x}, \omega) = \delta(x - x_0)\delta(z - z_0)e^{i\omega\tau_0(y)}. \tag{8.3.13}$$

We expect this source to give rise to a wave whose rays emanate from the line $x = x_0$, $z = z_0$, where the delta functions "act." We parameterize this line by a single parameter γ_2 by setting

$$x = x_0, \qquad y = \gamma_2, \qquad z = z_0, \qquad \tau = \tau_0(\gamma_2), \qquad \sigma = 0. \tag{8.3.14}$$

Differentiation of this data with respect to the parameter γ_2 [as described in (1.3.12) in two independent variables] leads to the result

$$p_2 = p_{20}(\gamma_2) = d\tau_0(\gamma_2)/d\gamma_2. \tag{8.3.15}$$

Since $\mathbf{p}$ must satisfy the eikonal equation (8.2.8), there will be no real solutions for the other two components of $\mathbf{p}$ unless $|p_{20}(\gamma_2)| \le p(x, \gamma_2, z)$, which we shall assume. In this case, the initial value of the magnitude, but not the direction of the two-vector (p_1, p_3), is determined by the eikonal equation.

Labeling the directions of these vectors will introduce the second ray parameter. The total vector (p_1, p_2, p_3) will make a fixed angle with the y axis, with tangential and normal components of fixed magnitude. Thus, these initial ray parameters will fill out the surface of a circular cone with the y axis as its axis.

We define

$$q_0(\gamma_2) = \sqrt{p^2(x_0, \gamma_2, z_0) - p_{20}^2(\gamma_2)}. \qquad (8.3.16)$$

We then express the initial values of p_1 and p_3 as

$$\begin{aligned} p_1 &= p_{10}(\gamma_1, \gamma_2) = q_0(\gamma_2) \sin \gamma_1, \\ p_3 &= p_{30}(\gamma_1, \gamma_2) = q_0(\gamma_2) \cos \gamma_1, \end{aligned} \qquad \sigma = 0. \qquad (8.3.17)$$

We remark that the cone angle of the initial ray directions has the ratio $p_{20}(\gamma_2)/q_0(\gamma_2)$ as its cosine.

By differentiating the ray data for $\mathbf{x}$ and $\mathbf{p}$ with respect to γ_1 and γ_2, we obtain the ray data for y_{jk} and z_{jk}. The results are

$$y_{11} = y_{21} = y_{31} = y_{12} = y_{32} = 0, \qquad y_{22} = 1,$$

$$z_{11} = q_0(\gamma_2) \cos \gamma_1, \qquad z_{21} = 0, \qquad z_{31} = -q_0(\gamma_2) \sin \gamma_1, \quad (8.3.18)$$

$$z_{12} = q_0'(\gamma_2) \sin \gamma_1, \qquad z_{22} = p_{20}'(\gamma_2), \qquad z_{32} = q_0'(\gamma_2) \cos \gamma_1.$$

As in the preceding example, we can see here that $J(\sigma)$, as defined by (8.2.9), will be zero initially, since its entire second row y_{11}, y_{21}, y_{31} is zero initially. Thus, again we must use the device of finding a constant $C(\gamma_1, \gamma_2)$, as defined by (8.3.5), in order to determine the solution A_0. We leave as an exercise the determination of the following approximate results for small σ:

$$x \approx x_0 + \sigma \lambda q_0(\gamma_2) \sin \gamma_1, \qquad y \approx \gamma_2 + \sigma \lambda p_{20}(\gamma_2),$$

$$z \approx z_0 + \sigma \lambda q_0(\gamma_2) \cos \gamma_1,$$

$$\tau \approx \tau_0(\gamma_2) + \lambda p^2 \sigma, \qquad (8.3.19)$$

$$y_{11} \approx \sigma \lambda q_0(\gamma_2) \cos \gamma_1, \qquad y_{21} \approx 0, \qquad y_{31} \approx -\sigma \lambda q_0(\gamma_2) \sin \gamma_1,$$

$$y_{12} \approx \sigma \lambda q_0'(\gamma_2) \sin \gamma_1, \qquad y_{22} \approx 1 + \sigma \lambda p_{20}'(\gamma_2), \qquad y_{32} \approx \sigma \lambda q_0'(\gamma_2) \cos \gamma_1;$$

$$\lambda p J \approx \sigma \lambda^2 q_0^2 \qquad (8.3.20)$$

and

$$A_0(\sigma) \approx C(\gamma_1, \gamma_2)/q_0 \sqrt{\lambda \sigma}. \qquad (8.3.21)$$

As in the preceding example, we can determine $C(\gamma_1, \gamma_2)$ by considering a *canonical problem* that is solvable asymptotically by other means. In particular, let us consider a problem for u with a source of the form of (8.3.13) but with p_{20} a constant and $p = 1/c$ equal to a constant as well. Therefore, we consider the problem

$$\nabla^2 u + (\omega^2/c^2)u = -\delta(x - x_0)\delta(z - z_0)e^{i\omega\alpha y}, \qquad (8.3.22)$$

for which

$$p_{20} = \alpha = \text{const}, \qquad q_0 = \sqrt{p^2 - \alpha^2}. \qquad (8.3.23)$$

The substitution

$$u(\mathbf{x}, \omega) = v(x, z, \omega)e^{i\omega\alpha y} \qquad (8.3.24)$$

leads to the following problem for v:

$$(\partial^2 v/\partial x^2) + (\partial^2 v/\partial z^2) + \omega^2 q_0^2 v = -\delta(x - x_0)\delta(z - z_0). \qquad (8.3.25)$$

We see here that v is just the *Green's function* for the Helmholtz equation in two spatial dimensions, with wave number given by ωq_0. Thus, the exact solution for v is given by (6.3.23) with $1/c$ replaced by q_0 and r measuring distance from the fixed point (x_0, z_0) to the point (x, z). We are interested in the asymptotic expansion of this solution. This can be obtained by using the asymptotic expansion of the Hankel function of order zero given by (6.3.18) with the argument z of that formula given by

$$z \to \omega q_0 \rho, \qquad \rho = \sqrt{(x - x_0)^2 + (z - z_0)^2}. \qquad (8.3.26)$$

We leave it to an exercise to confirm that these substitutions lead to the leading order asymptotic solution

$$u \sim \frac{\exp(i\omega\{q_0\rho + \alpha y\} + i\pi/4)}{2\sqrt{2\pi\omega q_0\rho}}. \qquad (8.3.27)$$

This solution is indeed of the form of the leading term in (8.2.2), this time with $\beta = -\frac{1}{2}$. We can identify this result with the ray solution more easily if we rewrite ρ in terms of σ through the solutions of the ray equations (8.3.19)–(8.3.21), using also (8.3.23). The result is

$$u \sim \frac{\exp(i\omega\{\lambda p^2\sigma + \alpha\gamma_2 y\} + i\pi/4)}{2\sqrt{2\pi\omega\lambda q_0(\gamma_2)p(x_0, \gamma_2, z_0)\sigma}}. \qquad (8.3.28)$$

The multiplier of ω in the phase of this result is exactly τ, as given in (8.3.19). We have already chosen $\beta = -\frac{1}{2}$ to give the right power of ω. By comparing the amplitude in (8.3.28) with the solution (8.3.21) for A_0, we conclude that

$$C(\gamma_1, \gamma_2) = \frac{e^{i\pi/4}}{2\sqrt{2\pi}}\sqrt{\frac{q_0(\gamma_2)}{p(x_0, \gamma_2, z_0)}}. \qquad (8.3.29)$$

With this constant determined, the solution for $A_0(\bar{\sigma})$ is given by

$$A_0(\sigma) = \frac{e^{i\pi/4}}{2\sqrt{2\pi p(\sigma)J(\sigma)}} \sqrt{\frac{q_0(\gamma_z)}{p(x_0, \gamma_2, z_0)}}. \qquad (8.3.30)$$

RAY DATA FOR REFLECTED AND TRANSMITTED WAVES

Let us now consider the case of a wave incident on a surface across which the velocity (and slowness) are discontinuous. We will denote the surface by S and the sound speed and slowness on the same side of S as the incident wave by $c_-(x, y, z) = 1/p_-(x, y, z)$; while on the opposite side of S from the incident wave, we will denote the sound speed and slowness by $c_+(x, y, z) = 1/p_+(x, y, z)$. Correspondingly, we will denote the two sides of S by D_- (including u_1) and D_+.

We will take as conditions at the interface that both the total solution and its normal derivative be continuous across S. We assume an incident wave of the form

$$u_1(\mathbf{x}, \omega) \approx \omega^{\beta_1} \sum_{j=0}^{\infty} \frac{A_j^I(\mathbf{x})e^{i\omega\tau_I(\mathbf{x})}}{(i\omega)^j}. \qquad (8.3.31)$$

This wave is of the form (8.2.2) with $\beta = \beta_1$. It is likely in practice that u_1 is known only in terms of ray parameters, as was the case in Exercise 8.15.

We assume that the incidence of u_1 on the boundary gives rise to a reflected wave u_R and a transmitted wave u_T having the same form

$$u_R(\mathbf{x}, \omega) \sim \omega^{\beta_R} \sum_{j=0}^{\infty} \frac{A_j^R(\mathbf{x})e^{i\omega\tau_R(\mathbf{x})}}{(i\omega)^j} \qquad (8.3.32)$$

and

$$u_T(\mathbf{x}, \omega) \sim \omega^{\beta_T} \sum_{j=0}^{\infty} \frac{A_j^T(\mathbf{x})e^{i\omega\tau_T(\mathbf{x})}}{(i\omega)^j}. \qquad (8.3.33)$$

The total solution u then is of the form

$$u = u_1 + u_R, \quad \mathbf{x} \text{ in } D_-, \quad u = u_T, \quad \mathbf{x} \text{ in } D_+. \qquad (8.3.34)$$

It would not be possible to balance these solutions on the surface S if their orders in ω were different. Therefore, we require that

$$\beta_R = \beta_T = \beta_1. \qquad (8.3.35)$$

With this identification made, we shall drop the multiplicative power of ω in all further discussion.

The condition that the solution be continuous on S leads to the equation

$$\sum_{j=0}^{\infty} \frac{A_j^I(\mathbf{x})e^{i\omega\tau_I(\mathbf{x})}}{(i\omega)^j} + \sum_{j=0}^{\infty} \frac{A_j^R(\mathbf{x})e^{i\omega\tau_R(\mathbf{x})}}{(i\omega)^j} = \sum_{j=0}^{\infty} \frac{A_j^T(\mathbf{x})e^{i\omega\tau_T(\mathbf{x})}}{(i\omega)^j}, \quad \mathbf{x} \text{ on } S. \quad (8.3.36)$$

These solutions could not agree asymptotically everywhere on S unless the phases agreed. Thus, we conclude that

$$\tau_I(\mathbf{x}) = \tau_R(\mathbf{x}) = \tau_T(\mathbf{x}), \qquad \mathbf{x} \quad \text{on} \quad S. \tag{8.3.37}$$

From these equations, we can conclude that the parts of the gradients of τ_I, τ_R, and τ_T that are in the surface S must be equal. To show this more precisely, let us introduce a parameterization of the surface

$$\mathbf{x} = \mathbf{x}(\gamma_1, \gamma_2), \qquad \mathbf{x} \quad \text{on} \quad S. \tag{8.3.38}$$

Now we differentiate the equations in (8.3.37) with respect to γ_1 and γ_2 to obtain

$$\nabla\tau_I \cdot \frac{d\mathbf{x}}{d\gamma_i} = \nabla\tau_R \cdot \frac{d\mathbf{x}}{d\gamma_i} = \nabla\tau_T \cdot \frac{d\mathbf{x}}{d\gamma_i}, \qquad i = 1, 2. \tag{8.3.39}$$

These equations state that the projection of the gradient on two linearly independent directions in the surface S are equal. Let us define

$$\nabla_\gamma = \nabla - \hat{\mathbf{n}}(\hat{\mathbf{n}} \cdot \nabla) = \nabla - \hat{\mathbf{n}}\frac{\partial}{\partial n}. \tag{8.3.40}$$

Here $\hat{\mathbf{n}}$ denotes a unit normal to the surface S, and ∇_γ is, therefore, the component of the gradient in the surface.

We conclude from (8.3.39) that

$$\nabla_\gamma\tau_I = \nabla_\gamma\tau_R = \nabla_\gamma\tau_T. \tag{8.3.41}$$

This result provides two equations for the three unknown initial values of the components of each of the gradients $\nabla\tau_R$ and $\nabla\tau_T$ in terms of the known values of $\nabla\tau_I$. The eikonal equation (8.2.8) provides a third equation for the initial values of each of the gradient vectors. In magnitude, $\hat{\mathbf{n}} \cdot \nabla\tau_R = \partial\tau_R/\partial n$ must agree with $\hat{\mathbf{n}} \cdot \nabla\tau_I$, since they satisfy the eikonal equation on the same side of S. The latter gradient must have the appropriate sign to ensure that the rays are directed toward S. If the former had the same sign as well, then $\nabla\tau_R$ would agree with $\nabla\tau_I$ completely, and the reflected rays would be directed toward the surface S as well, rather than away from it. Thus, we conclude that the two normal components of these gradients must be of opposite sign. The magnitude of $\partial\tau_T/\partial n$ is also determined by the eikonal equation, however; now p_- must be replaced by p_+. In order that the transmitted rays be directed away from the surface S, this normal component must have the same sign as $\partial\tau_I/\partial n$. Thus, we conclude that

$$\frac{\partial\tau_R}{\partial n} = -\operatorname{sign}\frac{\partial\tau_I}{\partial n}\sqrt{p_-^2 - (\nabla_\gamma\tau_I)^2} = -\frac{\partial\tau_I}{\partial n},$$

$$\tag{8.3.42}$$

$$\frac{\partial\tau_T}{\partial n} = \operatorname{sign}\frac{\partial\tau_I}{\partial n}\sqrt{p_+^2 - (\nabla_\gamma\tau_I)^2} = \operatorname{sign}\frac{\partial\tau_I}{\partial n}\sqrt{p_+^2 - p_-^2 + \left[\frac{\partial\tau_I}{\partial n}\right]^2}.$$

We shall assume for the present that the square roots appearing here are real. By using (8.3.39) and (8.3.42), we obtain initial data at the surface S for $\mathbf{p}_R = \nabla \tau_R$ and $\mathbf{p}_T = \nabla \tau_T$.

In order to determine ray data at the surface for the amplitude coefficients, we now consider the boundary condition that the normal derivatives across S be continuous. Thus, we return to the representation (8.3.31)–(8.3.34) and set the normal derivatives on the two sides of S equal to one another. The result is

$$i\omega \frac{\partial \tau_I}{\partial n} \sum_{j=0}^{\infty} \frac{A_j^I(\mathbf{x})}{(i\omega)^j} + \sum_{j=0}^{\infty} \frac{\partial A_j^I(\mathbf{x})}{\partial n} \frac{1}{(i\omega)^j} + i\omega \frac{\partial \tau_R}{\partial n} \sum_{j=0}^{\infty} \frac{A_j^R(\mathbf{x})}{(i\omega)^j} + \sum_{j=0}^{\infty} \frac{\partial A_j^R(\mathbf{x})}{\partial n} \frac{1}{(i\omega)^j}$$

$$= i\omega \frac{\partial \tau_T}{\partial n} \sum_{j=0}^{\infty} \frac{A_j^T(\mathbf{x})}{(i\omega)^j} + \sum_{j=0}^{\infty} \frac{\partial A_j^T(\mathbf{x})}{\partial n} \frac{1}{(i\omega)^j}, \qquad \mathbf{x} \quad \text{on} \quad S. \qquad (8.3.43)$$

In this equation, we have used (8.3.37) to eliminate the phases of the three waves. Here and in (8.3.36), we equate the coefficients of each power of ω in order to obtain a pair of equations for the ray data for A_j^R and A_j^T for each j. We leave it to an exercise to verify that those equations are

$$A_j^I + A_j^R = A_j^T,$$

$$\frac{\partial \tau_I}{\partial n} A_j^I + \frac{\partial \tau_R}{\partial n} A_j^R = \frac{\partial \tau_T}{\partial n} A_j^T + [A_{j-1}^T - A_{j-1}^I - A_{j-1}^R], \qquad j = 1, 2, \dots. \qquad (8.3.44)$$

In these equations, we have taken $A_{-1} = 0$. We shall solve explicitly here only the equations for the leading order coefficients. Those equations are

$$A_0^I + A_0^R = A_0^T, \qquad \frac{\partial \tau_I}{\partial n} A_0^I + \frac{\partial \tau_R}{\partial n} A_0^R = \frac{\partial \tau_T}{\partial n} A_0^T. \qquad (8.3.45)$$

The solution to this pair of equations is

$$A_0^R = R A_0^I, \qquad A_0^T = T A_0^I. \qquad (8.3.46)$$

The reflection and transmission coefficients are given by

$$R = \frac{\dfrac{\partial \tau_I}{\partial n} - \dfrac{\partial \tau_T}{\partial n}}{\dfrac{\partial \tau_I}{\partial n} - \dfrac{\partial \tau_R}{\partial n}} = \frac{\dfrac{\partial \tau_I}{\partial n} - \text{sign} \dfrac{\partial \tau_I}{\partial n} \sqrt{p_+^2 - p_-^2 + (\partial \tau_I/\partial n)^2}}{\dfrac{\partial \tau_I}{\partial n} + \text{sign} \dfrac{\partial \tau_I}{\partial n} \sqrt{p_+^2 - p_-^2 + (\partial \tau_I/\partial n)^2}}, \qquad (8.3.47)$$

and

$$T = \frac{2 \dfrac{\partial \tau_I}{\partial n}}{\dfrac{\partial \tau_I}{\partial n} - \dfrac{\partial \tau_R}{\partial n}} = \frac{2 \dfrac{\partial \tau_I}{\partial n}}{\dfrac{\partial \tau_I}{\partial n} + \text{sign} \dfrac{\partial \tau_I}{\partial n} \sqrt{p_+^2 - p_-^2 + (\partial \tau_I/\partial n)^2}}. \qquad (8.3.48)$$

In obtaining the final result in each of the latter two equations, we have used the results (8.3.42). We leave the determination of the higher-order reflection and transmission coefficients to the exercises.

The reader is cautioned that the initiation of only an ordinary reflected and an ordinary transmitted wave at an interface requires that the surface be smooth and that the square root in these results remain real. We have already seen in Section 8.2 the effect of a point (points) on S, where the square root is zero, bounding a region of real square root from a region of imaginary square root. That boundary gives rise to another kind of wave in the region D_-, a *head wave*, which is a form of diffracted wave. Yet another type of wave is initiated in the second region D_+, namely, an *evanescent wave*.

We have also seen in Chapter 1 that points on the surface at which the incident rays are tangent to the boundary also give rise to another type of ray family that propagates along the surface itself and provides characteristic data for the eikonal equation and generates yet another ray family that leaves the surface tangentially. The wave field associated with the former ray family is called the *creeping wave*, while the wave family associated with the latter set of rays is called the *smooth-body diffracted field*.

At edges of the surface S, the results are also suspect. First, the surface S is certainly discontinuous at an edge, so that the derivation of the ordinary reflected wave is suspect. (In fact, right along the *shadow boundary* of the reflected wave, the actual amplitude is half of the result predicted here.) Furthermore, as discussed in Chapter 1, the edge becomes the initial curve for an entirely new ray field and associated wave, namely, *the edge-diffracted wave*. A complete discussion of the extension of ray methods to the *geometrical theory of diffraction* is beyond the scope of this book. However, we shall briefly describe one example of this theory.

DIFFRACTION BY AN EDGE

The introduction of the geometrical theory of diffraction by J. B. Keller [1958] and the subsequent development of this method by him and his associates is one of the triumphs of ray methods. We shall qualitatively describe this theory here in the context of the problem of diffraction by an edge. Thus, let us suppose that there is a boundary surface S with an edge E. We suppose that the surface S gives rise to a reflected wave and a diffracted wave emanating from the edge. Thus, we require ray data for the family of rays emanating from the edge. We suppose that the diffracted wave also has a representation of the form (8.2.2), namely,

$$u^D(\mathbf{x}; \omega) \sim \omega^{\beta_D} e^{i\omega \tau_D(\mathbf{x})} \sum_{j=0}^{\infty} \frac{A_j^D(\mathbf{x})}{(i\omega)^j}. \tag{8.3.49}$$

We shall discuss only the determination of the initial data for the phase and the leading order coefficient A_0^D. Let us suppose that the edge is defined by

$$\mathbf{x} = \mathbf{x}(\gamma_2), \qquad \mathbf{x} \quad \text{on} \quad E. \tag{8.3.50}$$

We require that the phases of the incident and diffracted waves agree along the edge. Thus,

$$\tau_D(\mathbf{x}(\gamma_2)) = \tau_I(\mathbf{x}(\gamma_2)). \tag{8.3.51}$$

By differentiating this equation with respect to γ_2, we find that

$$\nabla\tau_D \cdot \dot{\mathbf{x}} = \nabla\tau_I \cdot \dot{\mathbf{x}}; \qquad \dot{\mathbf{x}} = \frac{(d\mathbf{x}(\gamma_2)/d\gamma_2)}{|d\mathbf{x}(\gamma_2)/d\gamma_2|}, \qquad \mathbf{x} \quad \text{on} \quad E. \tag{8.3.52}$$

This result is the analog of (8.3.15). That is, the projection of $\nabla\tau_D$ on the tangent to the edge is determined by differentiation of the ray data for τ_D itself. Since the total magnitude of $\nabla\tau_D$ is also known, $p(\mathbf{x}(\gamma_2))$, this condition also determines the initial angle that the diffracted rays must make with the tangent to the edge. Thus, let us define θ as the angle that the incident ray makes with the edge,

$$\cos\theta(\gamma_2) = (\nabla\tau_I \cdot \dot{\mathbf{x}})/p, \qquad \mathbf{x} \quad \text{on} \quad E; \tag{8.3.53}$$

then the diffracted rays satisfy the same equality. Indeed, as with the line source problem, there will be a whole family of such diffracted rays making a cone around the tangent. See Fig. 8.7. In that figure, we have also introduced

   (i)   the normal $\hat{\mathbf{n}}(\gamma_2)$ to the surface S at each point of the edge,
   (ii)   the vector $\dot{\mathbf{x}} \times \hat{\mathbf{n}}$, which together with $\dot{\mathbf{x}}$ and $\hat{\mathbf{n}}$ makes a right-handed vector triad at $\mathbf{x}(\gamma_2)$,
   (iii) ϕ, for which $\phi + \pi$ is the azimuthal angle of the incident ray in the plane of $\hat{\mathbf{n}}$ and $\dot{\mathbf{x}} \times \hat{\mathbf{n}}$, and
   (iv) γ_1, which is the azimuthal angle labeling the diffracted rays.

Using the discussion of the line source as a guide, we introduce the magnitude of the normal component of $\nabla\tau_P$ as a new variable; thus, we define

$$q_0(\gamma_2) = |\nabla\tau_D - \dot{\mathbf{x}}(\dot{\mathbf{x}} \cdot \nabla\tau_D)|$$

$$= \sqrt{p^2(\mathbf{x}(\gamma_2)) - [\nabla\tau_I(\mathbf{x}(\gamma_2)) \cdot \dot{\mathbf{x}}(\gamma_2)]^2}$$

$$= p(\mathbf{x}(\gamma_2))\sin\theta(\gamma_2) \tag{8.3.54}$$

and set

$$\nabla\tau_D - \dot{\mathbf{x}}(\dot{\mathbf{x}} \cdot \mathbf{V}_D) = q_0(\gamma_2)[\hat{\mathbf{n}}\cos\gamma_1 + \dot{\mathbf{x}} \times \hat{\mathbf{n}}\sin\gamma_1]. \tag{8.3.55}$$

The analysis then proceeds as in the problem for the line source earlier.

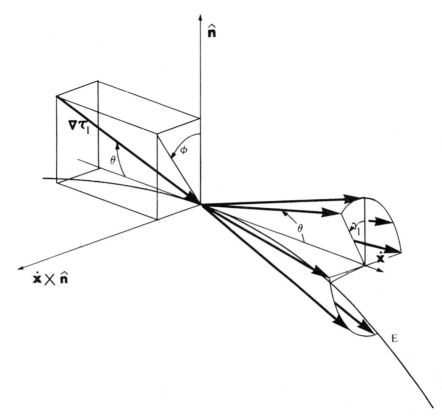

Fig. 8.7. The cone of diffracted rays.

In fact, we obtain an answer for A_0^D of the same form as (8.3.21). Now we make the assumption that the constant $C(\gamma_1, \gamma_2)$ is proportional to the incident wave. More precisely, let us rewrite the result (8.3.21) in the form

$$A_0^D(\sigma) \approx \frac{D(\gamma_1, \gamma_2)}{2\sqrt{2\pi p(\sigma)J(\sigma)}} \sqrt{\sin \theta(\gamma_2)} \, A_0^I(\mathbf{x}(\gamma_2)), \qquad (8.3.56)$$

with $D(\gamma_1, \gamma_2)$ a *diffraction coefficient* to be determined.

The diffraction coefficient is assumed to depend on the relative angle of incidence θ between the incident rays and the edge and on the direction of propagation away from the edge determined by θ and γ_1. Thus, D is indeed a function of γ_1 and γ_2.

As in the previous examples, the way to determine D is to consider the problem locally. Then, we might contemplate solving a problem in which the variables of the general problem are replaced by the constants of the local problem. Let us consider this approach to the problem of scattering in an inhomogeneous medium by a surface S on which the total field satisfies either the homogeneous Dirichlet boundary condition ($u = 0$) or the homogeneous Neumann boundary condition ($\partial u/\partial n = 0$). The corresponding local problem would then be scattering of a plane wave by a semi-infinite plane on which the same boundary condition is satisfied. This is the Sommerfeld problem [Sommerfeld, 1964], which admits a closed form solution from which we can determine the reflected and diffracted waves asymptotically. We leave it to the exercises to determine the ray solution for this problem. When the solution is compared to Sommerfeld's, we can conclude for this type of problem that

$$D(\gamma_1, \gamma_2) = -e^{i\pi/4}[\sec \tfrac{1}{2}(\phi - \gamma_1) \pm \csc \tfrac{1}{2}(\phi + \gamma_1)] \qquad (8.3.57)$$

for the Dirichlet ($+$) or Neumann ($-$) boundary condition. Also,

$$\beta_D = -\tfrac{1}{2}. \qquad (8.3.58)$$

We remark that the diffraction coefficient becomes unbounded when $\gamma_1 \to \phi + \pi$ or $-\phi$. The former angle defines the shadow boundary of the incident wave, while the latter defines the shadow boundary of the reflected wave. In these directions, the exact solution is well behaved, but the decomposition into the incident, reflected, and diffracted waves of geometrical optics is incorrect. The singular behavior of the diffraction coefficient is one manifestation of this invalidity.

In regions such as this, it is necessary to derive more exotic asymptotic expansions with respect to one large parameter (say, dimensionless frequency) that remains valid *uniformly* in a second parameter (say, $\gamma_1 - \phi$) that is near a critical value, such as π. This is a whole other area of current and ongoing research, both as regards ray methods and as regards other asymptotic expansion techniques. In particular, uniform asymptotic expansions for shadow boundaries of edge diffracted waves, as well as for many other cases of interest, have been derived.

Exercises

8.17 (a) Given the data (8.3.2) under the assumption of constant p, use (8.2.9) to find the rays for the conoidal solution to the eikonal equation, use (8.2.29) to determine J, and verify (8.3.7).
  (b) Repeat (a) for the data (8.3.14) and confirm (8.3.20).
8.18 Verify (8.3.44).
8.19 (a) Verify (8.3.46)–(8.3.48).

(b)   Show that the higher-order reflection and transmission coefficients corresponding to (8.3.46)–(8.3.48) are given by

$$A_j^R = RA_j^I + \frac{A_{j-1}^T - A_{j-1}^I - A_{j-1}^R}{(\partial \tau_1/\partial n) + \text{sign}(\partial \tau_1/\partial n)\sqrt{p_+^2 - p_-^2} + (\partial \tau_1/\partial n)^2},$$

$$A_j^T = TA_j^I - \frac{A_{j-1}^T - A_{j-1}^I - A_{j-1}^R}{(\partial \tau_1/\partial n) + \text{sign}(\partial \tau_1/\partial n)\sqrt{p_+^2 - p_-^2} + (\partial \tau_1/\partial n)^2}.$$

Here R and T are defined by (8.3.47) and (8.3.48), respectively.

8.20   The purpose of this problem is to provide some practice with the development of a ray solution to a problem with diffraction phenomena. The problem to be considered is exactly the canonical problem from which the diffraction coefficient (8.3.57) is derived. Thus, let us consider the problem of a plane wave

$$u_1(\mathbf{x}; \omega) = \exp(-i\omega\{x \sin \theta \sin \phi + y \cos \theta - z \sin \theta \cos \phi\}/c)$$

incident from the region $z < 0$ on the half plane $z = 0$, $x \geq 0$, on which u satisfies the homogeneous Dirichlet boundary condition $u = 0$.

(a)   Show that the continuation of the incident rays to the region $z > 0$ has a shadow boundary that is the plane through $x = 0$ making an angle ϕ with the z axis.

(b)   Find the reflected wave and show that it has a shadow boundary that makes an angle $\pi - \phi$ with the z axis.

(c)   Find the diffracted wave by using the diffraction coefficient (8.3.57) and by using (8.3.58). Verify that the diffraction coefficient becomes undefined on the two shadow boundaries determined in (a) and (b).

8.21   Consider now the problem of diffraction by an edge of the wave produced by a point source. Use the coordinate system of Fig. 8.8.

(a)   Show that

$$\tau_D(0, \gamma_1, \gamma_2) = p\sqrt{H^2 \sec^2 \phi + \gamma_1^2}, \qquad x = 0, \qquad y = \gamma_1, \qquad z = 0.$$

The parameter γ_2 is as defined in Fig. 8.7.

(b)   Define p_{20} by (8.3.15) and q_0 by (8.3.16). Show that

$$J(\sigma) = [1 - (q_0'p_{20}/q_0) + p_{02}'\sigma][\sigma q_0^2].$$

(c)   Show that back at the source point (backscattered signal),

$$A_0^D = D(\gamma_1, \gamma_2)/16\pi r_e\sqrt{\pi p r_e}$$

and that

$$u_D \sim (1/\sqrt{\omega})A_0^D e^{2i\omega r_e/c},$$

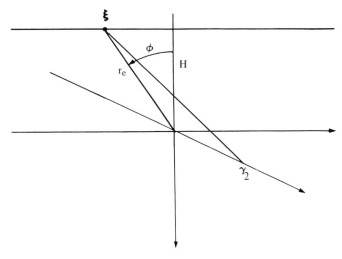

Fig. 8.8. Coordinate system for Exercise 8.21.

with r_e the normal distance from the edge to the observation point as shown in Fig. 8.8: $r_e = \sqrt{\xi^2 + H^2}$.

8.4 THE KIRCHHOFF APPROXIMATION

We consider in this section another high-frequency approximation for the solution of the Helmholtz equation for a class of scattering problems. We shall start from the Kirchhoff integral equation (6.4.14) for the solution to a scattering problem and make certain simplifying approximations in the integrand to turn that equation into a representation for the scattered field. Those simplifying assumptions are based on insights developed from the asymptotic analysis of exact solutions, such as in Section 8.1, and on other approximate solution techniques, such as the ray methods of Sections 8.2 and 8.3.

Originally, this approximation technique was applied to the field scattered by an aperture in a plane screen or its complement, the disk. The names of Rayleigh, Sommerfeld, Fresnel, and Huygens, among others, are associated with various refinements and alternatives to the approximation attributed to Kirchhoff. Strictly speaking, we are neither considering the problem that Kirchhoff considered nor making the approximation in the same form as he did. However, we refer to the result here as the Kirchhoff approximation in recognition of his fundamental role in its development. The distinctions between the various refinements can be found in Goodman [1968], Kuhn and Alhilali [1977], and Wolf and Marchand [1964], among others.

To begin, let us suppose that we have a scattering problem in a homogeneous medium (constant c) in which a wave is incident on a single, convex opaque scatterer with boundary surface denoted by B and Dirichlet or Neumann boundary condition on B. We shall represent the total field in terms of its values on the scattering surface by (6.4.14), with one change in the representation. In that formula, the normal derivative was a directional derivative directed *out* of the domain of the solution and therefore *into* the scatterer. Here we shall take the normal direction as the direction *out* of the scatterer and therefore *into* the domain of the wave field of interest. This change in interpretation only causes a change in sign in the surface integral in (6.4.14). Hence, we represent the total field by

$$u(\xi) + \int_{\mathscr{D}} G(\mathbf{x}; \xi) f(\mathbf{x}) \, dV = \int_B \left[u(\mathbf{x}) \frac{\partial G(\mathbf{x}; \xi)}{\partial n} - G(\mathbf{x}; \xi) \frac{\partial u(\mathbf{x})}{\partial n} \right] dS, \quad (8.4.1)$$

with $G(\mathbf{x}; \xi)$ a Green's function and $\mathscr{D}$ the domain exterior to B. We have suppressed the ω dependence in this representation but shall reintroduce it later in the discussion.

The incident wave is the response to the source in the absence of the scatterer. Let us choose the free-space Green's function (6.3.24), with $r = |\mathbf{x} - \xi|$. In this case, the incident wave is given by

$$u_I(\xi) = - \int_{\mathscr{D}} G(\mathbf{x}; \xi) f(\mathbf{x}) \, dV. \quad (8.4.2)$$

We remark that $u_I(\xi)$ and $G(\mathbf{x}; \xi)$ both have sources that are zero in the interior of the scatterer B, and therefore we conclude from (6.1.10) that

$$\int_B \left[u_I(\mathbf{x}) \frac{\partial G(\mathbf{x}; \xi)}{\partial n} - G(\mathbf{x}; \xi) \frac{\partial u_I(\mathbf{x})}{\partial n} \right] dS = 0. \quad (8.4.3)$$

Let us now set

$$u(\xi) = u_I(\xi) + u_S(\xi) \quad (8.4.4)$$

and use (8.4.2) and (8.4.3) in (8.4.1) to conclude that

$$u_S(\xi) = \int_B \left[u_S(\mathbf{x}) \frac{\partial G(\mathbf{x}; \xi)}{\partial n} - G(\mathbf{x}; \xi) \frac{\partial u_S(\mathbf{x})}{\partial n} \right] dS. \quad (8.4.5)$$

Up to this point, everything that we have done has been *exact*. We have obtained an integral equation that relates the scattered field to its values on the scattering surface. A Dirichlet or Neumann boundary condition on B would not suffice to give both of these boundary values directly, but only allow us to derive an integral equation for the determination of the remaining boundary values (Exercise 6.10, for example) and with both boundary values in hand, to recast (8.4.5) as a solution formula for $u_S(\xi)$.

Motivated by our experience with plane waves, geometrical optics, and ray methods, we propose to proceed now in another manner. We shall consider in particular the *backscatter problem* in which the source is a point source located at the same point ξ at which we seek the solution. In particular, then, when the time strucutre of the source is neglected, the incident wave is just the Green's function itself,

$$u_1(\mathbf{x}) = G(\mathbf{x}; \xi). \tag{8.4.6}$$

In the high-frequency limit, we think of this incident wave as *illuminating* a portion of B, which we shall denote by L, while leaving *dark* the remaining part of B, which we denote by D. See Fig. 8.9. The boundary between L and D on B is the curve along which the rays from the source point are tangent to B. Motivated by our ray theory, we introduce the following boundary values for $u_S(\mathbf{x})$:

$$u_S(\mathbf{x}) = Ru_1(\mathbf{x}), \qquad \frac{\partial u_S(\mathbf{x})}{\partial n} = -R\frac{\partial u_1(\mathbf{x})}{\partial n}, \qquad \mathbf{x} \quad \text{on} \quad L,$$

$$u_S(\mathbf{x}) = 0, \qquad \frac{\partial u_S(\mathbf{x})}{\partial n} = 0, \qquad \mathbf{x} \quad \text{on} \quad D. \tag{8.4.7}$$

In this equation, R is given by the reflection coefficient

$$R = -1 \qquad \text{(Dirichlet boundary condition)},$$
$$R = 1 \qquad \text{(Neumann boundary condition)}. \tag{8.4.8}$$

We use (8.4.7) and (8.4.6) in (8.4.5) to obtain the integral representation of the scattered field

$$u_S(\xi) = \int_L R\frac{\partial}{\partial n}\left[G^2(\mathbf{x}; \xi)\right] dS. \tag{8.4.9}$$

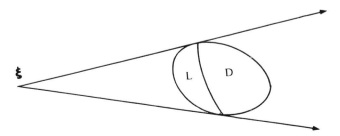

Fig. 8.9. Light (L) and dark (D) regions on the scatterer B.

In this equation and in the remainder of this section, *equality* should be taken in context to mean approximately equal in some not completely defined context. When we introduce the explicit form of the Green's function (6.3.24) into (8.4.9), we find that

$$u_S(\xi) = \frac{2i\omega}{c} \int_L R\hat{n} \cdot \hat{r} \frac{e^{2i\omega r/c}}{(4\pi r)^2} \, dS, \quad \mathbf{r} = \mathbf{x} - \xi, \quad r = |\mathbf{r}|, \quad \hat{r} = \frac{\mathbf{r}}{r}. \quad (8.4.10)$$

This approximate field representation is the main consequence of the Kirchhoff approximation (8.4.7). In practice, this result is used with a more general reflection coefficient deduced from a ray theoretic analysis of the particular scattering problem at hand. For example, suppose that the interior of B were another medium and that u and its normal derivative were required to be continuous across B. Then we might think of using for R the reflection coefficient given by the second expression in (8.3.47). Clearly, then, the backscattered field would, at best, represent the response due to the initial scattering from the front lighted part of B. We remark that the result (8.3.47) simplifies significantly for this case. The calculation is left to the exercises.

It has been shown that the boundary values for both u and $\partial u/\partial n$ imposed by (8.4.7) are inconsistent. It is known that prescribing either of these quantities suffices to determine the solution completely, so that in prescribing both we are *overdetermining* the solution. Therefore, our result is at best approximate in some sense. There is much discussion of this in the literature. Some references (Baker and Copson [1953] and Wolf and Marchand [1964]) are given at the end of the chapter, both for this problem and for related scattering problems to which the Kirchhoff approximation has been applied.

In the form (8.4.9), we might think to apply this approximation to the case of an *inhomogeneous* medium. In this case, we use for $G(\mathbf{x}; \xi)$ the ray method solution to the point source problem discussed in the preceding section. The light region L is then defined as the region lighted by the (no longer straight) rays of the conoidal problem for the eikonal equation.

As another generalization, we might consider the case in which the source and receiver are separated, say, with one of them at η. Then, in (8.4.9), the quadratic $G^2(\mathbf{x}; \xi)$ would be replaced by $G(\mathbf{x}; \xi)G(\eta; \mathbf{x})$, and the lighted region L would now be defined as the intersection of the lighted regions from the two points ξ and η. This generalization becomes seriously inaccurate as the angle of separation between source and receiver increases.

The Kirchhoff approximation has been successfully applied to surfaces with edges. *Successful* means that the approximate solution not only produces the primary response—the reflected wave—accurately, but also describes the edge diffracted field adequately. We shall see these separate components of the field arise from further asymptotic analysis of the representation (8.4.10) in the last part of this section, which discusses diffraction by a straight edge.

THE FAR-FIELD APPROXIMATION

It is possible to further simplify the representation (8.4.10) when the distance from L to the observation point ξ is large compared to the dimensions of the scatterer itself. In particular, we can use the expansions (6.4.9) and (6.4.10) in the phase and amplitude of the integrand in (8.4.10). In that expansion, we assume that the origin of the coordinate system is nearby or inside the scatterer B.

We shall use two terms of (6.4.9) in the phase and one term of (6.4.10) in the amplitude. In this manner, we obtain the result

$$u_S(\xi) = \frac{e^{2i\omega\rho/c}}{(4\pi\rho)^2} S(\hat{\xi};\omega), \tag{8.4.11}$$

where

$$S(\hat{\xi};\omega) = -\frac{2i\omega}{c} \int_L R\hat{n}\cdot\hat{\xi}e^{-2i\omega\hat{\xi}\cdot x/c}\, dS, \qquad \rho = |\xi|, \quad \hat{\xi} = \frac{\xi}{\rho}. \tag{8.4.12}$$

The factor $S(\hat{\xi};\omega)$ is called the *far-field (high-frequency) phase and range-normalized scattering amplitude*. We note that in this solution representation, the factor is multiplied by the square of the Green's function with argument measuring the range from the source/receiver point to the origin of the coordinate system. The fact that it depends only on $\hat{\xi}$ and ω results from the fact that, in this limit, R is at worst a function of the former of these, and hence the entire integrand only depends on these two arguments and x, over which the integral is to be calculated.

It is the factor $S(\hat{\xi};\omega)$ that distinguishes the far-field scattering patterns of different shapes and reflection coefficients from one another. Had we used a plane wave of unit amplitude as the incident wave, the multiplier of $S(\hat{\xi};\omega)$ would have been only one factor of the Green's function. We leave the verification of this to the exercises.

STATIONARY PHASE ANALYSIS

Let us now return to the representation (8.4.10) and apply the method of stationary phase to this integral. In order to do this, let us introduce a parameterization of the surface L in terms of two parameters σ_1 and σ_2. Thus,

$$\mathbf{x} = \mathbf{x}(\sigma_1,\sigma_2), \qquad dS = \sqrt{g}\, d\sigma_1\, d\sigma_2,$$

$$g = \det[g_{jk}], \qquad g_{jk} = \frac{d\mathbf{x}}{d\sigma_j}\cdot\frac{d\mathbf{x}}{d\sigma_k}, \qquad j,k = 1, 2. \tag{8.4.13}$$

The matrix $[g_{jk}]$ is a symmetric matrix of the elements of the *first funda-mental form* of differential geometry.

In (8.4.10), we use $2\omega/c$ as the large parameter. As in previous discussions, we consider ω to be positive and obtain the results for ω negative by observing that the integral in (8.4.10) is transformed into its complex conjugate when the sign of ω is changed. Alternatively, we could use $|\omega|$ in defining the large parameter and introduce a factor of sign ω in the phase to be analyzed. We choose the former method.

The phase of (8.4.10) whose stationary points we seek is

$$\Phi(\sigma_1, \sigma_2) = |\mathbf{x}(\sigma_1, \sigma_2) - \boldsymbol{\xi}|. \tag{8.4.14}$$

The first derivatives of this phase function are

$$\frac{\partial \Phi}{\partial \sigma_j} = \frac{\mathbf{x}(\sigma_1, \sigma_2) - \boldsymbol{\xi}}{|\mathbf{x}(\sigma_1, \sigma_2) - \boldsymbol{\xi}|} \cdot \frac{\partial \mathbf{x}}{\partial \sigma_j}, \qquad j = 1, 2. \tag{8.4.15}$$

The condition that this phase be stationary,

$$\frac{\mathbf{x}(\sigma_1, \sigma_2) - \boldsymbol{\xi}}{|\mathbf{x}(\sigma_1, \sigma_2) - \boldsymbol{\xi}|} \cdot \frac{\partial \mathbf{x}}{\partial \sigma_j} = 0, \qquad j = 1, 2, \tag{8.4.16}$$

states that the point $\mathbf{x}$ must be determined so that the vector $\mathbf{x} - \boldsymbol{\xi}$ is *ortho-gonal* to two (noncolinear) tangents to the surface L. Thus, at the stationary point, this vector must lie along the normal vector $\hat{\mathbf{n}}$. For this point, the direction of incidence and reflection make the same angle with the normal (zero), and, hence, this is the point of ordinary or *specular* reflection from L. For a convex scatterer, such a point, and only one such, always exists. Therefore, we shall proceed as though that point has been determined. In particular, at that point,

$$\mathbf{x} - \boldsymbol{\xi} = -|\mathbf{x} - \boldsymbol{\xi}|\hat{\mathbf{n}} = -r_n\hat{\mathbf{n}}; \tag{8.4.17}$$

that is, the vectors $\mathbf{x} - \boldsymbol{\xi}$ and $\hat{\mathbf{n}}$ are oppositely directed. We have also introduced the notation r_n for the (normal) distance from the stationary point to the observation point.

As in (2.8.5), we denote by A_{jk} the elements of the Hessian matrix for the phase Φ evaluated at the stationary point. In calculating this matrix from (8.4.15), we will exploit the condition of stationarity immediately to neglect the term in the differentiation that arises from differentiating $1/|\mathbf{x} - \boldsymbol{\xi}|$. Thus, we find that

$$A_{jk} = \frac{1}{r_n} \frac{\partial \mathbf{x}}{\partial \sigma_j} \cdot \frac{\partial \mathbf{x}}{\partial \sigma_k} - \hat{\mathbf{n}} \cdot \frac{\partial^2 \mathbf{x}}{\partial \sigma_j \, \partial \sigma_k}, \qquad j, k = 1, 2. \tag{8.4.18}$$

We remark that the dot product in the first term here is again the set of elements of the first fundamental form of differential geometry. In fact, we

can write the second term here in terms of the elements of the *second fundamental form* of differential geometry as

$$b_{jk} = -\hat{\mathbf{n}} \cdot \frac{\partial^2 \mathbf{x}}{\partial \sigma_j \, \partial \sigma_k}, \qquad j, k = 1, 2. \tag{8.4.19}$$

The minus sign in this definition comes from the fact that the normal direction is opposite to the one usually taken in the definition of these coefficients.
The matrix $[b_{jk}]$ is a symmetric matrix when the surface is smooth enough,
which we assume.

Now, we rewrite (8.4.18) as

$$A_{jk} = r^{-1}g_{jk} + b_{jk}, \qquad j, k = 1, 2. \tag{8.4.20}$$

In order to interpret this result, we need some further facts from differential
geometry. Also, to simplify our analysis, we shall specialize the (σ_1, σ_2) coordinate system in a manner that will not affect our final result but that will
make the intermediary computation easier.

Consider all of the curves on L formed by passing planes through the
surface L, all of which contain the normal vector $\hat{\mathbf{n}}$. For each of these, consider the *curvature* κ, defined by

$$|d^2\mathbf{x}(s)/ds^2| = \kappa, \tag{8.4.21}$$

where s is the arc-length parameterization of the curve. The reciprocal of
the curvature $\rho = 1/\kappa$ defines the *radius of curvature*, which is the radius of
a circle making second-order contact with the curve in question. The vector
$d^2\mathbf{x}/ds^2$, with initial point on L, has terminal point at the center of the circle
with second-order contact. That center is called the *center of curvature*.
Except for planes and spheres, the set of curvatures must have a minimum
and a maximum. These are called the *principal curvatures*. Their reciprocals
are called the *principal radii of curvature*. The directions of the tangents to
the curves in L along which the curvatures take on these extreme values are
called the *principal directions* at the point in question. These directions are
orthogonal.

Let us suppose now that we have chosen σ_1 and σ_2 so that they are coordinates along the principal directions at the stationary point and further
that they are arc-length variables along these curves, at least at the stationary
point. Then we can show that both the elements $g_{12} = 0$ and $b_{12} = 0$ while
$g_{11} = 1$ and $g_{22} = 1$. Finally, in this very special coordinate system, $b_{jj} = 1/\rho_j$,
$j = 1, 2$, where the ρ_j's are the principal radii of curvature. For this simplification, the matrix of elements A_{jk} is diagonal,

$$A_{jk} = \delta_{jk}\left[\frac{1}{r_n} + \frac{1}{\rho_j}\right] = \frac{\rho_j + r_n}{r_n \rho_j}, \qquad j, k = 1, 2, \tag{8.4.22}$$

and its signature is $+2$ because both eigenvalues are positive.

Had we not specialized the coordinate system, an equivalent result would have been obtained in the sense that the signature of the matrix would still be $+2$ and

$$\det(A_{jk}) = g \frac{[\rho_1 + r_n][\rho_2 + r_n]}{r^2 \rho_1 \rho_2}. \tag{8.4.23}$$

In this equation, g is as defined in (8.4.13). Thus, the quotient $\sqrt{g/A}$ is the same for the general coordinate system as it is for the specific coordinate system and, as we shall see, leads to an asymptotic expansion that is independent of the parameterization of the surface L.

We now have all of the information we need to write the leading term of the asymptotic expansion as given by (2.8.23). The result is

$$u_S(\xi) \sim \frac{R_n}{8\pi r_n} e^{2i\omega r_n/c} \sqrt{\frac{\rho_1 \rho_2}{[\rho_1 + r_n][\rho_2 + r_n]}}. \tag{8.4.24}$$

We have used the subscript n on R to note that for a more general reflection coefficient, such as the result derived in Exercise 8.22, the reflection coefficient at normal incidence is the value to appear in the asymptotic result here.

Let us now provide some interpretation of this result. First, let us consider the limit as we allow the principal radii of curvature to approach infinity. In this case, the surface L becomes flat, and the square root appearing in (8.4.24) approaches unity. In this limit, the solution formula reduces to

$$u_S(\xi) \sim (R_n/8\pi r_n)e^{2i\omega r_n/c}. \tag{8.4.25}$$

The functions $1/4\pi r_n$ and r_n/c are the geometrical optics amplitude and phase of the incident wave at the reflector. The function R_n is the reflection coefficient in the backscatter direction. The additional phase r_n/c accounts for propagation back from the scatterer to the observation point. The ratio $2r_n/r_n = 2$ is just the square root of the ratio of the Jacobians of the reflected wave evaluated at the observation point and on the scattering surface, respectively, and provides the extra factor of 2 in the denominator. Thus, in the absence of curvature of the reflecting surface, (8.4.25) is the reflected wave in the backscatter direction. Therefore, we conclude that the quantity under the square root appearing in (8.4.24) must be the ratio of Jacobians that accounts for the effect of the curvature of the scatterer.

To confirm that this is indeed the case, we exploit the interpretation of the ratio of Jacobians as a ratio of cross-sectional areas. In Fig. 8.10, we depict a differential surface element that has the same principal curvatures as L has at the stationary point. A plane through the normal to the surface in the

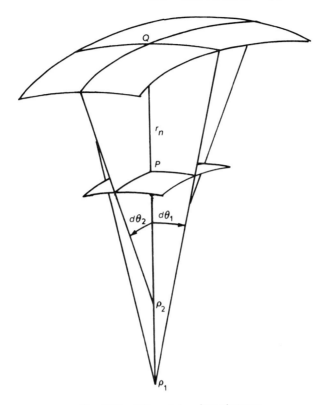

Fig. 8.10. Differential surface elements.

first principal direction cuts the surface in a differential element of a circle of radius ρ_1, while in the orthogonal direction, the curve is a circle of radius ρ_2. If the differential half angles of these surfaces are $d\theta_1$ and $d\theta_2$, as shown, then the differential surface element at P is just $4\rho_1\rho_2\,d\theta_1\,d\theta_2$. Performing the same calculation at the point Q in the figure requires only that we replace ρ_1 and ρ_2 by $\rho_1 + r_n$ and $\rho_2 + r_n$. Thus, the ratio of these quantities, which is the ratio under the square root in (8.4.24), is the ratio of cross-sectional areas in the normal direction for a family of rays that leaves this surface normally over the differential surface element. It is another matter entirely to confirm that the second-order contact of this surface with the surface L is sufficient to ensure that the same result holds true for the local family of specular rays on L when the central ray, at least, is normal to the surface and the incident rays introduce no spreading factor. We shall not discuss this last step here.

TIME DOMAIN RESPONSE

Let us suppose now that the original point source had in the time domain a time dependence $F(t)$, with Fourier transform $f(\omega)$. Let us suppose further that for the bandwidth of frequencies in $f(\omega)$, the asymptotics presented here are valid. This multiplier of the source term would then merely become a multiplier of all of the fields of interest and therefore would become a multiplier of solution formula (8.4.24) as well.

Let us now consider the inverse Fourier transform of our result. For this purpose, we reintroduce ω in u_S ($u_S(\xi) \equiv u_S(\xi; \omega)$) and take the inverse Fourier transform to find that

$$U_S(\xi, t) \sim \frac{R_n}{8\pi r_n} \sqrt{\frac{\rho_1 \rho_2}{[\rho_1 + r_n][\rho_2 + r_n]}} \, F(t - 2r_n/c); \qquad (8.4.26)$$

that is, the leading order asymptotic return is the time signal shifted by the two-way travel time—to the scatterer and back—scaled by factors that account for the geometrical spreading of the point source and the geometrical spreading due to the curvature of the scatterer at the specular reflection point.

A particular case of interest is that in which the source F is a *band-limited* Dirac delta function. That is, in some range of frequencies in which the asymptotics is valid, $f(\omega)$ is nearly equal to unity. This same band-limited delta function is reproduced at a delayed time that is the minimal travel time to the scatterer.

The simplest example of a band-limited delta function is provided by setting

$$f(\omega) = \begin{cases} 1, & \omega_0 = 2\pi f_0 \leq |\omega| \leq \omega_1 = 2\pi f_1, \\ 0, & \text{otherwise.} \end{cases} \qquad (8.4.27)$$

In this case,

$$F(t) = [\sin 2\pi f_1 t - \sin 2\pi f_0 t]/\pi t$$
$$= 2\{\sin[\pi(f_1 - f_0)t] \cos[\pi(f_1 + f_0)t]\}/\pi t. \qquad (8.4.28)$$

From the first form, we see that $F(t)$ is a difference of functions of the type (2.1.3) with n replaced by $2f_1$ or $2f_0$. If the latter of these were zero and f_1 in the former large enough, we would expect from the discussion of Chapter 2 that $F(t)$ was indeed a useful representation of the delta function. However, f_0 cannot be zero and the asymptotic analysis still be valid. For values of f_0 large enough for the asymptotics to be valid, (8.4.28) still can define a Dirac delta function with sufficient accuracy for many practical problems. We see from the second representation that

(i) $F(t)$ has a maximum at $t = 0$ equal to $2[f_1 - f_0]$;

(ii) its zeros nearest the origin are at $t = \pm 1/[2(f_1 + f_0)]$; the next pair are at $\pm t_1$, where

$$t_1 = \min[1/(f_1 - f_0), 3/[2(f_1 + f_0)]];$$

(iii) beyond the latter pair of points, $|F(t)| \leq (2/\pi t_1)$,

and $|F(t)/F(0)| \leq \pi$. The quotient $(f_1 - f_0)/(f_1 + f_0)$, when multiplied by 100, is called the *percentage bandwidth*. In a typical seismic application, the range of frequencies might be 6–24 Hz, in which case the percentage bandwidth is 60% and the quotient of function values is approximately 2. This extremely conservative estimate suggests that the center and shape of this band-limited Dirac delta function is sufficiently precise to characterize a delta function. As confirmation, see Fig. 8.11, which is a depiction of the function in (8.4.28) for the case of 60% bandwidth.

SCATTERERS OF INFINITE EXTENT

We shall now consider the extension of the result from the preceding discussion to the case in which the reflector is of infinite extent and separates two regions of constant but different velocities from one another. We shall assume that both u and its normal derivative must be continuous across this reflecting surface. Let us suppose that again we have a point source situated on the side of the reflector where the velocity is given by c_0. On the other side of the reflector, we assume that the velocity is c_1. We shall again consider only the backscatter problem in which the source point and observation point are the same.

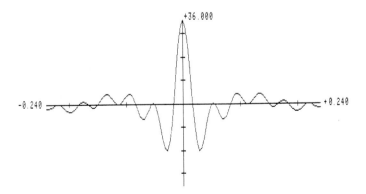

Fig. 8.11. Band-limited Dirac delta function, 60% bandwidth. (Program by Steven J. Bleistein.)

As regards surface shape, we are considering a generalization of the problem of Section 8.1. In this generalization, the planar reflector is replaced by an arbitrary smooth surface flat enough so that within some angular aperture about the normal to the surface that passes through the observation point, the surface has no *shadows*; that is, the rays from the source propagating toward the surface should touch every point of the surface in some finite aperture.

The total field is assumed to satisfy a Sommerfeld radiation condition (6.4.3), with c having its appropriate value on each side of the interface.

We shall again begin a Kirchhoff integral equation (8.4.1), with the domain $\mathcal{D}$ as shown in Fig. 8.12. The surface integral over the surface of the spherical portion of the boundary approaches zero with increasing radius of the cap by the Sommerfeld radiation condition. (See the discussion in Section 6.4.) Thus, in the surface integral, we are left in the limit with the integral over the reflector itself to consider.

We again choose $G(\mathbf{x}; \xi)$ as the free-space Green's function with the propagation speed of the upper domain. We define $u_1(\xi)$ as in (8.4.2). The result (8.4.3) must hold, but with different justification. Close a finite part of the reflector with a spherical cap in the lower half space. In the interior of this domain, neither $u_1(\xi)$ nor $G(\mathbf{x}; \xi)$ has any singularity when ξ is in the upper half space. Consequently, each of these functions satisfies the homogeneous Helmholtz equation in this lower domain but with propagation speed c_0. Therefore, from (6.1.10), the integral over the boundary surface of this spherical cap in the lower half space must be zero, since $\mathcal{L}u_1(\mathbf{x}) = \mathcal{L}G(\mathbf{x}; \xi) = 0$ inside of this domain. The integral over the spherical part is zero in the limit of increasing radius, because both functions also satisfy the appropriate Sommerfeld radiation condition. Consequently, the integral over the reflector itself must be zero.

Therefore, with $u_S(\xi)$ defined by (8.4.4), we again obtain the result (8.4.5). For this problem, we disregard the possible shadow zones, thus taking L to be all of B, and apply the first set of approximations over the entire surface B, with R the geometrical optics reflection coefficient. The result (8.4.10) again obtains for the case of constant propagation speed, although again the

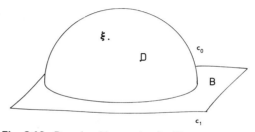

Fig. 8.12. Domain of integration for Kirchhoff integral.

form (8.4.9) is used in inhomogeneous media with $G(\mathbf{x}; \xi)$ the ray theory Green's function.

We *cannot* use the far-field approximation that leads to (8.4.11) and (8.4.12), since the transverse length scales of the reflector do not allow the approximations (6.4.9) and (6.4.10).

The asymptotic analysis described for the convex scatterer can proceed here in exactly the same way. However, one major difference occurs because we have not introduced any analog of convexity here. As a result of this difference, the centers of curvature need not be on the opposite side of the reflector from the source/receiver point. Let us define the parameters μ_1 and μ_2 by

$$\mu_j = \text{sgn}\left[-\hat{\mathbf{n}} \cdot \frac{d^2\mathbf{x}}{d\sigma_j}\right], \qquad j = 1, 2. \qquad (8.4.29)$$

The parameters in the surface (σ_1, σ_2) are again taken to be arc-length variables in the principal directions. Thus, $\mu_j = +1$ when the center of curvature in the jth direction is on the opposite side of the reflector from the observation point, and $\mu_j = -1$ when the center of curvature and the observation point are on the same side of the reflector. In the former case, the principal curve is convex *downward*, or *anticlinal*; in the latter case, the curve is convex *upward*, or *synclinal*.

In our earlier discussion of the matrix of second derivatives, we must now replace (8.4.22) by

$$A_{jk} = \delta_{jk}[(1/r_n) + (\mu_j/\rho_j)], \qquad j, k = 1, 2, \qquad (8.4.30)$$

and we must replace (8.4.23) by

$$\det[A_{jk}] = g\left|\frac{[\rho_1 + \mu_1 r_n][\rho_2 + \mu_2 r_n]}{r^2\rho_1\rho_2}\right|, \qquad (8.4.31)$$

$$\text{sgn}[A_{jk}] = \mu = \text{sign}[\rho_1 + \mu_1 r_n] + \text{sign}[\rho_2 + \mu_2 r_n].$$

We must now use this modification in the stationary phase formula (2.8.23) to obtain the asymptotic expansion of the integral (8.4.10). We must also take care here to account for the sign of ω. As noted earlier, this can be done by first stating the result for ω positive and then taking the complex conjugate of that result for ω negative. [In the case of the convex body, this was explicitly true for the result (8.4.24), and no special accommodation for negative ω was necessary.] The result is

$$u_s(\xi) \sim \frac{R_n}{8\pi r_n}\exp\left(\frac{2i\omega r_n}{c} + i\,\text{sign}\,\omega\left[\frac{\mu\pi}{4} - \frac{\pi}{2}\right]\right)$$

$$\cdot \sqrt{\frac{\rho_1\rho_2}{|[r_n + \mu_1\rho_1][r_n + \mu_2\rho_2]|}}. \qquad (8.4.32)$$

When μ_1 and μ_2 are positive, this formula is the same as (8.4.24). Now let us suppose that one of these parameters is negative, say, μ_1. In this case, for $r_n < \rho_1$, the factor $\rho_1 - r_n$ decreases toward zero as r_n increases toward ρ_1. The decrease of this factor causes the amplitude of the signal $u_s(\xi)$ to increase.

This increase in amplitude can be related to the nature of the back-scattered rays for this case. At the stationary point, the reflecting surface must be saddlelike, as shown in Fig. 8.13. The rays that reflect from points along the principal direction associated with ρ_1 must converge as they propagate back toward ξ, while rays that reflect from the principal direction associated with ρ_2 diverge as they propagate upward. At the center of curvature, P_1 in the figure, the cross-sectional area of the ray tube is equal to zero (as it is all along the caustic that touches P_1). The asymptotic expansion we have derived actually breaks down at this point. However, "sufficiently far away" from this point, where the method is valid, the cross-sectional area of the ray tube is nonetheless progressively decreasing with increasing r_n. Thus, from our ray method discussion, especially (8.2.11), we conclude that the amplitude must increase as r_n approaches ρ_1 from below.

The phase for this case is also worthy of attention. For $\rho_1 - r_n > 0$ (that is, for the observation point nearer the reflector than the center of curvature), $\mu = 2$ and the phase function is unchanged from what it was for the case of a convex scatterer. In particular, then, in this case, the discussion of the signal in the time domain is as earlier.

Let us suppose that $\rho_1 - r_n < 0$. Now the amplitude increases again with increasing r_n; the rays are diverging with increasing r_n. Let us consider the phase. We note that the eigenvalues now have opposite sign. Therefore, $\mu = 0$ and the total phase is given by $2i\omega r_n/c - i(\mathrm{sign}\ \omega)\pi/2$, and this is the

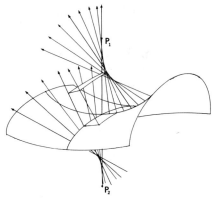

Fig. 8.13. A saddlelike reflection point.

only ω dependence of the result (8.4.32). Thus, after passing through the caustic, the phase must be adjusted by a factor $-(\text{sign } \omega)\pi/2$. This phase shift phenomenon is well known and has been observed experimentally.

Of all time dependences of the source, it is easiest to see the effect of this phase shift on the band-limited Dirac delta function source. Therefore, let us consider the source (8.4.26). Now the time signal will not be the inverse Fourier transform of unity but the inverse Fourier transform of the function

$$e^{-i(\text{sign}\omega)\pi/2} = -i \text{ sign } \omega.$$

This transform was discussed for full bandwidth in Section 2.3. In particular, the results are given by (2.3.9) if we identify x with ω and t with k. We must also multiply by $-i$. Thus, the band-limited delta function source will be imaged as a band-limited *doublet* $-2/t$.

The band-limited doublet that corresponds to the band-limited delta function (8.4.28) is given by

$$F_1(t) = [\cos 2\pi f_1 t - \cos 2\pi f_0 t]/\pi t. \tag{8.4.33}$$

In Fig. 8.14, we depict this doublet with the same 60% bandwidth as we used for our band-limited delta function in Fig. 8.11. When r_n is above the center of curvature, we say that the reflector has a *buried focus*. In seismological applications, the focus is indeed buried! For more realistic sources, analysis of the effect of a buried focus on the time trace is by no means as simple as it is for the delta function. This can cause great confusion in interpretation of returns as regards the nature of the reflecting mechanism that produced them.

We have yet another case to consider, namely, that for which the surface is completely synclinal, or convex upward, so that *both centers of curvature are above the reflector*. In this case, $\mu_1 = \mu_2 = -1$. Now if r_n is below both

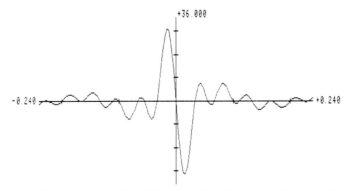

Fig. 8.14. Band-limited doublet, 60% bandwidth. (Program by Steven J. Bleistein.)

centers of curvature again, in (8.4.31), $\mu = 1$ and our discussion is much like that for the convex case as regards the source signature except that as in the saddle case, the amplitude increases with increasing r_n, now with both factors in the square root in (8.4.32) decreasing with increasing r_n.

For r_n between the two centers of curvature, one of the terms μ_1 or μ_2 changes sign and the phase shift $-(\text{sign } \omega)\pi/2$ is introduced into the asymptotic solution as earlier. Thus, for this configuration again, a delta function source will produce a doublet time signature in the reflected wave.

Finally, we consider the new case in which r_n is above both centers of curvature. Now both μ_1 and μ_2 are equal to -1 and in (8.4.31), $\mu = -2$. The total phase in (8.4.32) is now $2\omega r_n/c - \pi \text{ sign } \omega$. The latter term only introduces a multiplication by -1 for either choice of sgn ω. Thus, the formula for the backscattered field above both centers of curvature is analytically just the negative of the result obtained for the observation point below both centers of curvature.

This result again can be a source of great confusion. If we were to use the sign of the reflection strength alone to draw a conclusion about the reflection properties of the interface, we would likely draw the wrong conclusion. It is necessary to be aware of the two-dimensional geometry of the reflector in order to draw a correct conclusion as regards the increase or decrease in velocity across the interface.

DIFFRACTION BY A STRAIGHT EDGE

We shall close this section with a discussion of a simple problem with a straight edge. In fact, we shall consider the simplest of such problems. We shall assume that the reflecting surface is the half plane

$$x \geq 0, \qquad -\infty < y < \infty, \qquad z = H, \tag{8.4.34}$$

and that observations are to be made on the surface $\xi_3 = 0$. Furthermore, we shall consider only the Dirichlet or Neumann problem, so that $R = \pm 1$, as defined by (8.4.8). We shall assume that the result (8.4.10) is valid and proceed with asymptotic analysis as previously. We use for parameters on the surface

$$\sigma_1 = x, \qquad \sigma_2 = y, \qquad g \equiv 1, \qquad dS = dx \, dy. \tag{8.4.35}$$

The conditions of stationarity (8.4.16) now require that

$$x = \xi_1, \qquad y = \xi_2. \tag{8.4.36}$$

It is necessary that $\xi_1 \geq 0$ for there to be a stationary point, since $x \geq 0$ on the domain of integration. That is, for a flat half plane, the normal direction is vertical, and a reflected signal is observed only over the half plane itself.

We shall proceed with $\xi_1 > 0$, so that the stationary point is an interior point of the domain of integration. In this case, we evaluate the constituents of the integrand at the stationary point as

$$\Phi = r_n = H, \qquad A_{jk} = \delta_{jk}/r_n, \qquad j, k = 1, 2, \qquad \text{sgn}[A_{jk}] = 2. \quad (8.4.37)$$

Now we can apply the stationary phase formula (2.8.23) to the integral (8.4.10) to obtain the result

$$u_s(\xi) \sim Re^{i\omega H/c}/8\pi H. \quad (8.4.38)$$

This result can be seen to agree with (8.4.24) in the special case of a flat horizontal reflector.

Now let us suppose that $\xi_1 < 0$. We remark that the condition of stationarity in the variable $y = \sigma_2$ will still be satisfied. Therefore, let us view (8.4.10) as an interated integral and proceed to apply the method of stationary phase in the second variable. We begin by writing the phase explicitly in terms of x and y as

$$\Phi = r = \sqrt{(x - \xi_1)^2 + (y - \xi_2)^2 + H^2}, \quad (8.4.39)$$

with derivatives

$$\frac{d\Phi}{dy} = \frac{y - \xi_2}{r}, \qquad \frac{d^2\Phi}{d^2y} = \frac{1}{r} - \frac{(y - \xi_2)^2}{r^3}. \quad (8.4.40)$$

The stationary point is at $y = \xi_2$, in which case,

$$\Phi = r_0 = \sqrt{(x - \xi_1)^2 + H^2}, \qquad d^2\Phi/d^2y = 1/r_0, \quad (8.4.41)$$

and we can apply the stationary phase formula (2.7.18) to the integral (8.4.10). The result is

$$u_s(\xi) \sim \frac{RH}{8} \sqrt{\frac{|\omega|}{c\pi^3}} e^{-i(\text{sign}\,\omega)\pi/4} \int_0^\infty \frac{e^{2i\omega r_0/c}}{r_0^{5/2}} dx. \quad (8.4.42)$$

In this result, we have explicitly evaluated the dot product $\hat{n} \cdot \hat{r}$, used the fact that R is a constant to take it outside of the integral sign, and accounted for negative ω as well as positive ω.

We know that the phase has no stationary point in x. Therefore, we calculate the leading term of the asymptotic expansion by integration by parts through using the first term of (2.6.11). The result is

$$u_s(\xi) \sim \frac{RH}{\xi_1} \sqrt{\frac{c}{|\omega| r_e}} \frac{\exp(2i\omega r_e/c - i(\text{sign}\,\omega)3\pi/4)}{16\pi r_e}, \quad (8.4.43)$$

$$r_e = \sqrt{\xi_1^2 + H^2}.$$

This result should be compared to the result in Exercise 8.21. It does not contain the exact diffraction coefficient. The results will agree if we expand the result in (8.3.57) for small values of the ratio ξ_1/H, which is the offset of the observation point from the normal direction. That is, the amplitude of this result based on the Kirchhoff approximation will degrade with increasing offset from the normal direction. We remark that for this backscatter example,

$$\phi = \gamma_1, \qquad (\xi_s/H) = \tan \gamma_1 \sim \sin \gamma_1, \qquad D \sim \pm(H/\xi_1)e^{-i(\text{sign}\,\omega)3\pi/4}. \quad (8.4.44)$$

With this factor in place, the two results agree.

We remark that the result (8.4.43) also provides a second-order effect in the region $\xi_1 > 0$. The result fails in the limit $\xi_1 = 0$, which is the shadow boundary for the reflected wave in this backscatter analysis. Qualitatively, we see that the diffracted wave decays algebraically faster than the reflected wave (8.4.38) by a factor of $1/\sqrt{|\omega|r_e/c}$, in addition to the reciprocal range decay of the incident and reflected waves. Consequently, the diffracted wave falls off relatively rapidly with increasing offset angle γ_1. Thus, the preceding result deviates from the diffraction result of Section 8.3 in a region in which both results are small compared to the reflected wave anyway.

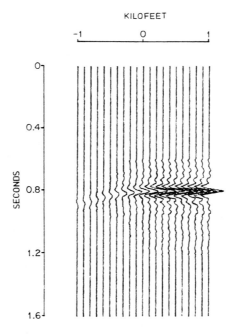

Fig. 8.15. Time record for backscatter from half plane. (Program by Sandra Bleistein.) [From Bleistein and Cohen, 1982.]

In Fig. 8.15, we depict a time record for backscattering by an edge. The horizontal coordinate is ξ_1, and the vertical coordinate is increasing time. In the region $\xi_1 > 0$, the backscattered signal is dominated by the reflected wave, which occurs at the same time on each time trace. In the region $\xi_1 < 0$, the diffracted wave dominates the signal. This wave is approximately the Fourier time transform of the result (8.4.43). The arrival times of the diffracted wave are seen to fall on the hyperbola given by $t = r_e$.

Exercises

8.22 (a) For the incident wave (6.3.24) used in (8.4.10), show that the reflection coefficient (8.3.47) is given by

$$R = \frac{p_-|\hat{n}\cdot\hat{r}| - \sqrt{p_+^2 - p_-^2[1 - (\hat{n}\cdot\hat{r})^2]}}{p_-|\hat{n}\cdot\hat{r}| + \sqrt{p_+^2 - p_-^2[1 - (\hat{n}\cdot\hat{r})^2]}}.$$

(b) Verify that $S(\hat{r}; \omega)$ defined by (8.4.12) is indeed a function of $\hat{\xi}$ and ω. That is, show that in the far-field limit, the reflection coefficient becomes

$$R = \frac{p_-\hat{n}\cdot\hat{\xi} - \sqrt{p_+^2 - p_-^2[1 - (\hat{n}\cdot\hat{\xi})^2]}}{p_-\hat{n}\cdot\hat{\xi} + \sqrt{p_+^2 - p_-^2[1 - (\hat{n}\cdot\hat{\xi})^2]}}.$$

8.23 Use both the Kirchhoff approximation (8.4.7) and the far-field approximations (6.4.9) and (6.4.10) in the integral (8.4.5) with the incident wave being the plane wave propagating in the direction $-\hat{\xi}$, and obtain the result

$$u_s(\hat{\xi}) = e^{i\omega\rho/c}S(\hat{\xi}; \omega)/4\pi\rho,$$

with $S(\hat{\xi}; \omega)$ again defined by (8.4.12).

8.24 Apply the method of stationary phase in two dimensions to the integral $S(\hat{\xi}; \omega)$ in (8.4.12). Show that at the stationary point, the normal points in the direction $\hat{\xi}$ and that

$$S(\hat{\xi}; \omega) \sim \frac{c_+ - c_-}{c_+ + c_-}\pi\sqrt{\rho_1\rho_2}\, e^{-2i\omega\hat{\xi}\cdot x_n/c}.$$

Here ρ_1 and ρ_2 are the principal radii of curvature at the stationary point, and x_n is the value of x at the stationary point.

References

Baker, B. B., and Copson, E. T. [1953]. "The Mathematical Theory of Huygens Principle," 2nd ed. Clarendon Press, Oxford.

Bleistein, M., and Cohen, J. K. [1982]. The velocity inversion problem—present status, new directions, *Geophysics* **47**, 1499–1511.

Brekhovskikh, L. M. [1980]. "Waves in Layered Media," 2nd ed. Academic Press, New York.

Červený, V., Molotkow, I. A., and Pšenčík, I. [1977]. "Ray Methods in Seismology." Univerzita Karlova, Prague, Czechoslovakia.

Cohen, J. K., and Bleistein, N. [1983]. The influence of out-of-plane surface properties on unmigrated time sections, *Geophysics* **48**, 125–132.

Ewing, W. M., Jardetzky, W. S., and Press, F. [1957]. "Elastic Waves in Layered Media." McGraw-Hill, New York.

Felsen, L. B., and Marcuvitz, N. [1973]. "Radiation and Scattering of Waves." Prentice-Hall, Englewood Cliffs, New Jersey.

Goodman, J. W. [1968]. "Introduction to Fourier Optics." McGraw-Hill, New York.

Grant, F. S., and West, G. F. [1965]. "Interpretation Theory in Applied Geophysics." McGraw-Hill, New York.

Hubral, P., and Krey, T. [1980]. "Interval Velocities from Seismic Reflection Time Measurements." Soc. Explor. Geophysicists, Tulsa, Oklahoma.

Keller, J. B. [1958]. A geometrical theory of diffraction, *in* "Calculus of Variations and Its Applications," pp. 27–52. McGraw-Hill, New York.

Keller, J. B. [1978]. Rays, waves, and asymptotics, *Bull. Am. Math. Soc.* **84**, 727–750.

Kuhn, M. J., and Alhilali, K. A. [1977]. Weighting factors in the construction and reconstruction of acoustical wave fields, *Geophysics* **42**, 1183–1198.

Lewis, R. M., and Keller, J. B. [1964]. Asymptotic Methods for Partial Differential Equations: The Reduced Wave Equation and Maxwell's Equations. Res. Rep. EM-194, Div. Electromag. Res., Courant Inst. Math. Sci. New York Univ., New York.

Pekeris, C. L. [1963]. Theory of propagation of explosive sound in shallow water, *in* "Propagation of Sound in the Ocean," Memoir 27. Geol. Soc. Am., New York.

Sommerfeld, A. [1964]. "Optics, Lectures on Theoretical Physics," Vol. 4. Academic Press, New York.

Wait, J. R. [1962]. "Electromagnetic Waves in Stratified Media." Pergamon, New York.

Wolf, E., and Marchand, E. W. [1964]. Comparison of the Kirchhoff and Rayleigh–Sommerfeld theories of diffraction at an aperture, *J. Opt. Soc. Am.* **54**, 587–594.

9 INVERSE METHODS
FOR REFLECTOR IMAGING

Inverse problems in the physical and biological sciences are a current source of ongoing research in applied mathematics and its related disciplines. Mathematically, the objective of inverse methods can be viewed as the determination of one or more parameters in the governing equation or system of equations of some process. A closely related problem is the determination of the size or shape of a scattering domain, which can be viewed as a domain with parameters that are different from those in the host medium. In this chapter, we shall discuss this subclass of inverse problems in which the objective is to describe the shape of a scatterer from a set of experiments that measure the backscattered field from a known source.

We shall show how the shape of a scatterer can be characterized mathematically. We shall then formulate mathematical problems for the determination of the *shape functions* and discuss the approximate solutions of these equations.

Delineation of a shape requires distinction between those points that are on a surface and those that are not. In one dimension, this would require simply the identification of the boundary points of an interval. A point is well defined by a Dirac delta function whose support (singularity) is at that point. Thus, in simplest form, shape discrimination requires identification of the support of a delta function.

In physical experiments, the ability to locate a point is limited by the bandwidth of the experiment at hand. In Section 8.4, we described the simplest form of a band-limited delta function in (8.4.28) and Fig. 8.11. In that example, the width of the main lobe of the band-limited delta function was $T = 1/[f_1 + f_0]$. On the other hand, the peak value of the band-limited delta function was $\max |F(t)| = 2[f_1 - f_0]$. The ratio of these two values, or twice the percentage bandwidth, $2[f_1 - f_0]/[f_1 + f_0]$ defines, in some sense, the *resolution* of the band-limited delta function. The objective in experiments, then, is to achieve a percentage bandwidth sufficiently large to provide a main lobe of an approximate delta function adequate for the discrimination desired.

In Exercise 2.2, we introduced the *singular function of a surface*. It is this function that will be the basis of the methods presented here for surface imaging. We shall discuss the singular function and a related function, the characteristic function, later in this introduction and in the following sections. We shall be concerned with the Fourier transforms of these functions, especially properties of these transforms for large values of the magnitude of the transform vector. We shall develop methods for the determination of these functions from backscattered data. It will be seen that these data are related to incomplete data about the Fourier transforms of the singular or characteristic functions.

It is important to emphasize that the inverse methods described in this chapter represent only one class of methods of current use and interest. They are applicable in situations in which a *backscattered* or *near-backscattered* signal is available. In some important applications, such as imaging human tissue, methods that deal with *transmitted* signals and employ *holographic imaging techniques* would seem to be more important. The difficulty with employing backscatter methods is that the x rays used for holograms produce weak backscattered signals from human tissue, while at the other end of the spectrum, ultrasonic signals do not penetrate human tissue to sufficient depth to image the human body.

On the other hand, acoustic or elastic waves are the signals of choice in seismic exploration and many types of nondestructive testing applications. The combination of penetration depths and resolution is adequate to the task at hand, although it seems to be a general rule that greater penetration and higher resolution always hold promise of enhanced useful information.

9.1 THE SINGULAR FUNCTION AND THE CHARACTERISTIC FUNCTION

THE SINGULAR FUNCTION

We shall begin by defining again the singular function $\gamma(\mathbf{x})$, $\mathbf{x} = (x_1, x_2, x_3)$, of a smooth surface S. For each point on S, we introduce the distribution $\delta(\sigma)$, where σ measures normal distance from S. We decompose a surface with corners, say, a "piecewise smooth" surface, into its smooth constituents and define $\gamma(\mathbf{x})$ separately for each of those. We shall assume that the surface of interest is finite but not necessarily closed.

A more rigorous definition of the singular function is provided by defining its effect under integration with test functions. Therefore, let us introduce the class of *infinitely differentiable* functions that vanish outside a finite domain. For $f(\mathbf{x})$ being any such function, we require that

$$\int_{-\infty}^{\infty} f(\mathbf{x})\gamma(\mathbf{x})\,dV = \int_{S} f(\mathbf{x})\,dS. \tag{9.1.1}$$

That is, the effect of the singular function on a volume integral is to reduce it to a surface integral over the support surface of the singular function.

From the point of view of inverse problems, we can see that the mathematical imaging of the surface S amounts to the determination of its singular function.

THE CHARACTERISTIC FUNCTION

The second function of interest is called the *characteristic function* of a domain D. This function $\Gamma(\mathbf{x})$ is defined by

$$\Gamma(\mathbf{x}) = \begin{cases} 1, & \mathbf{x} \quad \text{in} \quad D, \\ 0, & \mathbf{x} \quad \text{not in} \quad D. \end{cases} \tag{9.1.2}$$

FOURIER TRANSFORMS

As suggested in the introduction, we shall be interested in reconstructing singular or characteristic functions from information about their Fourier transforms. Therefore, let us introduce the two Fourier transforms

$$\bar{\gamma}(\mathbf{k}) = \int_{-\infty}^{\infty} \gamma(\mathbf{x})e^{-i\mathbf{k}\cdot\mathbf{x}}\,dV = \int_{S} e^{-i\mathbf{k}\cdot\mathbf{x}}\,dS, \qquad \mathbf{k} = (k_1, k_2, k_3), \tag{9.1.3}$$

and

$$\bar{\Gamma}(\mathbf{k}) = \int_{-\infty}^{\infty} \Gamma(\mathbf{x})e^{-i\mathbf{k}\cdot\mathbf{x}}\,dV = \int_{D} e^{-i\mathbf{k}\cdot\mathbf{x}}\,dV. \tag{9.1.4}$$

ASYMPTOTIC ANALYSIS OF THE SINGULAR FUNCTION

As stated in the introduction, the singular function is to be identified from band-limited data in the Fourier domain, that is, from data in which the magnitude of the Fourier transform vector variable is bounded above and below. In addition, in practice, the angular range will be limited as well. The combination of these effects constitutes a *limited aperture* in the Fourier domain.

In practice, the magnitude k of the transform vector variable $\mathbf{k}$ will be proportional to the frequency of some time signal $k = 2|\omega|/c$. Our interest will be in reconstruction of the singular function from *high-frequency data.* Thus, we shall proceed to study first the high-frequency behavior of the singular and characteristic functions and then the Fourier inversion of these aperture-limited functions.

We consider first the singular function and introduce a parameterization of the surface S, as in (8.4.13). We then rewrite the integral (9.1.3) as

$$\bar{\gamma}(\mathbf{k}) = \int_S e^{-ik\Phi(\mathbf{x};\hat{\mathbf{k}})} \sqrt{g} \, d\sigma_1 \, d\sigma_2, \tag{9.1.5}$$

$$\hat{\mathbf{k}} = \frac{\mathbf{k}}{k}, \qquad \Phi(\mathbf{x};\hat{\mathbf{k}}) = -\hat{\mathbf{k}} \cdot \mathbf{x}, \tag{9.1.6}$$

with g defined by (8.4.13).

We shall analyze this integral by the method of stationary phase for large values of k. This analysis was carried out in Exercise 8.24 in the context of the phase and range normalized scattering amplitude. We repeat that analysis here for completeness. The derivatives of the function $\Phi(\mathbf{x};\hat{\mathbf{k}})$ are

$$\frac{\partial\Phi(\mathbf{x};\hat{\mathbf{k}})}{\partial\sigma_j} = \hat{\mathbf{k}} \cdot \frac{\partial\mathbf{x}}{\partial\sigma_j}, \qquad \frac{\partial^2\Phi(\mathbf{x};\hat{\mathbf{k}})}{\partial\sigma_j \, \partial\sigma_k} = \hat{\mathbf{k}} \cdot \frac{\partial\mathbf{x}}{\partial\sigma_j \, \partial\sigma_k}, \qquad j,k = 1,2. \tag{9.1.7}$$

As in the discussion in Section 8.4, the condition that the first derivatives be zero will be satisfied at those points on S where $\hat{\mathbf{k}}$ is colinear or anticolinear with the normal vector to S. Let us assume that there is at least one such point. At that stationary point,

$$\hat{\mathbf{k}} = \hat{\mathbf{n}} \operatorname{sign} \hat{\mathbf{n}} \cdot \hat{\mathbf{k}} \tag{9.1.8}$$

We also introduce the principal curvature vectors at the stationary point $\mathbf{\kappa}_1$ and $\mathbf{\kappa}_2$ and define

$$\mu_j = \operatorname{sign} \hat{\mathbf{k}} \cdot \mathbf{\kappa}_j, \qquad \mathbf{\kappa}_j = d^2\mathbf{x}/ds_j^2, \qquad j = 1,2. \tag{9.1.9}$$

In this equation, s_j denotes an arc-length parameter in the jth principal direction. From the analysis carried out in Section 8.4, we know that

$$|\det A| = \sqrt{g}/\rho_1\rho_2, \qquad \operatorname{sgn} A = -\mu = -[\mu_1 + \mu_2]. \tag{9.1.10}$$

Here A is again the Hessian matrix, defined by (2.8.5), but for the function $\Phi(\mathbf{x};\hat{\mathbf{k}})$, $\rho_j = |\mathbf{\kappa}_j|^{-1}, j = 1,2$.

The asymptotic expansion of the integral in (9.1.5) is now determined by using the formula (2.8.23),

$$\bar{\gamma}(\mathbf{k}) \sim \frac{2\pi}{k} \sum \sqrt{\rho_1\rho_2} \, e^{-i\mathbf{k}\cdot\mathbf{x}_n - i\mu\pi/4}. \tag{9.1.11}$$

The summation must be carried out over all of the stationary points, that is, over all of the points $\mathbf{x}_n$ where (9.1.8) is true. For a closed convex scatterer, there will be two such points for each choice of $\mathbf{k}$. At one of these, $\hat{\mathbf{k}}$ and $\hat{\mathbf{n}}$ will be colinear; at the other point, they will be anticolinear. When S is not a smooth closed surface, this need not be the case. In fact, there need not be any stationary points at all. From the result (2.8.2) and (2.8.3), we can see that in this case the integral is of lower order asymptotically. The leading order term of the asymptotic expansion will then arise from applying the method of stationary phase in one dimension to the first term of the right side of (2.8.3). However, there is some disadvantage to applying the method of Section 2.8 directly to the integral (9.1.3). The reason is that the variables of integration are σ_1 and σ_2. Therefore, the boundary integral in (2.8.3) is an integral over the boundary in this σ domain rather than over the boundary of the actual surface S in the coordinates $\mathbf{x}$. Therefore, in Exercise 9.4, we specialize (2.8.1) to a surface integral, and we extend the integration by parts technique of (2.8.2) and (2.8.3) via the Stokes theorem to recast the surface integral in $\mathbf{x}$ directly to a line integral in $\mathbf{x}$. This amounts to accounting for the fact that the surface in $\mathbf{x}$ need not be planar while the surface in σ_1 and σ_2 is always planar.

Let us now consider the volume integral (9.1.4) defining $\bar{\Gamma}(\mathbf{k})$. Again, we seek an asymptotic expansion for large values of k, with the phase again defined by (9.1.6) except that $\mathbf{x}$ is not a function of surface variables in this case. The conditions of stationarity in $\mathbf{x}$ are that $\mathbf{k} = \mathbf{0}$. Clearly, then, there are no stationary points for the *volume* integral for large values of k.

Let us now denote by S the boundary surface of the domain D. Then, we can apply (2.8.2) and (2.8.3) to obtain a surface integral for the leading order approximation of $\bar{\Gamma}(\mathbf{k})$. The result is

$$\bar{\Gamma}(\mathbf{k}) \sim \frac{i}{k} \int_S \hat{\mathbf{n}} \cdot \mathbf{k} e^{-ik\Phi(\mathbf{x};\hat{\mathbf{k}})} \sqrt{g}\, d\sigma_1\, d\sigma_2. \tag{9.1.12}$$

In this equation, $\hat{\mathbf{n}}$ is an outward normal to the surface S.

Comparison of (9.1.12) and (9.1.5) reveals that the two integrands differ only in the factor $i\hat{\mathbf{n}} \cdot \hat{\mathbf{k}}/k$. Therefore, we evaluate the asymptotic expansion by adjusting (9.1.11) for this new amplitude factor. The result is

$$\bar{\Gamma}(\mathbf{k}) \sim \frac{2\pi i}{k^2} \sum [\text{sign } \hat{\mathbf{n}} \cdot \hat{\mathbf{k}}] \sqrt{\rho_1 \rho_2}\, e^{-i\mathbf{k} \cdot \mathbf{x}_n - i\mu\pi/4}. \tag{9.1.13}$$

Again, we sum over the solutions of (9.1.8). Let us denote the separate terms in the sums (9.1.11) and (9.1.13) by γ_n and Γ_n, respectively. Then, we can see that

$$\gamma_n \sim -ik[\text{sign } \hat{\mathbf{n}} \cdot \hat{\mathbf{k}}]\Gamma_n. \tag{9.1.14}$$

That is, the leading order asymptotic expansion of the elements of the Fourier transform of $\bar{\gamma}(\mathbf{k})$ is obtained from the same terms in $\bar{\Gamma}(\mathbf{k})$ by multiplication by $\pm ik$, $\pm 1 = \text{sign } \hat{\mathbf{n}} \cdot \hat{\mathbf{k}}$.

In some sense, this result is an extension, asymptotically, of the correspondence (2.2.18) between differentiation in the spatial domain and multiplication in the Fourier domain. For a directional derivative in a fixed direction in the spatial domain defined, say, by the unit vector $\hat{\alpha}$, we would have the results

$$\hat{\mathbf{n}}\gamma(\mathbf{x}) = -\nabla\Gamma(\mathbf{x}), \qquad \hat{\alpha} \cdot \hat{\mathbf{n}}\gamma(\mathbf{x}) = -\hat{\alpha} \cdot \nabla\Gamma(\mathbf{x}) \leftrightarrow -ik\hat{\alpha} \cdot \hat{\mathbf{k}}\bar{\Gamma}(\mathbf{k}). \quad (9.1.15)$$

For $\hat{\alpha} = \hat{\mathbf{n}}$, this result would be incorrect, since at the very least, $\hat{\mathbf{n}}$ is a function of $\mathbf{x}$ and the rightmost term in the second line makes no sense at all. However, at the stationary point, $\hat{\mathbf{k}}$ and $\hat{\mathbf{n}}$ coline and (9.1.14) imply that (9.1.15) is true for normal derivatives of $\Gamma(\mathbf{x})$ as well, at least asymptotically.

The significance of this result will become apparent in the inversion techniques in the following sections. In many problems, it is $\Gamma(\mathbf{x})$ that arises more naturally, but $\gamma(\mathbf{x})$ that we seek. The result (9.1.14) provides a means of obtaining the Fourier transform of the latter from the Fourier transform of the former, at least asymptotically.

SURFACES WITH DISTINGUISHED DIRECTIONS

Let us consider a surface S that is defined by

$$x_3 = h(x_1, x_2), \qquad (x_1, x_2) \quad \text{in} \quad X, \quad (9.1.16)$$

with X some (not necessarily finite) two-dimensional domain. We then define the singular and characteristic functions

$$\gamma(\mathbf{x}) = \sqrt{g}\,\delta(h(x_1, x_2) - x_3),$$

$$\Gamma(\mathbf{x}) = \begin{cases} 1, & x_3 > h(x_1, x_2), \\ 0, & x_3 < h(x_1, x_2), \end{cases} \qquad (x_1, x_2) \quad \text{in} \quad X. \quad (9.1.17)$$

The function g is defined by

$$g = [\partial h/\partial x_1]^2 + [\partial h/\partial x_2]^2 + 1. \quad (9.1.18)$$

We leave it as an exercise to verify that the function g as defined by (8.4.13) is given by this result in the special case of a surface defined as in (9.1.16), with surface parameters being two of the Cartesian variables. We have also used here the result (2.1.8) with the derivative there interpreted as a normal derivative, consistent with the first definition of the singular function at the beginning of this section.

We shall calculate the Fourier transform (9.1.4) as an interated integral by carrying out the x_3 integration first and using the result (2.3.11) and (2.3.12). The result is

$$\bar{\Gamma}(\mathbf{k}) = \left[\frac{1}{k_3} + \pi\delta(k_3) \right] \int_{-\infty}^{\infty} e^{-ik\Psi(x_1, x_2, \hat{k})} \, dx_1 \, x_2 . \qquad (9.1.19)$$

The phase in this equation is

$$\Psi(x_1, x_2, \hat{\mathbf{k}}) = -\hat{k}_1 x_1 - \hat{k}_2 x_2 - \hat{k}_3 h(x_1, x_2). \qquad (9.1.20)$$

Similarly, either by using the definition of $\gamma(\mathbf{x})$ in (9.1.17) in the first line of (9.1.3) or by directly using the second line of (9.1.3) with the special variables x_1 and x_2, we find that

$$\bar{\gamma}(\mathbf{k}) = \int_{-\infty}^{\infty} \sqrt{g} \, e^{-ik\Psi(x_1, x_2, \hat{k})} \, dx_1 \, x_2 . \qquad (9.1.21)$$

Let us suppose not only that k is large but also that k_3 is nonzero. Then, the delta function appearing in (9.1.19) is zero. Therefore, in our further asymptotic analysis here, we shall neglect this term. The conditions that the phase be stationary in either integral (9.1.19) or (9.1.21) are

$$\hat{k}_j/\hat{k}_3 = -(\partial h/\partial x_j), \qquad j = 1, 2, \qquad (9.1.22)$$

from which it follows that

$$\sqrt{g} = 1/|\hat{k}_3| = k/|k_3|. \qquad (9.1.23)$$

We can now evaluate both integrals by the method of stationary phase (2.8.23). The results are again of the form (9.1.11) and (9.1.13). The value of $\mathbf{x}_n$ is determined by (9.1.22). The contribution to each integral from any stationary point is now given by

$$\bar{\gamma}_n \sim (2\pi/k)\sqrt{\rho_1 \rho_2} \exp(-ik\Psi(\mathbf{x}_n; \hat{\mathbf{k}}) - i\mu\pi/4); \qquad (9.1.24)$$

$$\bar{\Gamma}_n \sim (2\pi/ikk_3)\sqrt{\rho_1 \rho_2/g} \exp(-ik\Psi(\mathbf{x}_n; \hat{\mathbf{k}}) - i\mu\pi/4)$$

$$\sim (2\pi/ik^2)(-\operatorname{sign} k_3)\sqrt{\rho_1 \rho_2} \exp(-ik\Psi(\mathbf{x}_n; \hat{\mathbf{k}}) - i\mu\pi/4). \qquad (9.1.25)$$

We conclude for this case that

$$\bar{\gamma}_n \sim ik(\operatorname{sign} k_3)\bar{\Gamma}_n. \qquad (9.1.26)$$

To compare this result with (9.1.14), let us define

$$\hat{\mathbf{n}} = \left[\frac{\partial h}{\partial x_1}, \frac{\partial h}{\partial x_2}, -1 \right] \Big/ \sqrt{g}. \qquad (9.1.27)$$

This definition is consistent with the definition of $\hat{\mathbf{n}}$ as an outward normal, used above, since *outward* for a bounded domain is directed from the region where $\Gamma(\mathbf{x}) = 1$ to the region where $\Gamma(\mathbf{x}) = 0$. The identification with the result (9.1.14) is complete when we use (9.1.22) and (9.1.27) to verify that

$$\text{sgn } \hat{\mathbf{n}} \cdot \hat{\mathbf{k}} = -\text{sign } k_3. \qquad (9.1.28)$$

We remark that for surfaces in which one coordinate is distinguished, as was x_3 here, the task of determining a multiplier on $\bar{\Gamma}(\mathbf{k})$ to produce the asymptotic value of $\bar{\gamma}(\mathbf{k})$ has become much easier; it has been simply reduced to multiplication of the former Fourier transform by a variable sign k_3 that is known a priori no matter what the direction of the normal to S is at the stationary point.

THE APERTURE–LIMITED SINGULAR FUNCTION

We shall now consider the effects of limited aperture on the Fourier inversion of $\bar{\gamma}(\mathbf{k})$. In order to this, we introduce the function $\gamma_B(\mathbf{y})$, defined by

$$\gamma_B(\mathbf{y}) = \frac{1}{(2\pi)^3} \int_K dk_1 \, dk_2 \, dk_3 \, e^{i\mathbf{k}\cdot\mathbf{y}} \int_S e^{-i\mathbf{k}\cdot\mathbf{x}} \, dS. \qquad (9.1.29)$$

In this equation, K connotes the limited aperture in the Fourier domain over which the data $\bar{\gamma}(\mathbf{k})$ are known. This domain has the form

$$K : k_- \leq k \leq k_+, \qquad \hat{\mathbf{k}} \text{ in } \Omega; \qquad (9.1.30)$$

that is, the magnitude of $\mathbf{k}$ is restricted to some finite domain while the direction of $\mathbf{k}$ is restricted to some solid angle on the unit sphere.

We rewrite this result as

$$\gamma_B(\mathbf{y}) = \frac{1}{(2\pi)^3} \int_{k_-}^{k_+} dk \int_\Omega \sin\theta \, d\theta \, d\phi \int_S \sqrt{g} \, e^{ik\Phi(\mathbf{y},\mathbf{x},\hat{\mathbf{k}})} \, d\sigma_1 \, d\sigma_2. \qquad (9.1.31)$$

In this equation,

$$\Phi(\mathbf{y}, \mathbf{x}, \hat{\mathbf{k}}) = \hat{\mathbf{k}} \cdot [\mathbf{y} - \mathbf{x}(\sigma_1, \sigma_2)],$$
$$\hat{\mathbf{k}} = (\sin\theta \cos\phi, \sin\theta \sin\phi, \cos\theta). \qquad (9.1.32)$$

Also, we have used S and Ω to denote the domains of integration in the specific parameterization for the surface S and for the angular aperture Ω.

We shall apply the method of stationary phase to $\Phi(\mathbf{y}, \mathbf{x}, \hat{\mathbf{k}})$ in the four variables $\sigma_1, \sigma_2, \theta = \sigma_3$, and $\phi = \sigma_4$, under the assumption that this expansion will be valid for all values of k in the interval (k_-, k_+). We shall use σ_3 and σ_4 or θ and ϕ as is convenient in context. The four first derivatives of

$\Phi(\mathbf{y}, \mathbf{x}, \hat{\mathbf{k}})$ are given by

$$\frac{\partial \Phi}{\partial \sigma_j} = -\mathbf{k} \cdot \frac{\partial \mathbf{x}}{\partial \sigma_j}, \qquad j = 1, 2,$$

(9.1.33)

$$\frac{\partial \Phi}{\partial \sigma_3} = \hat{\boldsymbol{\theta}} \cdot [\mathbf{y} - \mathbf{x}], \qquad \frac{\partial \Phi}{\partial \sigma_4} = \hat{\boldsymbol{\phi}} \cdot [\mathbf{y} - \mathbf{x}] \sin \theta.$$

Here, $\hat{\boldsymbol{\theta}}$ and $\hat{\boldsymbol{\phi}}$ are the two other unit vectors on the sphere that together with $\hat{\mathbf{k}}$ form a right-handed triple of vectors

$$\hat{\boldsymbol{\theta}} = (\cos \theta \cos \phi, \cos \theta \sin \phi, -\sin \theta), \qquad \hat{\boldsymbol{\phi}} = (-\sin \phi, \cos \phi, 0). \quad (9.1.34)$$

The condition that the first pair of derivatives of $\Phi(\mathbf{y}, \mathbf{x}, \hat{\mathbf{k}})$ in (9.1.33) be zero leads us to conclude that at a stationary point, $\hat{\mathbf{k}}$ must be orthogonal to the surface S at the stationary point. The condition that the second pair of derivatives be zero leads to the conclusion that $\hat{\mathbf{k}}$ must also be orthogonal to $\mathbf{y} - \mathbf{x}$ at the stationary point. Thus, given a point $\mathbf{y}$, we find a point on S, $\mathbf{x} = \mathbf{x}_n$, for which $\mathbf{y} - \mathbf{x}_n$ is orthogonal to S. This determines σ_1 and σ_2 at the stationary point. We then find $\hat{\mathbf{k}}$ so that this vector is colinear or anticolinear with the normal at that point as well. At a stationary point, then,

$$\mathbf{y} - \mathbf{x}_n = r_n \hat{\mathbf{n}} \mu_r, \qquad r_n = |\mathbf{y} - \mathbf{x}(\sigma_1, \sigma_2)|, \qquad \mu_r = \operatorname{sign} \hat{\mathbf{n}} \cdot [\mathbf{y} - \mathbf{x}_n],$$

(9.1.35)

$$\hat{\mathbf{k}} = \hat{\mathbf{n}} \mu_k, \qquad \mu_k = \operatorname{sign} \hat{\mathbf{n}} \cdot \hat{\mathbf{k}}.$$

We leave it to the exercises to verify the following results about the Hessian matrix for $\Phi(\mathbf{y}, \mathbf{x}, \hat{\mathbf{k}})$:

$$\frac{\partial^2 \Phi}{\partial \sigma_j \partial \sigma_k} = -\hat{\mathbf{k}} \cdot \frac{\partial^2 \mathbf{x}}{\partial \sigma_j \partial \sigma_k}, \qquad j, k = 1, 2,$$

$$\frac{\partial^2 \Phi}{\partial \sigma_j \partial \sigma_3} = -\hat{\boldsymbol{\theta}} \cdot \frac{\partial \mathbf{x}}{\partial \sigma_j}, \qquad \frac{\partial^2 \Phi}{\partial \sigma_j \partial \sigma_4} = -\hat{\boldsymbol{\phi}} \cdot \frac{\partial \mathbf{x}}{\partial \sigma_j} \sin \theta, \qquad j = 1, 2,$$

(9.1.36)

$$\frac{\partial^2 \Phi}{\partial \sigma_3^2} = -r_n \mu_k \mu_r, \qquad \frac{\partial^2 \Phi}{\partial \sigma_3 \partial \sigma_4} = -r_n \mu_r \hat{\boldsymbol{\phi}} \cdot \hat{\mathbf{n}} \cos \theta,$$

$$\frac{\partial^2 \Phi}{\partial \sigma_4^2} = -r_n \mu_r (\cos \phi, \sin \phi, 0) \cdot \hat{\mathbf{n}} \sin \theta$$

and

$$\det \left[\frac{\partial^2 \Phi}{\partial \sigma_j \partial \sigma_k} \right] = \det A = \sin^2 \theta (1 - \hat{\mathbf{n}} \cdot \boldsymbol{\kappa}_1 r_n \mu_r)(1 - \hat{\mathbf{n}} \cdot \boldsymbol{\kappa}_2 r_n \mu_r),$$

(9.1.37)

$$\operatorname{sgn} A = 0, \qquad r_n < \min(\rho_1, \rho_2).$$

We now use the results (9.1.35) and (9.1.37) to obtain the leading order asymptotic expansion of (9.1.31) by using the multidimensional stationary phase formula (2.8.23). We state the result only for r_n satisfying the inequality of (9.1.37),

$$\gamma_B(\mathbf{y}) \sim \frac{1}{2\pi} \sum \frac{1}{\sqrt{|(1 - \hat{\mathbf{n}} \cdot \boldsymbol{\kappa}_1 r_n \mu_r)(1 - \hat{\mathbf{n}} \cdot \boldsymbol{\kappa}_2 r_n \mu_r)|}} \int_{k_-}^{k_+} \exp(ikr_n \mu_k \mu_r)\, dk$$

$$\sim \frac{1}{2\pi} \sum \frac{1}{\sqrt{|(1 - \hat{\mathbf{n}} \cdot \boldsymbol{\kappa}_1 r_n \mu_r)(1 - \hat{\mathbf{n}} \cdot \boldsymbol{\kappa}_2 r_n \mu_r)|}}$$

$$\cdot \left[\frac{\sin kr_n}{r_n} - \frac{i \cos kr_n}{r_n \mu_k \mu_r} \right] \Bigg|_{k_-}^{k_+} . \tag{9.1.38}$$

The sum here must be carried out over all of the stationary points as defined by (9.1.35). We remark that the value of the integral in the first line here when $r_n = 0$ is correctly given by the limit in the next two lines. (The difference of cosines is quadratic in r_n.)

For $\mathbf{y}$ near the surface S, the sum in (9.1.38) will be dominated by the contribution from the point on S that defines the minimal distance from $\mathbf{y}$ to S. Furthermore, it is typical that the angular aperture contains a range of $\hat{\mathbf{k}}$ that is a subdomain of a hemisphere plus the symmetric range on the other side of the sphere. Thus, there will be two choices of $\hat{\mathbf{k}}$ for each choice of $\mathbf{y}$, and the stationary points occur in pairs for such an angular aperture. The values of μ_k at these two stationary points will be of opposite sign, ± 1, while all other constituents of the stationary phase formula remain unchanged. Therefore, in the neighborhood of the surface S,

$$\gamma_B(\mathbf{y}) \sim \frac{\sin k_+ r_n - \sin k_- r_n}{\pi r_n \sqrt{|(1 - \hat{\mathbf{n}} \cdot \boldsymbol{\kappa}_1 r_n \mu_r)(1 - \hat{\mathbf{n}} \cdot \boldsymbol{\kappa}_2 r_n \mu_r)|}} . \tag{9.1.39}$$

We see here that asymptotically the aperture-limited singular function in three dimensions behaves in the same manner as the one-dimensional, band-limited Dirac delta function. The surface S can then be depicted by its band-limited singular function when the percentage bandwidth is sufficiently large to distinguish the peaks of the Dirac delta function. In particular, we remark that the peak value is given by

$$\max |\gamma_B(\mathbf{y})| \sim (k_+ - k_-)/\pi . \tag{9.1.40}$$

As in one dimension, 60% bandwidth seems to be more than sufficient for peak discrimination.

If the aperture in $\hat{\mathbf{k}}$ does not contain points satisfying (9.1.35), then the asymptotic expansion will be of lower order in k, and the final k integration will not produce a function that peaks with the amplitude in (9.1.40). In fact, Mager and Bleistein [1978] have shown that the peak value of the asymptotic expansion in this case decays to zero with increasing minimum angle between the aperture in $\hat{\mathbf{k}}$ and the range of the normal directions of S.

Exercises

9.1 (a) Let S be a spherical cap of radius a centered at the origin, with polar angle range $0 \le \theta \le \theta_0$. Define the normal $\hat{\mathbf{n}}$ for S to be the unit radial vector at each point $(\cos \phi \sin \theta, \sin \phi \sin \theta, \cos \theta)$. Show that

$$\bar{\gamma}(\mathbf{k}) \sim (2\pi i \mu a/k)e^{-i\mu ka}, \qquad \mu = \text{sign } \hat{k}_3,$$

when either $\pm \hat{\mathbf{k}}$ is in the aperture of the cap.

(b) Find the leading term of the asymptotic expansion of $\bar{\gamma}(\mathbf{k})$ when $\hat{\mathbf{k}}$ is not in the aperture of the cap. Confirm that it is of lower order in k.

(c) Let $\theta_0 = \pi$ so that S is closed. Let $\Gamma(\mathbf{x})$ be the characteristic function of the interior of the sphere. Find the asymptotic expansion of $\bar{\Gamma}(\mathbf{k})$ and confirm (9.1.4).

9.2 Consider the surface S defined by (9.1.16). For this special parameterization $x_1 = \sigma_1$ and $x_2 = \sigma_2$, show that (8.4.13) agrees with (9.1.18).

9.3 (a) Verify (9.1.36).

(b) Introduce the notation

$$\cos \alpha = \hat{\boldsymbol{\theta}} \cdot \frac{\partial \mathbf{x}}{\partial \sigma_1}, \qquad \sin \alpha = \hat{\boldsymbol{\theta}} \cdot \frac{\partial \mathbf{x}}{\partial \sigma_2},$$

and show that

$$\cos \alpha = \hat{\boldsymbol{\phi}} \cdot \frac{\partial \mathbf{x}}{\partial \sigma_2}, \qquad \sin \alpha = -\hat{\boldsymbol{\phi}} \cdot \frac{\partial \mathbf{x}}{\partial \sigma_1}.$$

(c) Calculate $\det[A - \lambda I]$, where I is the identity matrix. Show that for $\alpha = 0$, this quartic in λ factors into a pair of quadratics, each of which has two roots of opposite sign for r_n small enough. Thus, conclude that sgn $A = 0$ in this case.

(d) Now verify the first line in (9.1.37) for any α.

(e) Explain why sgn $A = 0$ for any α and r_n small enough.

9.4 (a) Let $I(\lambda)$ be a surface integral of the form

$$I(\lambda) = \int_S f(\mathbf{x})e^{i\lambda\phi(\mathbf{x})} \, dS, \qquad \mathbf{x} = (x_1, x_2, x_3).$$

We have seen in this section that for $\phi(\mathbf{x})$ to have a stationary point, $\nabla\phi(\mathbf{x})$

must be normal to S at that point. Suppose that $\phi(\mathbf{x})$ has no stationary points on S. Denote the boundary curve to S by C and apply Stokes's theorem

$$\int_S (\mathbf{V} \times \mathbf{B}) \cdot \hat{\mathbf{n}} \, dS = \int_C \mathbf{B} \cdot d\mathbf{x}$$

to the vector function

$$\mathbf{B} = \frac{\hat{\mathbf{n}} \times \mathbf{V}\phi(\mathbf{x})}{i\lambda [(\mathbf{V}\phi(\mathbf{x}))^2 - (\hat{\mathbf{n}} \cdot \mathbf{V}\phi(\mathbf{x}))^2]} f(\mathbf{x}) e^{i\lambda\phi(\mathbf{x})}$$

to conclude that to leading order

$$I(\lambda) \sim \frac{1}{i\lambda} \int_C \frac{\hat{\mathbf{t}} \cdot (\hat{\mathbf{n}} \times \mathbf{V}\phi(\mathbf{x}))}{(\mathbf{V}\phi(\mathbf{x}))2 - (\hat{\mathbf{n}} \cdot \mathbf{V}\phi(\mathbf{x}))^2} f(\mathbf{x}) e^{i\lambda\phi(\mathbf{x})} \, ds.$$

In this equation, ds is differential arc length around the boundary of the surface S, and $\hat{\mathbf{t}} = d\mathbf{x}(s)/ds$ is the unit tangent vector to the boundary curve. Note that the denominators of the integrands here are nonzero under the assumption that $\phi(\mathbf{x})$ has no stationary points on S.

(b) Let the boundary curve C be parameterized by an arc-length parameter $s:\mathbf{x} = \mathbf{x}(s)$ on C. Show that the stationary points are determined by $\mathbf{V}\phi(\mathbf{x}) \cdot \hat{\mathbf{t}} = 0$ and that

$$I(\lambda) \sim \frac{1}{i\lambda} \sqrt{\frac{2\pi}{\lambda}} \sum \frac{\hat{\mathbf{t}} \cdot (\hat{\mathbf{n}} \times \mathbf{V}\phi(\mathbf{x}))}{\sqrt{|\mathbf{V}\phi(\mathbf{x}) \cdot \boldsymbol{\kappa}|} [(\mathbf{V}\phi(\mathbf{x}))^2 - (\hat{\mathbf{n}} \cdot \mathbf{V}\phi(\mathbf{x}))^2]} f(\mathbf{x}) e^{i\lambda\phi(\mathbf{x}) + i\mu\pi/4}.$$

In this equation, $\boldsymbol{\kappa}$ again represents the curvature vector at the stationary point and $\mu = \text{sign } \mathbf{V}\phi(\mathbf{x}) \cdot \boldsymbol{\kappa}$.

(c) Apply this result to the second integral in (9.1.3) under the assumption of no interior stationary points. Show that the assumption of no stationary points assures that $\hat{\mathbf{k}} \cdot \hat{\mathbf{n}} \neq \pm 1$ and that if all of the boundary stationary points are simple,

$$\bar{\gamma}(\mathbf{k}) \sim -\frac{1}{ik} \sqrt{\frac{2\pi}{k}} \sum \frac{\hat{\mathbf{t}} \cdot (\hat{\mathbf{n}} \times \hat{\mathbf{k}})}{\sqrt{|\hat{\mathbf{k}} \cdot \boldsymbol{\kappa}|} [1 - (\hat{\mathbf{n}} \cdot \hat{\mathbf{k}})^2]} e^{-i\mathbf{k} \cdot \mathbf{x} + i\mu\pi/4}.$$

9.2 PHYSICAL OPTICS FAR-FIELD INVERSE SCATTERING (POFFIS)

The physical optics far-field inverse scattering (POFFIS) method was originally developed by Bojarski [1967]. The approach presented here has

evolved from that original method while still retaining some of the fundamental features of the original derivation. Bojarskī's original derivation is
outlined in Exercise 9.9.

The objective of the method is to image the surface of a scatterer that is
remote enough to be considered to be in the far field. The method has found
application in nondestructive testing, where the objective is to image flaws
of size a few hundred micrometers (10^{-6} m) at a range on the order of a few
centimeters (10^{-2} m). The propagation speed of acoustic waves in solids is
on the order of thousands of meters per second (10^3 m/s). Consequently,
high frequency requires frequencies in the range of megahertz (10^6 Hz).
Typical numbers in an experiment might be a length scale H equal to
4×10^{-4} m, a sound speed c equal to 6×10^3 m/s, and a minimum frequency
f_0 equal to 3×10^6 Hz for which we obtain a dimensionless large parameter

$$\lambda = 4\pi f_0 H/c \approx 7,$$

which is certainly large enough for asymptotics. (The extra factor of 2 here
is due to the two-way travel time in backscatter experiments. See, for example,
(8.4.12) or (9.2.9) below.)

In nondestructive testing applications, the propagating signal is really an
elastic wave. However, in the backscatter direction, the signal is dominated
by the compressional or acoustic wave. Thus, our model based on the
acoustic wave equation is reasonable as an elemental model of this direct and
inverse scattering problem, among others.

Similarly, in electromagnetic applications, polarization effects are of
relatively minor importance in an implementation based on the backscattered
specular signal, for which the polarization of the incident wave is well
preserved.

We shall begin our discussion by assuming that we are observing the
backscattered signal from a closed scatterer. We assume that by *time gating*
the response, that is, by setting the time record equal to zero at times beyond
the return of the specularly reflected wave, we can assume that we have only
these primary specular returns as our observations.

The POFFIS method is based on an analysis of the Kirchhoff and far-field
approximate backscattered field discussed in Section 8.4. In particular, the
result (8.4.11) shows that when both of these approximations are valid, the
backscattered field from a point source depends on a phase and amplitude
scaling that is common to all scatterers and a factor $S(\hat{\xi}; \omega)$, the phase and
range normalized far-field scattering amplitude, which depends on the
properties of the specific scatterer. The result of Exercise 8.23 is that for the
model in which it is assumed that the incident signal is a plane wave, the same
factor $S(\hat{\xi}; \omega)$ arises but with a different phase and range normalization.

ASYMPTOTIC ANALYSIS OF $S(\hat{\xi}; \omega)$

The asymptotic analysis of the integral representation (8.4.12) for $S(\hat{\xi}; \omega)$ follows along the same lines as the asymptotic analysis of $\bar{\gamma}(\mathbf{k})$ in the preceding section. Indeed, we need only make the following identifications between variables in (8.4.12) and variables in the second equation of (9.1.3) in order to obtain the asymptotic expansion for $S(\hat{\mathbf{k}}; \omega)$ from the results of Section 9.1:

$$k = (2|\omega|)/c, \qquad \hat{\mathbf{k}} = \hat{\xi} \operatorname{sign} \omega. \qquad (9.2.1)$$

Thus, under the assumption of simple stationary points, we can deduce the leading order asymptotic expansion of $S(\hat{\xi}; \omega)$ from the result (9.1.11). To do so, we must evaluate the amplitude in (8.4.12) at the stationary point and use the special form of k and $\hat{\mathbf{k}}$. The result is

$$S(\hat{\xi}; \omega) \sim -2\pi i \operatorname{sign} \omega \sum R_n \sqrt{\rho_1 \rho_2} \, e^{-2i\omega \hat{\xi} \cdot \mathbf{x}_n - i\mu\pi/4}. \qquad (9.2.2)$$

In this equation, ρ_1, ρ_2, and μ are as defined in the discussion in Section 9.1: see, in particular, (9.1.9) and (9.1.10). The summation is to be carried out over the stationary points of the integral. Those points are defined by (9.1.8), but for the special choice of k and $\hat{\mathbf{k}}$ in (9.2.1), that condition becomes

$$\hat{\mathbf{n}} = \hat{\xi}. \qquad (9.2.3)$$

There is no choice of sign here, because on the lighted side of the scatterer, the normal at the stationary point must point back toward the observation point. The reflection coefficient, given in Exercise 8.22b, must also be evaluated at the stationary point. This yields the normal reflection strength R_n, as in Exercise 8.24,

$$R_n = (c_+ - c_-)/(c_+ + c_-), \qquad (9.2.4)$$

where $c_\pm$ are the propagation speeds on the two sides of the scatterer. We remark also that reversing the sign of ω in the result (9.2.2) simply transforms $S(\hat{\xi}; \omega)$ into its complex conjugate, which is equally true for the integral (8.4.12) from which this asymptotic expansion arose.

THE REFLECTIVITY FUNCTION

Let us define a new function, the *reflectivity function*, of a surface S to be the normal reflection strength multiplied by the singular function and to be denoted by $R_n \gamma(\mathbf{x})$. The reflectivity function locates the boundary of the scattering object and characterizes the change in medium properties through the normal reflection coefficient. For this function, the asymptotic expansion is readily obtained from the result (9.1.11) as

$$R_n \bar{\gamma}(\mathbf{k}) \sim \frac{2\pi}{k} \sum R_n \sqrt{\rho_1 \rho_2} \, e^{-i\mathbf{k} \cdot \mathbf{x}_n - i\mu\pi/4}. \qquad (9.2.5)$$

For a fixed value of $\mathbf{k}$, the sum is over all points where $\hat{\mathbf{k}}$ and $\hat{\mathbf{n}}$ are colinear or anticolinear.

THE POFFIS IDENTITY FOR THE REFLECTIVITY FUNCTION

Rather than consider the entire sum here, let us consider a single term, namely, the contribution from a particular stationary point, say, $\mathbf{x}_0$. If this point contributes to the sum at a particular choice of $\mathbf{k}$, then it also contributes to the sum at $-\mathbf{k}$, with the contribution in the latter case being the complex conjugate of the result in the former case. Now let us suppose that $\mathbf{x}_0$ is in the aperture of the observations for which (9.2.2) is valid. Then the contribution to the sum in (9.2.2) is proportional to the corresponding contribution to the sum in (9.2.5) when sign $\omega = +1$, while the result at sign $\omega = -1$ produces the complex conjugate result. In these results, the constant of proportionality is $i[\text{sign } \omega]/k = -c/2i\omega$. What is true for each term of the sum must be true for the entire sum; that is,

$$R_n \bar{y}(\mathbf{k}) \sim (i/k)[\text{sign } \omega]S(\hat{\xi};\omega) = -[2i\omega/c]^{-1}S(\hat{\xi};\omega), \qquad \mathbf{k} = (2\omega/c)\hat{\xi}, \quad (9.2.6)$$

with this result holding for $\hat{\xi}$ in the aperture of observations and negative ω given the results in the complement of that aperture in k-space. This is the POFFIS identity for the reflectivity function.

FOURIER INVERSION OF THE POFFIS IDENTITY

We have already discussed the aperture-limited Fourier inversion of the singular function in the preceding section. For the reflectivity function, we need only multiply those results, in particular, (9.1.38)–(9.1.40), by the normal reflection coefficient. Also, this is a case in which we sum over pairs of values of $\hat{\mathbf{k}}$ of opposite sign in (9.2.2). Therefore, in the neighborhood of the scattering surface, the Fourier inversion is dominated by a term of the type (9.1.39). That is,

$$R_n \gamma_B(\mathbf{y}) \sim R_n \frac{\sin[4\pi f_1 r_n/c] - \sin[4\pi f_0 r_n/c]}{\pi r_n \sqrt{(1 - \hat{\mathbf{n}} \cdot \mathbf{\kappa}_1 r_n \mu_r)(1 - \hat{\mathbf{n}} \cdot \mathbf{\kappa}_2 r_n \mu_r)}}. \qquad (9.2.7)$$

In this result, we have replaced the limits of integration in k by limits in frequency f, because in practice, the band limits are given in hertz. In particular, at the peak, this result becomes

$$\text{peak } R_n \gamma_B(\mathbf{y}) \sim R_n(4[f_1 - f_0]/c). \qquad (9.2.8)$$

Thus, the location of the peaks of $\gamma_B(\mathbf{y})$ delineates the scattering surface, while the magnitude of $\gamma_B(\mathbf{y})$ at the peak provides a means of determining the reflection strength.

Let us now express the result (9.2.6) and (9.2.7) explicitly in terms of the integrations to be performed. First, from (9.2.6), we obtain the following result for the reflectivity function:

$$R_n\gamma_B(\mathbf{y}) \sim \frac{2i}{\pi c^2}\left[\int_{f_0}^{f_1} + \int_{-f_1}^{-f_0}\right] f\, df \int_\Omega d\Omega\, S(\hat{\boldsymbol{\xi}}; 2\pi f)e^{4\pi i f\hat{\boldsymbol{\xi}}\cdot\mathbf{y}/c}. \tag{9.2.9}$$

The notation $d\Omega$ connotes the differential element of the solid angle in $\hat{\boldsymbol{\xi}}$, while the domain of integration Ω is determined by the range of observation directions $\hat{\boldsymbol{\xi}}$. The domain of integration f in hertz is both the positive and negative frequency ranges in the bandwidth of the signal to provide the full aperture of information in k-space that is available from (9.2.6) from positive and negative frequencies. We have expressed this final result in f rather than in ω because in implementation on real data, the fast Fourier transform (FFT) routine would be used to process the data; this routine uses frequencies f in hertz. The result of Exercise 9.5 exploits the fact that negative frequencies merely transform the integrand into its complex conjugate to rewrite (9.2.9) as an integral over positive frequencies alone. This is also more practical for numerical implementation.

As a check on self-consistency, let us consider the problem of back-scattering by a sphere of radius a and reflection coefficient 1. In particular, we apply (8.4.12) to determine $S(\hat{\boldsymbol{\xi}}; \omega)$. For the domain L, we use the hemisphere with line of symmetry alone $\hat{\boldsymbol{\xi}}$. The leading order asymptotic expansion of $S(\hat{\boldsymbol{\xi}}: \omega)$ is

$$S(\hat{\boldsymbol{\xi}}; \omega) \sim 2\pi ae^{-2i\omega a/c}, \tag{9.2.10}$$

independent of $\hat{\boldsymbol{\xi}}$. Let us assume that the data have been imaged over the entire 4π aperture in $\hat{\boldsymbol{\xi}}$. We substitute this result into (9.2.9) and carry out the straightforward integrations to obtain the result

$$R_n\gamma_B(\mathbf{y}) \sim \frac{a}{\pi y}\left[\frac{\sin[4\pi(y-a)/c]}{y-a} - \frac{\sin[4\pi(y+a)/c]}{y+a}\right]\Bigg|_{f_0}^{f_1}, \quad y = |\mathbf{y}|. \tag{9.2.11}$$

To compare this result with (9.2.7), we focus our attention on the first term in the brackets and note first that $r_n = y - a$. Then, we note also that both principal radii of curvature for a sphere are equal to the radius of the sphere at every point on the sphere. Thus,

$$a/y = 1/\sqrt{(1 - \hat{\mathbf{n}}\cdot\boldsymbol{\kappa}_1 r_n \mu_r)(1 - \hat{\mathbf{n}}\cdot\boldsymbol{\kappa}_2 r_n \mu_r)}.$$

Finally, we note that the peak value of this first term at $y = a$ agrees with the prediction (9.2.7).

For this same example, a series representation of the exact backscattered wave can be obtained by specializing (6.5.58) to the backscatter case $\cos\gamma = 1$. Since that is the solution for plane wave incidence, we can obtain $S(\hat{\boldsymbol{\xi}}; \omega)$

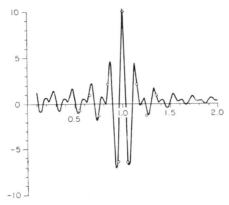

Fig. 9.1. Inversion of sphere data in units a/r. $31.5 \leq 4\pi fa/c \leq 63$. [From Bleistein, 1976.]

by using the phase and range normalization implicit in the result in Exercise 8.23; that is, we multiply by $4\pi r \exp\{-2i\omega r/c\}$. The angular integrations can then be carried out analytically, leaving only the integral in f to be computed numerically. Figure 9.1 depicts the output for such a numerical integration. The indicated parameter range is given in a dimensionless variable $4\pi fa/c$. For this range, $\lambda = 4\pi f_0 a/c \approx 31.5$ is well within the range defined as large for the purposes of asymptotics. On the other hand, the percentage bandwidth for this example is only 33%, significantly less that the recommended 60%. Nonetheless, the peak is seen to be well delineated.

In Fig. 9.2, we depict the output for a case in which the lower limit is zero. Our theory should not apply in this case. Yet, the band-limited Dirac delta function is clearly discernible. We remark that the dimensionless large parameter λ defined earlier fails to be large over only a very small portion of the range for this example. Thus, it is reasonable to expect relatively little damage to the final output as a result of inclusion of the range of values of f that are too small. The output confirms this expectation.

In both of these figures, the circles represent sample values of the asymptotic formulas (9.2.7) and (9.2.8). The agreement between leading order asymptotics and numerical integration is apparent. In particular, the error at the peak was about 0.3% in the first example and 1.6% in the second example.[†] In fact, this size error is significantly less than the errors due to noise in any real data experiment. Thus, experimental error becomes a more significant limitation on estimation of reflection strength than does all of the approximations made to obtain the POFFIS identity.

[†] The larger error in the second example is likely due to the inclusion of the low frequencies in the numerical integration.

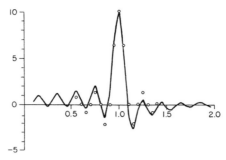

Fig. 9.2. Inversion of sphere data in units of a/r. $0 \leq 4\pi fa/c \leq 31.5$. [From Bleistein, 1976.]

We remark that variations in the amplitude of $S(\hat{\xi}; \omega)$ in (9.2.2) will affect the estimate of the reflection strength in processing (9.2.9), while variations in the phase in (9.2.2) affect the *structure* of the scattering surface. This is an important consideration when we are seeking only an image of the surface, with less interest in reflection strength. Within limits of variation that allow us to view the amplitude of $S(\hat{\xi}; \omega)$ as "slowly varying" and, hence, not affecting the phase, we can allow errors in the amplitude without seriously affecting the image of the surface S.

To obtain this result, we have assumed that the source has unit strength over the entire bandwidth. True sources do not behave at all like this. Therefore, as a practical matter, we must strip away the structure of the source in order to use the result as stated. That is not always an easy task and leads to the whole class of problems that come under the heading of *deconvolution*. This is the subject of a volume in itself and will not be addressed here. We only remark that small errors in deconvolution produce small errors in surface location. In particular, we leave to Exercise 9.6 the verification of the following result. Suppose that the source in the experiments that produced $u_S(\xi)$ had the form $\bar{F}(2\pi f)\delta(\xi - x)$. That is, the entire experiment was scaled by $\bar{F}(2\pi f)$. Assume further that $\bar{F}(2\pi f)$ had its support only in the bandwidth $(\pm f_0, \pm f_1)$ or, equivalently, that $\bar{F}(2\pi f)$ represents the consequences of truncation and deconvolution of the Fourier transform of the source on this interval and that this transform is *slowly varying* over this interval. We can show that in this case, (9.2.7) and (9.2.8) are replaced by

$$R_n \gamma_B(y) \sim R_n \frac{F(-2r_n/c)}{\sqrt{(1 - \hat{k} \cdot \kappa_1 r_n \mu_r \mu_k)(1 - \hat{k} \cdot \kappa_2 r_n \mu_r \mu_k)}}, \qquad r_n \neq 0, \quad (9.2.12)$$

and

$$\text{peak } R_n \gamma_B(y) \sim R_n F(0) = R_n \int_{f_0}^{f_1} + \int_{f_1}^{-f_0} \bar{F}(2\pi f)\, df. \qquad (9.2.13)$$

The case $\bar{F}(2\pi f) \equiv 1$ on the bandwidth is the extreme case of slow variation. Other slowly varying functions will produce alternative band-limited representations of the delta function and, consequently, alternative band-limited representations of the band-limited reflectivity function. See Exercise 9.7, for example.

While errors in amplitude of the backscattered signal will not seriously affect imaging, errors in the phase of the backscattered signal will produce significant errors in surface imaging. In particular, if we use (8.4.11) in (9.2.9), the following result can be derived:

$$R_{n\gamma_B}(\mathbf{y}) \sim -16 \int_\Omega d\Omega\, \rho^2\, \frac{\partial}{\partial t}\, U_S(\boldsymbol{\xi}, 2(\rho - \hat{\boldsymbol{\xi}} \cdot \mathbf{y})/c). \qquad (9.2.14)$$

Here $U_S(\boldsymbol{\xi}, t)$ is the observed data in the time domain. We see from this result that the phase in frequency domain arose from the arrival time of the signal in the time domain. Let us think of carrying out a set of experiments over a discrete set of values of $\hat{\boldsymbol{\xi}}$. It is assumed that the distance to the center of the coordinate system ρ is known for each experiment, that is, for each choice of $\hat{\boldsymbol{\xi}}$. Thus, errors in recording the arrival time of the response relative to the initial time, the time of onset of the source, will produce errors in the contribution $\hat{\boldsymbol{\xi}} \cdot \mathbf{y}$ for that contribution to the integral. Thus, an accurate record of time of occurrence of the return signal relative to time of initiation of the incident wave is essential.

While the POFFIS identity was originally derived for convex scatterers, it has also been applied to nonconvex scatterers. The key to the success of the method is that concavity should not be so severe as to shadow the specular points in any experiment. This is a limit on concavity but does not preclude it.

BOJARSKI'S RESULT

Bojarski's original result provides a POFFIS identity for the characteristic function rather than for the singular function. However, it is more difficult to depict the scatterer through band-limited Fourier inversion of a one–zero function than for a Dirac delta function. Thus, in implementation, Bojarski's result requires multiplication by k_1, k_3, or k_3 before inversion to produce a band-limited derivative of a characteristic function. This type of directional derivative has the disadvantage of image fading when the direction of the directional derivative is tangent to the surface, that is, in the direction of y_1, y_3, or y_3. Indeed, it was this disadvantage of the Bojarski implementation that motivated a search for a "pseudonormal derivative" and led ultimately to the POFFIS identity for singular functions.

Exercises

9.5 Show that the result (9.2.9) can be rewritten as

$$R_n \gamma_B(\mathbf{y}) \sim \frac{4}{\pi c^2} Re\left[\int_{f_0}^{f_1} if\, df \int_\Omega d\Omega\, S(\hat{\xi}; 2\pi f) e^{4\pi i f \hat{\xi} \cdot \mathbf{y}/c} \right].$$

9.6 Verify (9.2.12) and (9.2.13). [*Hint*: Use the stationary phase result in the first line of (9.1.38) to calculate the angular integral in (9.2.9) and then interpret the integral in frequency domain as an Fourier transform back to the "time" domain with time properly interpreted.]

9.7 (a) Define $F_\lambda(t) = \lambda^2 t \exp(-\lambda t)$. Use Watson's lemma to show that

$$\lim_{\lambda \to \infty} F_\lambda(t) = \delta(t).$$

That is, for any fixed but large value of λ, $F_\lambda(t)$ is an approximate Dirac delta function.

(b) Show that the Fourier transform of $F_\lambda(t)$ is

$$\bar{F}(2\pi f) = \lambda^2 \int te^{-(\lambda - 2\pi i f)t}\, df = \frac{\lambda^2}{(\lambda - 2\pi i f)^2}.$$

(c) Suppose that $f_0 = 0.25\, \lambda/2\pi$ and $f_1 = \lambda/2\pi$. Show that the modulus of $\bar{F}(2\pi f)$ varies from 0.5 to 0.96 over the bandwidth.

9.8 Verify (9.2.14).

9.9 The purpose of this exercise is to derive Bojarski's original POFFIS identity for the characteristic function. Suppose that the scattering obstacle to be imaged is convex and that the reflection coefficient is equal to ± 1. Suppose that for each experiment from the direction $\hat{\xi}$, with scattering amplitude defined in (8.4.12), the corresponding experiment from the direction $-\hat{\xi}$ is also carried out.

(a) Explain why this latter experiment produces the output

$$S(-\hat{\xi}; \omega) \frac{2i\omega R}{c} = \int_D \hat{n} \cdot \hat{\xi} e^{2i\omega \hat{\xi} \cdot \mathbf{x}/c}\, dS.$$

That is, the experiment from the opposite direction produces an integral over the part of the scatterer that was dark for the original experiment.

(b) Use the divergence theorem to show that

$$S(\hat{\xi}; \omega) + S^*(-\hat{\xi}; \omega) = -\frac{2i\omega R}{c} \int_B \hat{n} \cdot \hat{\xi} e^{2i\omega \hat{\xi} \cdot \mathbf{x}/c}\, dS$$

$$= -\frac{4\omega^2 R}{c^2} \int_{\mathbf{x} \text{ in } B} e^{-i\omega \hat{\xi} \cdot \mathbf{x}}\, dV = -\frac{4\omega^2 R}{c^2} \Gamma\left[\frac{2\omega\hat{\xi}}{c}\right].$$

This is Bojarski's POFFIS identity for the characteristic function of a scattering obstacle.

9.3   THE SEISMIC INVERSE PROBLEM

In this section, we shall describe inverse methods for the problem of imaging the layers in the earth from backscattered acoustic signals. In this seismic experiment, a source is set off at a point on the surface of the earth, and the upward propagating wave is then measured at an array of receivers nearby the source. The process is repeated with the source/receiver array displaced along a line or in a plane. Thus, the actual data-gathering process does not produce a backscattered signal. Furthermore, the source is often an explosion, so that the source shape is not entirely repeatable, nor is the time shape of the source signal recorded as a rule. Much preprocessing of the data must be done in order to create from this array of experiments a set of pseudobackscattered impulse responses, that is, the backscatter response to a three-dimensional spatial delta function multiplied by a delta function in time. Nonetheless, the preprocessing produces usable data for the backscatter model, and imaging techniques based on an acoustic backscatter model produce usable subsurface images. Here *usable* means that the images of the interior of the earth based on these techniques are a useful tool for geological interpretation, either for theoretical purposes or for the purpose of interpretation by a geologist for the identification of likely subsurface regions for resource extraction, such as oil or natural gas.

In order to describe our model, we begin by introducing a right-handed coordinate system $\mathbf{x} = (x_1, x_2, x_3)$, with x_3 being positive in the *downward* direction into the earth. We assume that we know the backscatter response from acoustic point sources set off at every point $x_1 = \xi_1, x_2 = \xi_2, x_3 = 0$ on the surface of the earth. We assume that the total field $u(\mathbf{x}; \omega)$ is a solution of the Helmholtz equation, with point source at $\boldsymbol{\xi} = (\xi_1, \xi_2, 0)$:

$$\nabla^2 u(\mathbf{x}; \omega) + (\omega^2/\nabla^2) u(\mathbf{x}; \omega) = -\delta(x_1 - \xi_1)\delta(x_2 - \xi_2)\delta(x_3). \quad (9.3.1)$$

In this equation, $v = v(\mathbf{x})$ is the variable reference speed that we seek.

We have modeled the problem as though the medium extends to negative infinity in x_3. Alternatively, we could have considered the problem on $0 \le x_3 < \infty$ and modeled the source through the boundary condition that the normal derivative be equal to $-\delta(x_1 - \xi_1)\delta(x_2 - \xi_2)$. The effect of this would be to rescale the incident wave to be defined later by a factor of 2 and to rescale the final answer by a factor of 4. Thus, there is no essential difference between these two models.

We introduce a reference velocity c_0 and a *perturbation* $\alpha(\mathbf{x})$ defined by the equation

$$v^{-2} = c_0^{-2}[1 + \alpha(\mathbf{x})]. \qquad (9.3.2)$$

We now decompose the total field into an incident and scattered field

$$u(\mathbf{x};\omega) = u_1(\mathbf{x};\xi;\omega) + u_S(\mathbf{x};\xi;\omega) \qquad (9.3.3)$$

in which $u_1(\mathbf{x};\xi;\omega)$ is the response to the source in the unperturbed medium

$$\nabla^2 u_1(\mathbf{x};\xi;\omega) + (\omega^2/c_0^2)u_1(\mathbf{x};\xi;\omega) = -\delta(x_1 - \xi_1)\delta(x_2 - \xi_2)\delta(x_3), \qquad (9.3.4)$$

and $u_S(\mathbf{x};\xi;\omega)$ must then satisfy

$$\nabla^2 u_S(\mathbf{x};\xi;\omega) + (\omega^2/c_0^2)u_S(\mathbf{x};\xi;\omega) = -\alpha(\mathbf{x})(\omega^2/c_0^2)[u_1(\mathbf{x};\xi;\omega)$$
$$+ u_S(\mathbf{x};\xi;\omega)]. \qquad (9.3.5)$$

We shall write down an integral equation relating the backscattered values of $u_S(\mathbf{x};\xi;\omega)$—$u_S(\xi;\xi;\omega)$—to the values in the interior. To do this, we use the Green's function representation (6.4.14) with no finite scatterers D, since we have replaced the scattering domain in this problem by a source term on the right side of the equation. The Green's function we shall use for (9.3.5) is just the free-space Green's function. Furthermore, for the backscattered signal, the Green's function is just the solution to (9.3.4); that is, the Green's function and $u_1(\mathbf{x};\xi;\omega)$ itself are the same function. Therefore, the equation for $u_S(\xi;\xi;\omega)$ is

$$u_S(\xi;\xi;\omega) = \omega^2 \int_{x_3 > 0} \frac{\alpha(\mathbf{x})}{c_0^2} u_1(\mathbf{x};\xi;\omega)[u_1(\mathbf{x};\xi;\omega) + u_S(\mathbf{x};\xi;\omega)]\, dV. \qquad (9.3.6)$$

This is an exact integral equation. Unfortunately, this equation has two unknown functions, $\alpha(\mathbf{x})$ and $u_S(\mathbf{x};\xi;\omega)$, and therefore does not directly lend itself to exact solution techniques. The equation is also *nonlinear* in the unknowns through the product $\alpha(\mathbf{x})u_S(\mathbf{x};\xi;\omega)$.

LINEARIZATION: AN INTEGRAL EQUATION FOR $\alpha(\mathbf{x})$

We now introduce our first approximation. Let us assume that the variation $\alpha(\mathbf{x})$ is small. Then, the source term in (9.3.5) is also small. It would be reasonable to expect that the solution to that problem is small as well, on the same order, $O(\alpha(\mathbf{x}))$, as the source term itself. Consequently, it is reasonable to expect that the product $\alpha(\mathbf{x})u_S(\mathbf{x};\xi;\omega)$ on the right side is *quadratic* in the small-scale measuring $\alpha(\mathbf{x})$—$O(\alpha^2(\mathbf{x}))$—while the product $\alpha(\mathbf{x})u_1(\mathbf{x};\xi;\omega)$ is *linear* in $\alpha(\mathbf{x})$. As a first-order approximation then, we neglect the higher-order term in $\alpha(\mathbf{x})$ in the source; that is, we replace the sum $u_1(\mathbf{x};\xi;\omega) + u_S(\mathbf{x};\xi;\omega)$ by the first term $u_1(\mathbf{x};\xi;\omega)$ alone. Mathematically, we are assuming that the solution $u(\mathbf{x}; \omega)$ can be derived as a *regular perturbation series* in $\alpha(\mathbf{x})$. This technique is equivalent to the Born approximation for potential scattering in theoretical physics, and that name has been attached to this application of the regular perturbation method as well.

The effect of modifying the source in this manner is to recast the integral equation (9.3.6) as

$$u_S(\boldsymbol{\xi}; \boldsymbol{\xi}; \omega) = \omega^2 \int \frac{\alpha(\mathbf{x})}{c_0^2} u_I^2(\mathbf{x}; \boldsymbol{\xi}; \omega) \, dV. \qquad (9.3.7)$$

This is a *linear* integral equation relating the (assumed known) surface values of the backscattered field $u_S(\boldsymbol{\xi}; \boldsymbol{\xi}; \omega)$ and $\alpha(\mathbf{x})$, which is now the only unknown in the equation. The latter is a function of the three variables $\mathbf{x}$, while the former is a function of the three variables $\boldsymbol{\xi}$ and ω; or back in the time domain, the upward scattered field would be a function of the two transverse coordinates of the source/receiver point and the time. In either case, the count of degrees of freedom agrees.

Equation (9.3.7) is a *Fredholm integral equation of the first kind* for $\alpha(\mathbf{x})$. When the kernel of such an integral equation—in this case, $u_I^2(\mathbf{x}; \boldsymbol{\xi}; \omega)$—is such that its modulus has a bounded square integral in all of its variables $\mathbf{x}, \boldsymbol{\xi}, \omega$, then it is known that the solution to this type of integral equation is *ill conditioned*, with eigenvalues that have a limit point at zero. However, this kernel is not square integrable in all of its variables, and that theory does not apply. Indeed, a prototypical one-dimensional analog of (9.3.7) has as its kernel the square of the Fourier kernel $\exp\{2i\omega z/c\}$ (Exercise 9.10), which is known to have all of its (complex) eigenvalues on a circle of nonzero radius. If we think of the kernel as always being the ray method generalization of this one-dimensional kernel, then it is reasonable to expect that the kernel *never* has a bounded square integral in all of its variables.

REFLECTOR IMAGING

For an arbitrary reference speed, we cannot solve (9.3.7) in closed form analytically. Therefore, we further specialize this result to the case in which c_0 is a constant. In this case, $u_I(\mathbf{x}; \boldsymbol{\xi}; \omega)$ is just the free-space Green's function (6.4.8), and the integral equation (9.3.7) takes the form

$$u_S(\boldsymbol{\xi}; \boldsymbol{\xi}; \omega) = \frac{\omega^2}{(4\pi c_0)^2} \int \alpha(\mathbf{x}) \frac{e^{2i\omega|\mathbf{x}-\boldsymbol{\xi}|/c}}{|\mathbf{x}-\boldsymbol{\xi}|^2} \, dV. \qquad (9.3.8)$$

We shall solve (9.3.8) for $\alpha(\mathbf{x})$ by Fourier methods. We observe, first, that in the variables x_1 and x_2, this equation is in convolution form (2.2.19) and (2.2.20). However, we do not have a closed analytical form for the Fourier transform of the square of the Green's function. To overcome this, let us introduce the function

$$\Theta(\xi; \omega) = -i \frac{\partial}{\partial \omega} \left[\frac{u_S(\xi; \xi; \omega)}{\omega^2} \right]. \tag{9.3.9}$$

We remark that the operator $-i(\partial/\partial\omega)$ applied to $u_S(\xi; \xi; \omega)$ simply requires multiplication by t before transforming the time domain data. Hence, differentiation of the observed data is not required by the operations in this equation.

The integral equation for $\Theta(\xi; \omega)$ deduced from (8.3.7) is

$$\Theta(\xi; \omega) = \frac{1}{2\pi c_0^3} \int \alpha(\mathbf{x}) \frac{e^{2i\omega|\mathbf{x} - \xi|/c}}{4\pi|\mathbf{x} - \xi|} dV. \tag{9.3.10}$$

[This integral equation, which, as we shall see, is amendable to solution by transform technique, is the consequence of the linearization process—the application of regular perturbation or Born approximation. The equation is written in the form of a convolution of the Green's function with another function. In order to solve this equation, we shall introduce the transverse Fourier transform in the variables ξ_1 and ξ_2. In this equation and later, we shall not introduce a new symbol for the Fourier transform of a function but only indicate the transform by the arguments of the function. (The reader will see later that too many "levels" of transform arise in this problem to create a new symbol for each of them.) Thus, we define the transform by

$$\Theta(k_1, k_2; \omega) = \int_{-\infty}^{\infty} \Theta(\xi; \omega) \exp(-2i\{k_1\xi_1 + k_2\xi_2\}) d\xi_1 d\xi_2. \tag{9.3.11}$$

The extra factor of 2 in the phase here will prove to be a convenience.

We now apply this Fourier transform to (9.3.10) to obtain an integral equation for $\alpha(k_1, k_2, x_3)$. We remark that the transverse Fourier transform of the Green's function is a solution of the one-dimensional Green's function problem (8.1.4), with the appropriate wave number. We leave it to the exercises to verify that the Fourier transform of (9.3.10) is the equation

$$-8\pi i c_0^3 k_3 \Theta(k_1, k_2; \omega) = \int_{-\infty}^{\infty} \alpha(k_1, k_2, x_3) e^{2ik_3|x_3|} dx_3; \tag{9.3.12}$$

$$k_3 = \begin{cases} \text{sign } \omega \sqrt{(\omega^2/c_0^2) - k_1^2 - k_2^2}, & (\omega^2/c_0^2) \geq k_1^2 + k_2^2, \\ i\sqrt{k_1^2 + k_2^2 - (\omega^2/c_0^2)}, & (\omega^2/c_0^2) < k_1^2 + k_2^2. \end{cases} \tag{9.3.13}$$

The function $\alpha(\mathbf{x})$ and, therefore, also $\alpha(k_1, k_2, x_3)$ are assumed to be nonzero only for x_3 positive. Therefore, there is no need for the absolute value sign in (9.3.12). Then, if ω is restricted to the upper range in (9.3.13), the integral in (9.3.12) is the three-dimensional spatial Fourier transform of

$\alpha(\mathbf{x})$; that is,

$$\alpha(\mathbf{k}) = -8\pi i c_0^3 k_3 \Theta(k_1, k_2; \omega), \qquad \mathbf{k} = (k_1, k_2, k_3), \qquad (9.3.14)$$

with k_3 defined by the *dispersion relation*, which is the upper choice in (9.3.13).

Fourier inversion of this result yields $\alpha(\mathbf{x})$. However, for field data, the solution is not quite so straightforward. A typical frequency range for field data might be 8–40 Hz; a typical sound speed 1500–6000 m/s; a typical length scale to the first significant variations in $\alpha(\mathbf{x})$, 500 m. For these values, we obtain a dimensionless parameter

$$\lambda = 4\pi f_0 H/c \approx 9$$

at minimum frequency and maximum sound speed, large enough for asymptotics. Thus, we contemplate not attempting to find $\alpha(\mathbf{x})$ but only its *discontinuities*; that is, we seek the *reflectors* in the subsurface. We remark that the principal radii of curvature of the interfaces are typically on the order of hundreds of meters as well. Therefore, asymptotics are valid as regards this parameter, also.

A weakness in the method arises when we consider the effect of many (more than one) layers. The incident wave refracts at the first interface, so that the representation for $u_I(\mathbf{x}; \boldsymbol{\xi}; \omega)$ is no longer valid. In addition, there are multiple reflections for which our model does not account at all. Indeed, implicit in the linearization process in which we neglected the product $\alpha(\mathbf{x})u_S(\mathbf{x}; \boldsymbol{\xi}; \omega)$ is the assumption that each point in the subsurface acts as an independent point scatterer producing only a singly reflected upward scattered signal. Nonetheless, we argue that consistent with the linearization process itself, we can assume that the reflections are of small enough amplitude that multiply reflected waves are small—at least, $O(\alpha^2(\mathbf{x}))$—and that, similarly, refraction effects are negligible. Of course, these assumptions will not always be justified. However, they *are* justified often enough for us to continue.

Let us assume, therefore, that $v(\mathbf{x})$ is a piecewise constant function, so that the same is true for $\alpha(\mathbf{x})$. In this case, $\alpha(\mathbf{x})$ is a sum of characteristic functions, each nonzero on some domain of constant $v(\mathbf{x})$ and each weighted by the constant value of $\alpha(\mathbf{x})$ on that domain. From our discussion of Section 9.1, we know, then, how to delineate the boundaries of the regions of constant $\alpha(\mathbf{x})$ from the band-limited Fourier data: we deduce from (9.1.26) that we must multiply the Fourier transform $\alpha(\mathbf{k})$ by $2ik$ sign k_3 in order to produce a scaled singular function from the scaled characteristic function. The extra multiple of 2 here arises from the extra factor of 2 in the definition of the spatial Fourier transform (9.3.11). After Fourier inversion, this will produce an array of band-limited Dirac delta functions with support on the interfaces between the layers of constant velocity.

In summary, we define

$$\beta(\mathbf{k}) = 2ik(\operatorname{sign} k_3)\alpha(\mathbf{k}) = (2i\omega/c_0)\alpha(\mathbf{k}). \tag{9.3.15}$$

The inverse transform of $\beta(\mathbf{k})$ is then an array of singular functions, each with support on an interface of the piecewise constant function $\alpha(\mathbf{x})$. The weighting of the singular function on each interface is the linearized jump in $\alpha(\mathbf{x})$ across the interface in the direction of increasing x_3.

Let us define v_+ (v_-) to be the value of $v(\mathbf{x})$ below (above) an interface and define α_+ correspondingly. Then the linearized jump in $\alpha(\mathbf{x})$ is determined by linearizing the difference in the values of $\alpha(\mathbf{x})$ through the definition (9.3.2). That is,

$$\alpha_- - \alpha_+ = \frac{c_0^2}{v_-^2} - 1 - \left[\frac{c_0^2}{v_+^2} - 1\right] = c_0^2 \frac{v_+^2 - v_-^2}{v_+^2 V_-^2} \approx 2\frac{v_+ - v_-}{c_0}. \tag{9.3.16}$$

To the same linear order, the reflection coefficient at the interface agrees with this result except that the factor of 2 must be replaced by $\frac{1}{2}$. That is, the change in $\alpha(\mathbf{x})$ across the interface is just 4 times the reflection coefficient, and therefore the inverse Fourier transform of $\beta(\mathbf{k})$ is just 4 times the *reflectivity function*.

Let us define $\beta(\mathbf{y})$ to be the *band-limited* Fourier inversion of $\beta(\mathbf{k})$ and $R_n\gamma_B(\mathbf{y})$ to be the entire array of reflectivity functions of the interfaces in the subsurface. Then we can summarize our discussion as

$$4R_n\gamma_B(\mathbf{y}) = \beta(\mathbf{y}) = \frac{16c_0^2}{\pi^2}\int \omega k_3\, dk_1\, dk_2\, dk_3\, \Theta(k_1, k_2; \omega)$$

$$\cdot \exp(2i\{k_1 y_1 + k_2 y_2 - k_3 y_3\}) \tag{9.3.17}$$

In this Fourier inversion, we have accounted for the multiple of 2 in the transform variables by dividing by π^3 rather than $(2\pi)^3$; we have also properly accounted for the reversal in sign in the Fourier inversion in k_3 as compared to k_1 and k_2. The domain of integration is all real values of k_1, k_2, k_3 available in the bandwidth of frequencies available in the data. For $\alpha(\mathbf{k})$, we have used (9.3.14).

Since we have specialized our result to high frequency and we intend to interpret our results in terms of the leading order asymptotic expansion of the singular function, we should, as well, only use the leading order high-frequency approximation of $\Theta(k_1, k_2; \omega)$. To do this, we must retain only the leading order term in ω in (9.3.9); that is,

$$\Theta(\xi; \omega) \sim -\frac{i}{\omega^2}\frac{\partial}{\partial\omega}u_S(\mathbf{x}; \xi; \omega) = \int_0^\infty tU_S(\xi, t)e^{i\omega t}\, d\omega. \tag{9.3.18}$$

In the last part of the equations we have used the notation $U_S(\xi, t)$ to denote the time domain backscatter response. We use this result and (9.3.11) in (9.3.17) to obtain

$$4R_n\gamma_B(\mathbf{y}) = \beta(\mathbf{y}) = \frac{16c_0^2}{\pi^2} \int \frac{k_3}{\omega} \, dk_1 \, dk_2 \, dk_3 \int d\xi_1 \, d\xi_2 \, dt \, t U_S(\xi, t)$$

$$\cdot \exp(2i[k_1(y_1 - \xi_1) + k_2(y_2 - \xi_2) - k_3 y_3] + i\omega t).$$

$$(9.3.19)$$

This formula provides a basis for processing band-limited, aperture-limited backscatter observations for imaging of reflectors in the earth.

As a simple check on this result, let us consider the case in which the subsurface has only one planar discontinuity at depth H across which the velocity changes from c_0 to c_1. From our ray theory, we expect that the leading order response to an impulsive point source for this problem to be just an impulse that has propagated down to the interface and back up again a distance $2H$ but has been modified by the normal reflection coefficient at the interface. That is,

$$U_S(\xi, t) = R[\delta(t - 2H/c)/8\pi H], \qquad R = (c_1 - c_0)/(c_1 + c_0), \quad (9.3.20)$$

which is independent of ξ. We leave it to the exercises to confirm that the calculation of $\beta(\mathbf{y})$ proceeds as follows. The t integration is carried out by exploiting the delta function. The integrations in ξ_1 and ξ_2 yield delta functions in k_1 and k_2, which, in turn, allow the integrations in those two variables to be carried out explicitly. The result after these computations is

$$4R_n\gamma_B(\mathbf{y}) \sim 4R \frac{1}{\pi} \int dk_3 \, e^{2ik_3\{H - y_3\}}. \qquad (9.3.21)$$

The multiplier of $4R$ on the right is recognized as a band-limited Dirac delta function. Since the delta functions in k_1 and k_2 evaluated those variables at zero, the dispersion relation (9.3.13) implies that $k_3 = \omega/c$, and therefore this is exactly the band-limited delta function predicted by the theory.

In the exercises, another example is presented in which it is possible to carry out the integrations analytically.

TWO AND ONE-HALF DIMENSIONS

In seismic exploration, it is often the case that data are not gathered over an entire planar array on the upper surface but only along a line. To compensate for the lack of data in the direction orthogonal to that line, it is assumed that the velocity variations in the subsurface are independent of that orthogonal variable. The experimental configuration for this situation is often referred to as *two-and-one-half-dimensional*, since the variations in the velocity are assumed to be two-dimensional while the propagation model is three-dimensional.

A formula for the two-and-one-half-dimensional case is easily deduced from the result (9.3.14). Let us suppose that the velocity variations do not depend on ξ_2. In this case, if experiments were carried in the direction ξ_2, the backscattered signal, or time traces, U_S would be independent of ξ_2. Consequently, the only ξ_2 dependence in the integrand in (9.3.14) is in the phase. Thus, the ξ_2 integration yields the result $\pi\delta(k_2)$, in which case the integration in k_2 can be carried out. The result for two and one-half dimensions, therefore, is

$$4R_n\gamma_B(y_1, y_3) = \beta(y_1, y_3) = \frac{16c_0^2}{\pi} \int \frac{k_3}{\omega} dk_1 \, dk_3 \int d\xi_1 \, dt$$

$$\cdot t U_S(\xi_1, t) \exp(2i[k_1(y_1 - \xi_1) - k_3 y_3] + i\omega t),$$

$$\omega = c_0 \operatorname{sign} k_3 \sqrt{k_1^2 + k_3^2}. \tag{9.3.22}$$

Computer algorithms to implement (9.3.19) and (9.3.22) have been developed. Both have been used on *synthetic data* (data generated numerically) and have proved quite accurate, both for reflector imaging and reflection strength estimation. The two-and-one-half-dimensional algorithm has also been used on field data and imaged reflectors from that data.

For the field-data examples, the data were not "true amplitude" data whose source spectrum was consistent and well defined. However, as in the preceding section, we remark that amplitude inaccuracies largely affect only the estimation of reflection strength but do not have a significant effect on the location of reflectors.

For the field data, it quickly becomes apparent that a model with a *constant* reference speed throughout the earth is not adequate. However, we can exploit the *linearity* of our model to consider the subsurface to be a superposition of domains for each of which the reference speed is constant although not necessarily the *same* constant in all of them. In practice, this is what is done.

The question then arises of how one chooses this "local" reference speed. Quite often, there is a priori information about the velocity. The time record itself also supplies a means of doing this. Consider, for example, the time record shown in Fig. 8.15 as a prototype of a time record from a discontinuous interface. In particular, the locus of peaks on the *diffraction tail* of that figure lies on a hyperbola in the space–time plot. This is verified for a point scatterer in the exercises. The edge of the half plane is indeed a point scatterer. The slope of the asymptote to that diffraction tail is a function of the sound speed. Thus, a best guess at a reference speed is provided by the sound speed that best fits the diffraction tails locally in the time plot, under the assumption that the depth at which such a sound speed c is to be used is approximately $ct/2$.

MIGRATION

The dominant method for reflector imaging from seismic data in geophysics today is known as *migration*. This term comes from the idea that the space–time record, such as in Fig. 8.14, is itself a subsurface plot with the reflectors imaged at their *temporal* location below a source/receiver point rather than at their *spatial* location. Thus, the objective of any subsurface imaging method might be viewed as moving or *migrating* the reflectors from their temporal location to their spatial location.

Migration based on the wave equation was introduced by Claerbout [1971, 1976] and his associates. The method is based on the observation that the scattering due to a reflector behaves much like a source distribution on the reflecting surface in the frequency domain or like Cauchy data restricted to the reflecting surface in the time domain. The former of these is suggested by the integral equation (8.4.5), while the latter is somewhat more obscure, from exact results. A feature common to both interpretations is that the propagation speed of this source distribution or initial data is $v/2$, half of the propagation speed of the medium. The method further assumes that the ensemble of backscatter responses is itself a wave whose propagation is governed by the wave equation in the time domain or the Helmholtz equation in the frequency domain. The reflectors in the subsurface are, therefore, imaged by tracing the data back to their origin; the support of the source in the frequency domain or the support of the initial data in the time domain is the reflecting surface(s). As with the discussion of inversion earlier, we start with a linearized theory with a constant reference speed and then "bootstrap" to variable reference speed through various devices.

The first premise of wave equation migration is strongly supported by the representations arrived at via the Kirchhoff approximation (8.4.10) or the Born approximation (9.3.10). The second premise will be confirmed later as a *leading order high-frequency* asymptotic result.

More recently, migration techniques based on Fourier analysis [Stolt, 1978] and the Kirchhoff representation [Schneider, 1978] have been introduced.

All three migration methods successfully image the subsurface. However, in these methods, while it is assumed that the support of the source distribution is the reflector, the relationship between the source strength and the reflection strength is not established. Nonetheless, these methods are all demonstrably reflector imaging methods.

MIGRATION DEDUCED FROM THE KIRCHHOFF APPROXIMATION

We shall show now that the basic premises of wave equation migration can be deduced from the Kirchhoff representation of the backscattered field in a constant background medium. To do this, we begin from that Kirchhoff representation (8.4.10). We assume that L is the reflecting surface we wish to image. We introduce

$$w(\boldsymbol{\xi}; \omega) = -i\omega \frac{\partial}{\partial \omega} \left[\frac{u_S(\mathbf{x}; \boldsymbol{\xi}; \omega)}{\omega} \right] \sim \frac{1}{i} \frac{\partial}{\partial \omega} u_S(\mathbf{x}; \boldsymbol{\xi}; \omega). \qquad (9.3.23)$$

We remark that the kernel of the Kirchhoff approximate integral representation for this function, deduced from (8.4.10), very nearly has the free-space Green's function *with sound speed* $c/2$ as integrand:

$$w(\boldsymbol{\xi}; \omega) = \frac{i\omega}{\pi c^2} \int_L R\hat{\mathbf{n}} \cdot \hat{\mathbf{r}} \frac{e^{2i\omega r/c}}{4\pi r} \, dS. \qquad (9.3.24)$$

We take the point of view now that $\boldsymbol{\xi}$ is a backscatter point in three-space—$\boldsymbol{\xi} = (\xi_1, \xi_2, \xi_3)$—and that $w(\boldsymbol{\xi}; \omega)$ is known on the surface $\xi_3 = 0$. We shall apply the Helmholtz operator with sound speed $c/2$ to this representation. Before doing so, however, let us consider the asymptotic order in ω of the function $w(\boldsymbol{\xi}; \omega)$. If we think of applying the method of stationary phase to the integral in (9.3.24), then we find that $w(\boldsymbol{\xi}; \omega) = O(1)$ in ω. Consequently, thinking of $w(\boldsymbol{\xi}; \omega)$ as represented by a wave series as in (8.2.2), we know that the leading order terms after applying the Helmholtz operator to $w(\boldsymbol{\xi}; \omega)$ will be $O(\omega^2)$ and $O(\omega)$, as in (8.2.3). By retaining these two orders, we shall have enough information to estimate the phase and leading order amplitude of $w(\boldsymbol{\xi}; \omega)$ or, equivalently, a wave equation that will produce this phase and leading order amplitude. Therefore, in the remainder of this discussion, we neglect terms that do not contribute to the two orders indicated and find that

$$\left[\nabla_\xi^2 + \left[\frac{2\omega}{c} \right]^2 \right] w(\boldsymbol{\xi}; \omega)$$

$$= \frac{i\omega}{\pi c^2} \int_L \left[-\delta(\mathbf{x} - \boldsymbol{\xi}) R\hat{\mathbf{n}} \cdot \hat{\mathbf{r}} - \frac{2i\omega}{c} \hat{\mathbf{r}} \cdot \mathbf{V}_\xi [R\hat{\mathbf{n}} \cdot \hat{\mathbf{r}}] \frac{e^{2i\omega r/c}}{r} \right] dS. \qquad (9.3.25)$$

The subscript ξ on the differential operators is to remind the reader that the differentiations are to be carried out with respect to $\boldsymbol{\xi}$.

The expression $\hat{\mathbf{r}} \cdot \mathbf{V}_3 [R\hat{\mathbf{n}} \cdot \hat{\mathbf{r}}]$ is equal to zero. To see why, note first, from Exercise 8.22a, that R is a function of $\hat{\mathbf{n}} \cdot \hat{\mathbf{r}}$. The vector $\hat{\mathbf{n}}$ is independent of $\boldsymbol{\xi}$; it is a function of $\mathbf{x}$. The vector $\hat{\mathbf{r}}$ is a unit vector whose gradient is orthogonal to $\hat{\mathbf{r}}$. Thus, $\hat{\mathbf{r}} \cdot \mathbf{V}_\xi \hat{\mathbf{r}} = 0$ and so does $\hat{\mathbf{r}} \cdot \mathbf{V}_\xi [\hat{\mathbf{n}} \cdot \hat{\mathbf{r}}]$. Since R depends on $\boldsymbol{\xi}$ totally through $\hat{\mathbf{n}} \cdot \hat{\mathbf{r}}$, the bracketed expression in the second line has zero derivative.

Thus, let us focus our attention on the first line in (9.3.25). We need only concern ourselves with points near the surface S, where we may introduce a local coordinate system to simplify our analysis. Let $\boldsymbol{\xi}$ be close enough to S so that we can uniquely identify a point P on S for which the normal through

P also passes through ξ. Introduce a local coordinate system in which one coordinate is distance along that normal and the other two coordinates are parameters on S along lines of principal curvature from P. The three-dimensional Dirac delta function is equally a three-dimensional delta function in these new coordinates or a product of three one-dimensional delta functions in these coordinates.

The pair of delta functions in the surface variables simply evaluates the surface integrand at the point P. [Had we not made the coordinates arc-length variables at P, we would have had $dS = \sqrt{g}\, d\sigma_1\, d\sigma_2$, but the surface product of delta functions would have had a compensating factor of $1/\sqrt{g}$ arising from the rule (2.1.16) for transformation of coordinates for delta functions.]

At P, $\hat{\mathbf{r}} = -\hat{\mathbf{n}}$ and $R = R_n$, the normal reflection coefficient given by $[c_1 - c]/[c_1 + c]$ if we define c_1 as the velocity below the interface. Thus, we conclude that

$$\left[\nabla_\xi^2 + \left[\frac{2\omega}{c}\right]^2\right] w(\xi; \omega) = \frac{i\omega}{\pi c^2} R_n \delta(r_n) = \frac{i\omega}{\pi c^2} R_n \gamma(\xi). \tag{9.3.26}$$

The product $R_n \delta(r_n)$ is just the *reflectivity function* Thus, we have verified that the function $w(\xi; \omega)$, related to $u_S(\mathbf{x}; \xi; \omega)$ through (9.3.23), does, indeed, satisfy the Helmholtz equation with a source distribution with support on the reflecting surface. Furthermore, we have quantified the relationship between the source strength and the reflection strength.

Let us define $W(\xi, t)$ as the inverse Fourier transform of $w(\xi; \omega)$. Then we can think of the problem for $w(\xi; \omega)$ as having been deduced from the problem for $W(\xi, t)$, similar to the result (4.3.2). By comparing the right side of (4.3.2) with the right side of (9.3.26), we conclude that a time domain problem for $W(\xi, t)$ equivalent to the problem for $w(\xi; \omega)$ is

$$\left[\nabla_\xi^2 - \frac{1}{c^2}\frac{\partial^2}{\partial t^2}\right] W(\xi, t) = 0, \qquad W(\xi, 0) = 0, \qquad W_t(\xi, 0) = \frac{R_n \gamma(\xi)}{\pi}. \tag{9.3.27}$$

In the frequency domain, we find the reflectivity function by solving an *inverse source problem* with data prescribed on the surface $\xi_3 = 0$. In the time domain, we find the reflectivity function by solving an *inverse initial value problem*, again with data prescribed on the surface $\xi_3 = 0$.

In Exercise 6.22, we demonstrated that, in general, inverse source problems for the Helmholtz equation have nonunique solutions. However, that demonstration depended on the fact that the ω dependence—$k = \omega/c$ in that exercise—in the source was completely general and unknown. In the present case, the ω dependence is known and multiplicative. By using that fact and the fact that the source is located below the observation surface, we outline in the exercises a solution technique. The solution formula is exactly the result (9.3.19).

The inverse initial value problem is a *Cauchy problem* for the wave equation, with only one function $W(\xi_1, \xi_2, 0, t)$ given on the upper surface. In Sections 3.2 and 5.1, we have shown that Cauchy problems, with data given on such *timelike* surfaces, have *ill-conditioned* solutions that grow exponentially. The solution formula (9.3.19) does not exhibit such behavior. There are two reasons for this. First, by restricting the dispersion relation (9.3.13) to the upper choice, we have *regularized* the solution and eliminated the exponentially growing components of the solution. Of course, in that formulation, where we were not tracing a wave, we justified such regularization by the fact that $\gamma(\mathbf{x})$ depends on its Fourier superposition over real values of $\mathbf{k}$ only. In addition, we replaced $|\mathbf{x}_3|$ in (9.3.12) by x_3 below that equation. We argued that $\alpha(\mathbf{x})$ was zero for x_3 negative. That is equivalent in the present analysis to assuming that the wave $W(\xi, t)$ is *upward propagating* only. In this case, with $\alpha(\mathbf{x})$ zero near $x_3 = 0$, we could argue that in the Fourier domain,

$$w(k_1, k_2, x_3; \omega) = A(k_1, k_2; \omega)e^{-ik_3 x_3} \qquad (9.3.28)$$

for x_3 near zero and, therefore, that

$$\frac{\partial w(k_1, k_2, 0; \omega)}{\partial x_3} = -ik_3 w(k_1, k_2, 0; \omega). \qquad (9.3.29)$$

Once this derivative is known in the Fourier domain, the inverse transform provides the value in the space–time domain. Thus, our assumptions about the nature of the solution $\alpha(\mathbf{x})$ are equivalent to completing the Cauchy data and regularizing the solution of the migration formulation.

CONSEQUENCES OF A SOLUTION BASED ON THE
KIRCHHOFF APPROXIMATION

We now have a solution technique based on the Kirchhoff approximation rather than on the Born approximation. That result has important consequences. In particular, we are no longer so strongly tied to a small reflection coefficient at the reflector, since that constraint does not apply to the Kirchhoff approximation. We are, however, still constrained by the requirement that the reference velocity be a good guess to account adequately for the effects of layers above the reflector in question. This is a matter of ongoing research. However, given that caveat, for a reflection coefficient of any size, we can now say that the processing (9.3.19) or (9.3.22) yields an estimate of the true reflection strength

$$R_n \gamma_B(\mathbf{y}) = [(v_+ - v_-)/(v_+ + v_-)]\gamma_B(\mathbf{y}), \qquad (9.3.30)$$

with v_+ being the relevant velocities on either side of each reflector where the theory is applied.

We derived the result (9.3.19) by starting with a linearized integral representation (9.3.8) of the backscattered field based on the Born approximation. The question arises then of whether one can deduce such a formula starting from the Kirchhoff approximation. The answer is that this cannot be done quite so simply, although our derivation via the mechanism of first introducing a differential equation (9.3.26) certainly is equivalent to this process.

An examination of the Kirchhoff representation of the backscattered wave (8.4.10) reveals the source of the difficulty: the dependence of the product $R\hat{n} \cdot \hat{r}$ on the variable ξ precludes a straightforward Fourier inversion of that equation, as was carried out for (8.3.9). However, we note that at the specular points, which dominate the backscatter response, $\hat{n} \cdot \hat{r} = -1$ and $R\hat{n} \cdot \hat{r} = -R_n$. Indeed, the Green's function representation of the solution of (9.3.26) reflects this approximation. In neglecting lower-order terms when deriving (9.3.25) and, hence, (9.3.26), we have effectively made the asymptotic approximation of replacing $R\hat{n} \cdot \hat{r}$ by $-R_n$.

We can, nonetheless, deduce a result such as (9.3.19) directly from the Kirchhoff integral representation (8.4.10). The derivation is carried out in the exercises.

Exercises

9.10 Consider the one-dimensional inverse problem

$$u''(z; \omega) + \frac{\omega^2}{v^2(z)} u(z; \omega) = -\delta(z), \qquad \text{where prime means } \frac{d}{dz},$$

$$\frac{1}{v^2(z)} = \frac{1}{c_0^2}, \qquad z \leq 0,$$

$$\frac{1}{v^2(z)} = \frac{1}{c_0^2}[1 + \alpha(z)], \qquad z > 0.$$

(a) Set

$$u(z; \omega) = u_1(z; \omega) + u_S(z; \omega),$$

with

$$u_1''(z; \omega) + (\omega^2/c_0^2)u_1(z; \omega) = -\delta(z).$$

Show that

$$u_S''(z; \omega) + (\omega^2/c_0^2)u_S(z; \omega) = -(\omega^2/c_0^2)\alpha(z)[u_1(z; \omega) + u_S(z; \omega)]$$

and that the linearized equation for $u_S(0; \omega)$ has the solution

$$u_S(0; \omega) = \omega^2 \int_0^\infty \frac{\alpha(z)}{c_0^2} u_I^2(z; \omega) \, dz.$$

(b) For $c_0 = \text{const}$, show that

$$u_S(0; \omega) = -\frac{1}{4} \int_0^\infty \alpha(z) e^{2i\omega z/c_0} \, dz = -\frac{\bar{\alpha}(2i\omega/c)}{4},$$

with solution

$$\alpha(z) = -\frac{4}{\pi c_0} \int u_S(0; \omega) e^{-2i\omega z/c_0} \, d\omega = -4U_S\left(0, \frac{2z}{c_0}\right),$$

with $U_S(0, t)$ the inverse transform to the time domain of $u_S(0; \omega)$.

(c) Show that for a one-dimensional reflectivity function, defined as in this section,

$$R_{\gamma B}(z) = -\frac{2i}{\pi c_0^2} \int \omega u_S(0; \omega) e^{-2i\omega z/c_0} \, d\omega.$$

9.11   For the same problem as in Exercise 9.10, introduce the new independent variable τ defined by

$$\tau = \int_0^z \frac{dz'}{v(z')},$$

and set

$$u(z(\tau); \omega) = w(\tau; \omega), \qquad v(z(\tau)) = c(\tau).$$

(a) Show that the problem for $w(\tau; \omega)$ is

$$\ddot{w}(\tau; \omega) + \omega^2 w(\tau; \omega) - \Gamma(\tau)\dot{w}(\tau; \omega) = -v(0)\delta(\tau),$$

$$\Gamma(\tau) = \frac{\dot{c}(\tau)}{c(\tau)}, \qquad \text{where the overdot means } \frac{d}{d\tau}.$$

(b) Set

$$w(\tau; \omega) = w_1(\tau; \omega) + w_S(\tau; \omega), \qquad \ddot{w}_1(\tau; \omega) + \omega^2 w_1(\tau; \omega) = -v(0)\delta(\tau).$$

Show that the linearized equation for $w_S(\tau; \omega)$ has the solution

$$w_S(0; \omega) = \frac{iv(0)}{4\omega} \int_0^\infty \Gamma(\tau) e^{2i\omega\tau} \, d\tau,$$

with inversion

$$\Gamma(\tau) = -\frac{4i}{\pi v(0)} \int_{-\infty}^\infty \omega w_S(0; \omega) e^{-2i\omega\tau} \, d\omega.$$

The implicit solution for $v(z)$ is then given by

$$c(\tau) = v(0)e^{\Gamma(\tau)}, \qquad z = \int_0^\tau c(\tau')\,d\tau', \qquad v(z(\tau)) = c(\tau).$$

The solution $\Gamma(\tau)$ has been shown to produce a more accurate reconstruction of the velocity than the solution $\alpha(z)$.

9.12 In Section 8.4, it was shown that the width of the main lobe of a band-limited delta function is $t = 1/[f_1 + f_0]$. Suppose two nearby reflectors are such that the zeros of their main lobes just touch. Show that this corresponds to a layer width between the the reflectors of $c/[f_1 + f_0]$ and that for the values $f_0 = 10$ Hz, $f_1 = 40$ Hz, and $c = 3000$ m/s, this yields a layer width of 60 m. This is an example of the constraints on resolution of nearby reflectors.

9.13 Follow the discussion after (9.3.20) to carry out the calculations that yield (9.3.21).

9.14 Suppose that the backscattered data from an array of experiments are given by

$$U_S(\xi, t) = R\frac{a\delta(t - 2[\rho - a]/c_0)}{8\pi[\rho - a]\rho},$$

$$\xi = (\xi_1, \xi_2, 0), \qquad \xi^2 = \xi_1^2 + \xi_2^2, \qquad \rho = \sqrt{H^2 + \xi^2}.$$

(a) Use the ray method to show that this is the leading order back-scatter response from a spherical domain of radius a, centered at depth H, with reflection coefficient R.

(b) Let Ra remain finite while $a \to 0$. Show that in this limit, the arrivals or *events* at the upper surface occur at the times $t = 2\rho/c$ at the distance ρ from the origin on the upper surface. Plot the surface of arrivals in a space–time plot in which the vertical axis is $ct/2$ and the horizontal axes are the coordinates ξ_1 and ξ_2. Confirm that the surface is one sheet of a two-sheeted hyperboloid of revolution, which has as asymptote the right circular cone with opening angle whose tangent is $2\rho/ct$.

(c) Return to nonzero a Substitute the data for $U_S(\xi, t)$ into the right side of (9.3.19) and obtain the left side asymptotically. Proceed to carry out the calculation as follows. Use the Dirac delta function to compute the time integral. Scale the variables $\mathbf{k}$ by $k = |\omega|/c_0$, and introduce the polar coordinates θ and ϕ:

$$k_1 = k\cos\phi\sin\theta, \qquad k_2 = k\sin\phi\sin\theta, \qquad k_3 = k\cos\theta.$$

Now carry out the integrations in θ, ϕ, and ξ by the method of stationary phase. The remaining integral in k will be recognized as the band-limited reflectivity function multiplied by 4.

9.15 In (9.3.19), use the dispersion relation to rewrite that integral as an integral with respect to ω.

(a) Interpret the multiplication by t as a derivative with respect to ω and integrate by parts. Rewrite the result as an integral in k_3 to obtain

$$\beta(\mathbf{y}) = \int U_S(\boldsymbol{\xi}, t) B(\mathbf{y} - \boldsymbol{\xi}, t) \, d\xi_1 \, d\xi_2 \, dt,$$

with

$$B(\mathbf{y}, t) = \frac{32 y_3}{\pi^2} \int dk_1 \, dk_2 \, dk_3 \, \exp(2i \{ k_1 y_1 + k_2 y_2 - k_3 y_3 \} + i\omega t).$$

In this form, the reflectivity is seen to be an operator on U_S itself rather than on $t U_S$.

(b) Consider the representation (2.4.6) of $U(\mathbf{x}, t)$, which was shown in (2.4.13) to be the three-dimensional Green's function $G(\mathbf{x}, t)$. Rewrite the denominator of the integrand as $(ck \, \text{sign} \, k_3)^2 - \omega^2$, and calculate the integral as a residue sum at the zeros of this expression to obtain

$$G(\mathbf{x}, t) = G_+(\mathbf{x}, t) + G_-(\mathbf{x}, t),$$

$$G_\pm(\mathbf{x}, t) = \pm \frac{ic}{16\pi^3} \int \frac{dk_1 \, dk_2 \, dk_3}{k \, \text{sign} \, k_3} \, e^{i\{\mathbf{k} \cdot \mathbf{x} \mp ckt \, \text{sign} \, k_3\}}, \qquad t > 0.$$

[*Remark*: The k_3 dependence in the phase is $[|k_3| x_3 \mp ckt] \, \text{sign} \, k_3$. That is, the Green's function is decomposed by this device into a down-going wave G_+ and an up-going wave G_-.]

(c) In G_+, differentiate with respect to t and make the change of variable of integration from k_3 to $-k_3$ to obtain the result

$$\frac{\pi}{c^2} \frac{\partial G_+(2\mathbf{y}, t)}{\partial t} = \frac{2}{y_3} B(\mathbf{y}, t).$$

9.16 (a) Write the Green's function representation of the solution to (9.3.26) under the assumption that the medium is homogeneous, with constant sound speed c. Set $\xi_3 = 0$ in that solution, and take the Fourier transform in the transverse variables, as in (9.3.11). Obtain the result

$$w(k_1, k_2, 0; \omega) = (\omega/2\pi c^2 k_3) R_n \bar{\gamma}(\mathbf{k}),$$

with k_3 defined by (9.3.13). In this result, it should also have been necessary to identify $|y_3| = y_3$ as in this section.

(b) Show that this solution for $R_n \gamma_B(\mathbf{y})$ agrees with (9.3.17).

9.17 Rewrite the result of the preceding exercise as

$$\beta(\mathbf{y}) = \int u_S(\mathbf{x}; \boldsymbol{\xi}; \omega) b(\mathbf{y} - \boldsymbol{\xi}) \, d\xi_1 \, d\xi_2,$$

with

$$b(\mathbf{y}) = \frac{32y_3}{\pi^2} \int dk_1\, dk_2\, dk_3 \exp(2i\{k_1 y_1 + k_2 y_2 - k_3 y_3\}).$$

Use for $u_S(\mathbf{x}; \xi; \omega)$ the Kirchhoff approximate backscatter field (8.4.10), thereby obtaining a sevenfold integral in ξ, $\mathbf{k}$, and $\boldsymbol{\sigma}$ for $\beta(\mathbf{y})$. Introduce the polar transformation of Exercise 9.14. Calculate the integrals in θ and ϕ by stationary phase. Now calculate the integrals in ξ_1 and ξ_2 by the method of stationary phase. Finally, calculate the integrals in σ_1 and σ_2 by the method of stationary phase. In this last stationary phase, the previous conditions of stationarity make the chain rule differentiations of the phase with respect to ξ at the stationary point easier to calculate. The stationarity conditions require that $\hat{\mathbf{k}}$, $\mathbf{y} - \mathbf{x}$, and $\xi - \mathbf{x}$ all coalign along the normal from the scattering surface through the output point $\mathbf{y}$. The actual output is $\beta(\mathbf{y})$ scaled by the ratio of distances along the normal from $\mathbf{x}$ to ξ and from $\mathbf{x}$ to $\mathbf{y}$. Note that this ratio is equal to unity on the scattering surface.

References

Berkhout, A. J. [1980]. "Seismic Migration: Imaging of Acoustic Energy by Wave Field Extrapolation." Developments in Solid Earth Geophysics, Vol. 12. Elsevier, New York.

Bleistein, N. [1976]. Physical optics farfield inverse scattering in the time domain, J. Acoust. Soc. Am. 60, 1249–1255.

Bojarski, N. N. [1967]. Three Dimensional Electromagnetic Short Pulse Inverse Scattering. Spec. Proj. Lab. Rep. Syracuse Univ. Res. Corp., Syracuse, New York.

Bojarski, N. N. [1968]. Electromagnetic Inverse Scattering Theory. Spec. Proj. Lab. Rep. Syracuse Univ. Res. Corp., Syracuse, New York.

Bojarski, N. N. [1982]. A survey of the physical optics inverse scattering identity, IEEE Trans. Antennas Propag. AP-30, 980–989.

Claerbout, J. F. [1971]. Toward a unified theory of reflector mapping, Geophysics 36, 467–481.

Claerbout, J. F. [1976]. "Fundamentals of Geophysical Data Processing." McGraw-Hill, New York.

Cohen, J. K., and Bleistein, N. [1979]. The singular function of a surface and physical optics inverse scattering, Wave Motion 1, 153–161.

Cohen, J. K., and Bleistein, N. [1979]. Velocity inversion procedure for acoustic waves, Geophysics 44, 1077–1087.

Lewis, R. M. [1970]. Physical optics inverse diffraction, IEEE Trans. Antennas Propag. AP-17, 308–314. [Correction, AP-18, 194.]

Mager, R. D., and Bleistein, N. [1978]. An examination of the limited aperture problem of physical optics inverse scattering, IEEE Trans. Antennas Propag. AP-26, 695–699.

Schneider, W. A. [1978]. Integral formulation for migration in two and three dimensions, Geophysics 43, 49–76.

Stolt, R. H. [1978]. Migration by Fourier transform, Geophysics 43, 23–48.

INDEX

A

Adjoint operator, 121
Anticlinal, 293
Ascent, direction, path, 212
Asymptotic sequence, 70
Asymptotically equal, 69
Auxiliary sequence, 71

B

Bessel functions, 196, 198
Bicharacteristics, 101
Bojarski's POFFIS identity, 319
Boundary curves, between hill and valley, 213
Buried focus, 295

C

Cauchy data, 91
Cauchy–Kowaleski theorem, 91
Cauchy principal value, 59
Cauchy problem, 91
Causality, 104
Caustic, 25
Center of curvature, 287
Characteristic
    base curves, 2
    curves, 2
    function, 303
    initial data, 5
    point, 5
    speed, 101
    strip, 13

Characteristics, 101
    method of, 3, 16
Complete set of functions, 112
Conoidal solution, 20, 269
Creeping wave, 30, 276
Critical angle, 249
Critical points, 74, 83
Critically reflected, refracted, transmitted ray,
    29
Curvature, 287
Cylindrical wave, 31, 194

D

d'Alembert solution, 103
Delta function
    band-limited. 290
    cylindrical coordinates, 49
    spherical coordinates, 50
Descent, direction, path, 212
Diffusion equation, 93
Digamma function, 208
Dirac delta function, 46
Direct scattering problems, 160
Dirichlet boundary condition, 165
Dispersion relation, 134
Domain of dependence, 5, 104
Doublet, band-limited, 295

E

Edge-diffracted wave, 276
Edge diffraction, 34
    coefficient, 279

Edge of regression, 25
Eigenfunction expansions, 112
Eigenfunctions, 112
Eigenvalues, 112
Eikonal equation, 18, 27, 146, 258
Elliptic partial, differential equation, 93
Envelope solution, 7
Evanescent wave, 254, 276

F

Far-field scattering amplitude, 184
Fermat's principle, 34
Fourier transform, 53
    half, 54
    multidimensional, 61

G

General solution, 19
Geometrical optics, 33
Geometrical theory of diffraction, 33, 276
Goursat problem, 107
    characteristic, 108
Green's first identity, 166
Green's function, 130
    Helmholtz equation, 178
    Klein–Gordon equation, 157
    Laplace's equation, 172
Green's second identity, 166
Green's theorem, 166
Group speed, 136

H

Hadamard well-posedness, 91
Hankel function, 236, 239
Head wave, 30, 249
Heat equation, 93
Heaviside function, 48
Helmholtz equation, 92
    cylindrical coordinates, 194
    spherical coordinates, 197
Hill, 213
Hyperbolic partial differential equation, 93

I

Impedance boundary condition, 168
Incident wave, 21, 158

Index of refraction, 18
Initial boundary value problem, 94
Inverse scattering problem, 160
Inverse source problem, 160

J

Jordan's lemma, 64, 220

K

Kirchhoff approximation, 283
Klein–Gordon equation, 125

L

Laplace operator, 92
Laplace's equation, 92
Large O, 67
Lateral wave, 30, 249
Legendre functions, associated, 199
Legendre polynomials, 198
Limited aperture, 303

M

Migration, 329
Mixed boundary condition, 165
Monkey saddle, 216

N

Neumann boundary condition, 165
Neutralizer, van der Corput, 75
Nonradiating sources, 202

O

Order estimates, 68
Orthonormal set of functions, 112

P

Parabolic partial differential equation, 93
Phase speed, 134
Physical optics far-field inverse scattering, 312
Plane wave(s), 31, 192
POFFIS identity, 315
Potential equation, 92
Principal curvatures, 287

Principal directions, 287
Principal radii of curvature, 287

R

Radius of curvature, 287
Range of influence, 5, 104
Ray data, 268
  line source, 272
  point source, 270
  reflected wave, 275
  transmitted wave, 275
Ray equations, 259
Rays, 19
Reduced wave equation, 92
Reflected wave, 21, 248, 273
Reflection coefficient, 275, 280, 299
Reflectivity function, 314
Reflector imaging, seismic inverse problem, 323
Refracted wave, 253
Resolution, 302
Riemann function, 121, 124, 128

S

Saddle point, 216
  method, 228
Scattered wave, 158
Scattering, 31
  by a circular cylinder, 196
  by a half space, 241
  by a sphere, 200
Scattering amplitude, 285
Self-adjoint, 124
Sifting property, 46
Sign(x), 59
Singular function, 51, 302
  aperture-limited, 308
Slowness, 259
Small O, 68
Smooth-body diffracted wave, 276
Smooth-body diffraction, 30
Snell's law, 253
Sommerfeld radiation condition, 182

Spacelike initial surface, 148
Spectral density, 54
Spherical harmonics, 199
Spherical waves, 197
Stationary phase formula, 79, 80, 81
  multidimensional, 88
Stationary point, 77, 83
  simple, 84
Steepest descent, ascent
  directions, 212
  formulas, 226, 227, 228
  paths, 213
Steepest descents, method of, 230
Stokes lines, 139
Stokes phenomenon, 139, 230
Synclinal, 293

T

Timelike initial surface, 148
Totally reflected wave, 29
Transmission coefficient, 275, 280
Transmitted ray, 29
  wave, 28, 252, 273
Transport equation, 259, 261
Two and one-half dimensions, 327

U

Ultrahyperbolic partial differential equation, 93
Uniqueness
  Helmholtz equation, 169
  wave equation, 152

V

Valley, 213

W

Watson's lemma, 205
Wave equation, 92
Well-posedness, 91
WKBJ connection formulas, 142

Computer Science and Applied Mathematics

A SERIES OF MONOGRAPHS AND TEXTBOOKS

Editor
Werner Rheinboldt
University of Pittsburgh

HANS P. KÜNZI, H. G. TZSCHACH, AND C. A. ZEHNDER. Numerical Methods of Mathematical Optimization: With ALGOL and FORTRAN Programs, Corrected and Augmented Edition

AZRIEL ROSENFELD. Picture Processing by Computer

JAMES ORTEGA AND WERNER RHEINBOLDT. Iterative Solution of Nonlinear Equations in Several Variables

AZARIA PAZ. Introduction to Probabilistic Automata

DAVID YOUNG. Iterative Solution of Large Linear Systems

ANN YASUHARA. Recursive Function Theory and Logic

JAMES M. ORTEGA. Numerical Analysis: A Second Course

G. W. STEWART. Introduction to Matrix Computations

CHIN-LIANG CHANG AND RICHARD CHAR-TUNG LEE. Symbolic Logic and Mechanical Theorem Proving

C. C. GOTLIEB AND A. BORODIN. Social Issues in Computing

ERWIN ENGELER. Introduction to the Theory of Computation

F. W. J. OLVER. Asymptotics and Special Functions

DIONYSIOS C. TSICHRITZIS AND PHILIP A. BERNSTEIN. Operating Systems

A. T. BERZTISS. Data Structures: Theory and Practice, Second Edition

N. CHRISTOPHIDES. Graph Theory: An Algorithmic Approach

SAKTI P. GHOSH. Data Base Organization for Data Management

DIONYSIOS C. TSICHRITZIS AND FREDERICK H. LOCHOVSKY. Data Base Management Systems

JAMES L. PETERSON. Computer Organization and Assembly Language Programming

WILLIAM F. AMES. Numerical Methods for Partial Differential Equations, Second Edition

ARNOLD O. ALLEN. Probability, Statistics, and Queueing Theory: With Computer Science Applications

ELLIOTT I. ORGANICK, ALEXANDRA I. FORSYTHE, AND ROBERT P. PLUMMER. Programming Language Structures

ALBERT NIJENHUIS AND HERBERT S. WILF. Combinatorial Algorithms, Second Edition

AZRIEL ROSENFELD. Picture Languages, Formal Models for Picture Recognition

ISAAC FRIED. Numerical Solution of Differential Equations

ABRAHAM BERMAN AND ROBERT J. PLEMMONS. Nonnegative Matrices in the Mathematical Sciences

BERNARD KOLMAN AND ROBERT E. BECK. Elementary Linear Programming with Applications

CLIVE L. DYM AND ELIZABETH S. IVEY. Principles of Mathematical Modeling

ERNEST L. HALL. Computer Image Processing and Recognition

ALLEN B. TUCKER, JR. Text Processing: Algorithms, Languages, and Applications

MARTIN CHARLES GOLUMBIC. Algorithmic Graph Theory and Perfect Graphs

GABOR T. HERMAN. Image Reconstruction from Projections: The Fundamentals of Computerized Tomography

WEBB MILLER AND CELIA WRATHALL. Software for Roundoff Analysis of Matrix Algorithms

ULRICH W. KULISCH AND WILLARD L. MIRANKER. Computer Arithmetic in Theory and Practice

LOUIS A. HAGEMAN AND DAVID M. YOUNG. Applied Iterative Methods

I. GOHBERG, P. LANCASTER, AND L. RODMAN. Matrix Polynomials

AZRIEL ROSENFELD AND AVINASH C. KAK. Digital Picture Processing, Second Edition, Vol. 1, Vol. 2

DIMITRI P. BERTSEKAS. Constrained Optimization and Lagrange Multiplier Methods

JAMES S. VANDERGRAFT. Introduction to Numerical Computations, Second Edition

FRANÇOISE CHATELIN. Spectral Approximation of Linear Operators

GÖTZ ALEFELD AND JÜRGEN HERZBERGER. Introduction to Interval Computations. Translated by Jon Rokne

ROBERT R. KORFHAGE. Discrete Computational Structures, Second Edition

MARTIN D. DAVIS AND ELAINE J. WEYUKER. Computability, Complexity, and Languages: Fundamentals of Theoretical Computer Science

LEONARD UHR. Algorithm-Structured Computer Arrays and Networks: Architectures and Processes for Images, Percepts, Models, Information

O. AXELSSON AND V. A. BARKER. Finite Element Solution of Boundary Value Problems: Theory and Computation

PHILIP J. DAVIS AND PHILIP RABINOWITZ. Methods of Numerical Integration, Second Edition

NORMAN BLEISTEIN. Mathematical Methods for Wave Phenomena

In preparation

ROBERT H. BONCZEK, CLYDE W. HOLSAPPLE, AND ANDREW B. WHINSTON. Micro Data Base Management: Practical Techniques for Application Development

PETER LANCASTER AND MIRON TISMENETSKY. The Theory of Matrices: With Applications, Second Edition